INTERNATIONAL

REVIEW OF CYTOLOGY

VOLUME 65

INTERNATIONAL

Review of Cytology

EDITED BY

G. H. BOURNE
St. George's University School of Medicine
St. George's, Grenada
West Indies

J. F. DANIELLI
Worcester Polytechnic Institute
Worcester, Massachusetts

ASSISTANT EDITOR
K. W. Jeon
Department of Zoology
University of Tennessee
Knoxville, Tennessee

Volume 65

1980

ACADEMIC PRESS   *A Subsidiary of Harcourt Brace Jovanovich, Publishers*
New York   London   Toronto   Sydney   San Francisco

ACADEMIC PRESS, INC.
111 Fifth Avenue, New York, New York 10003

United Kingdom Edition published by
ACADEMIC PRESS, INC. (LONDON) LTD.
24/28 Oval Road, London NW1 7DX

LIBRARY OF CONGRESS CATALOG CARD NUMBER: 52–5203

ISBN 0–12–364465–8

PRINTED IN THE UNITED STATES OF AMERICA

80 81 82 83   9 8 7 6 5 4 3 2 1

Contents

Cell Surface Glycosyltransferase Activities

MICHAEL PIERCE, EVA A. TURLEY, AND STEPHEN ROTH

The Transport of Steroid Hormones into Animal Cells

ELEONORA P. GIORGI

Structural Aspects of Brain Barriers, with Special Reference to the Permeability of the Cerebral Endothelium and Choroidal Epithelium

B. VAN DEURS

Immunochemistry of Cytoplasmic Contractile Proteins

UTE GRÖSCHEL-STEWART

The Ultrastructural Visualization of Nucleolar and Extranucleolar RNA Synthesis and Distribution

S. FAKAN AND E. PUVION

Cytological Mechanisms of Calcium Carbonate Excavation by Boring Sponges

SHIRLEY A. POMPONI

Neuromuscular Disorders with Abnormal Muscle Mitochondria

Z. KAMIENIECKA AND H. SCHMALBRUCH

List of Contributors

Numbers in parentheses indicate the pages on which the authors' contributions begin.

S. Fakan (255), *Swiss Institute for Experimental Cancer Research, Lausanne, Switzerland*

Eleonora P. Giorgi* (49), *Department of Biological Chemistry, The Hebrew University, Jerusalem, Israel*

Ute Gröschel-Stewart (193), *Institut für Zoologie der Technischen Hochschule, Darmstadt, Federal Republic of Germany*

Z. Kamieniecka (321), *Institute of Neurophysiology, University of Copenhagen, Copenhagen, Denmark*

Michael Pierce (1), *Department of Biology, Johns Hopkins University, Baltimore, Maryland 21218*

Shirley A. Pomponi† (301), *University of Miami, Rosenstiel School of Marine and Atmospheric Science, Miami, Florida 33149*

E. Puvion (255), *Institut de Recherches Scientifiques sur le Cancer, Villejuif, France*

Stephen Roth (1), *Department of Biology, Johns Hopkins University, Baltimore, Maryland 21218*

H. Schmalbruch (321), *Institute of Neurophysiology, University of Copenhagen, Copenhagen, Denmark*

Eva A. Turley‡ (1), *Department of Biology, Johns Hopkins University, Baltimore, Maryland 21218*

B. van Deurs (117), *Institute of Anatomy, The Panum Institute, University of Copenhagen, Copenhagen, Denmark*

*Present address: Hormone Department, The National Institute for Biological Standards and Controls, Holly Hill, Hampstead, London NW3, England.

†Present address: Horn Point Environmental Laboratories, University of Maryland, Center for Environmental and Estuarine Studies, P.O. Box 775, Cambridge, Maryland 21613.

‡Present address: Department of Pathology, Queen's University, Kingston, Ontario, Canada.

INTERNATIONAL REVIEW OF CYTOLOGY, VOL. 65

Cell Surface Glycosyltransferase Activities

Michael Pierce, Eva A. Turley, and Stephen Roth

Department of Biology, Johns Hopkins University, Baltimore, Maryland

I. Introduction

There are three types of biological polymers, and each of these contains information in the linear sequence of its units. The nucleic acids are generally composed of four bases, proteins are composed of 21 amino acids, and the complex carbohydrates are made, in animals, from about ten monosaccharides. Unlike nucleic acids and proteins, however, the synthesis of the complex carbohydrates is not directed by a molecular template. Sugars are attached singly or, occasionally, as groups to the nonreducing ends of oligosaccharides, polysaccharides, proteins, and lipids by enzymes called glycosyltransferases. Most of the cellular glycosyltransferases are bound to membranes, especially those of the smooth endoplasmic reticulum and Golgi apparatus. The enzymes are specific for a particular class of sugar donors (e.g., uridine diphosphate galactose (UDP-Gal)[1] and uridine diphosphate N-acetylgalactosamine (UDP-GalNAc) are in different donor classes) and acceptors. For example, a galactosyltransferase that will recognize and galactosylate a serum glycoprotein with an exposed, N-acetylglucosaminide may also utilize free N-acetylglucosamine as an acceptor, but it will not utilize N-acetylgalactosamine, free or bound. In addition to the catalytic specificity, the synthesis of a complete oligosaccharide sequence depends upon the sequential encounters between the acceptor molecule and the appropriate transferases. If a glycoprotein fails to receive, say, its fourth monosaccharide, then none of the subsequent sugars will be attached because the subsequent transferases will not recognize the incomplete oligosaccharide chain. The synthesis of the nonrepeating polysaccharides requires, therefore, enzymes that are substrate specific *and* enzymes that have a specific cellular localization.

Because polysaccharide biosynthesis is directed primarily by the specificity and position of glycosyltransferases relative to one another, the "template" for

[1]Abbreviations: UDP, uridine 5'-diphosphate; UMP, uridine 5'-monophosphate; GDP, guanosine 5'-diphosphate; CMP, cytidine 5'-monophosphate; AMP, adenosine 5'-monophosphate; Gal, galactose; Glc, glucose; GalNAc, N-acetylgalactosamine; GlcNAc, N-acetylglucosamine; NeuNAc, N-acetylneuraminic acid; Man, mannose; Fuc, fucose, GA, glucuronic Acid; UDP-Gal, uridine 5'-diphosphate galactose; UDP-Glc, uridine 5'-diphosphate glucose; UDP-GalNAc, uridine 5'-diphosphate N-acetylgalactosamine; UDP-GlcNAc, uridine 5'-diphosphate N-acetylglucosamine; UDP-GA, uridine 5'-diphosphate glucuronic acid; CMP-NeuNAc, cytidine 5'-monophosphate N-acetylneuraminic acid; GDP-Man, guanosine 5'-diphosphate mannose; GDP-Fuc, guanosine 5'-diphosphate fucose; TCA, trichloroacetic acid; UDP-Xyl, uridine 5'-diphosphate xylose; GM_2, Svennerholm notation for the ganglioside with the following structure: GalNAc—Gal—Glu— | NeuNac Ceramide; GM_1, ganglioside with the following structure: Gal—GalNAc—Gal—Glu—Ceramide; CDH, | NeuNAc ceramide dihexoside; AS-, asialo; AG-, agalacto; OSM, ovine submaxillary mucin; BSM, bovine submaxillary mucin; cAMP, 3':5'-cyclic adenosine monophosphate.

polysaccharide biosynthesis resides in the structure of the cell itself. Thus, disruption of cell structure, a common first step in transferase analyses, is probably one of the reasons for our relatively poor understanding of these enzymes compared to those that synthesize DNA, RNA, and protein. Another reason is that, until recently, very few functions of the carbohydrate portions of proteins and lipids were known. At present, it is clear that sugar sequences on glycoproteins can be signals to membrane receptors on liver cells and fibroblasts. These signals may instruct a liver cell to remove a glycoprotein from the blood and to destroy the glycoprotein or, alternatively, they may cause a fibroblast to internalize and utilize a glycoprotein with enzymatic activity. Sugar sequences on cell surface gangliosides are also toxin and hormone receptors and may be involved in intercellular recognition as well.

Contemporary investigations on the glycosyltransferases deal with their subcellular localizations, the varieties of their activated sugar donors, differences between the membrane-bound and soluble forms of the transferases, and the possibility that some transferases may act directly as carbohydrate-binding proteins instead of synthetic enzymes. This article will deal with research that has occurred within the past few years on the topic of cell surface glycosyltransferases. Criteria for assigning a surface localization will be delineated and the stress will be on the potential functions for surface transferases in systems for which some of these criteria have been met.

II. The Glycosyltransferases

Three basic components are involved in the reactions catalyzed by glycosyltransferases. These are the enzyme itself, a sugar donor, and a sugar acceptor. Often, but not always, divalent cations are needed for catalytic activity. Another component, α-lactalbumin also participates in the reactions of some galactosyltransferases (discussed in Section IV,B).

A. Sugar Donor

The glycosyltransferases are named according to the sugars they transfer. As shown in Fig. 1, a galactosyltransferase transfers galactose from UDP-galactose to an acceptor. In the more familiar reactions, sugar nucleotides serve as the sugar donor. These donors consist of a nucleoside di- or monophosphate linked

$$\text{UDP-Galactose} + N\text{-Acetylglucosamine} \xrightarrow[\text{enzyme}]{\text{Mn}^{2+}} \begin{array}{c}\text{Galactose-}\\ N\text{-acetylglucos-}\\ \text{aminegalactose}\end{array} + \text{UDP}$$

FIG. 1. Reaction of N-acetylglucosamine:UDP-galactosyltransferase.

through its terminal phosphate ester to the reducing terminus of the carbohydrate. Recently, another class of sugar donors has been discovered in animal systems, although they have been known for some time in bacteria and yeasts. These are the lipid-sugar intermediates and consist of a mono- or oligosaccharide linked through a phosphate or pyrophosphate ester to a polyisoprenoid lipid such as dolichol or retinol. The possible role of these intermediates in cell surface glycosyltransferase reactions will be discussed in Section IX.

B. Sugar Acceptor

The glycosyltransferases within each class may be further defined according to the particular acceptors they utilize. The degree of acceptor specificity of these enzymes can be known only for the few that have been purified. In the case of the UDP-GalNAc:mucin N-acetylgalactosaminyltransferase purified from porcine submaxiallary gland by Schwaezer and Hill (1977), both asialo-porcine sub-maxillary mucin and cytolipin H, a glycolipid with a similar terminal carbohydrate sequence, serve well as acceptors. Other molecules with slightly different terminal carbohydrate sequences are relatively inactive as acceptors.

Often when cell surface glycosyltransferase activity is measured, no exogenous acceptor is added and the cells are allowed to transfer sugars to native acceptors on their own surfaces. These endogenous acceptors have been partly characterized in only a few cases (Patt and Grimes, 1975; Roth and White, 1972).

C. Cation

The majority of the glycosyltransferase activities require millimolar levels of divalent cations for optimal activity. Depending on the enzyme, Mn^{2+}, Mg^{2+}, and Ca^{2+} most often stimulate activity to the greatest extent. Some cell surface transferase activities require lower levels of cation for maximal stimulation than do the same activities of homogenates of cells (Patt and Grimes, 1974; Roth et al., 1977).

D. Glycosyltransferases

The enzymes have been found in soluble form in many vertebrate body fluids, but are generally in membrane-bound form when associated with cells. Many of the membrane-bound enzymes studied thus far are considered to be intrinsic proteins; that is, they are not released from the membranes by sonication and require detergents for solubilization. Before 1971, glycosyltransferase activities were generally thought to be localized in the Golgi-rich fractions of cells, since that was the finding in rat liver. Roseman (1970) suggested that the enzymes function as a concerted system in the synthesis of carbohydrate side chains of glycoproteins and glycolipids as these acceptor molecules travel from the endoplasmic reticulum through the Golgi (reviewed by Jentoft *et al.*, 1976). Golgi vesicles that have pinched off from their former location are thought to travel to and fuse with the inner surface of the plasma membrane (Whaley *et al.*, 1972). It is easy to envision, therefore, that some glycosyltransferases would eventually be present on the exterior of the cell as Roseman (1970) postulated. Strictly speaking, these enzymes are partitioned from the cytoplasm by membrane and are always on the topologic "outside" of the cell.

Figure 2 depicts four types of surface transferase reactions. Each reaction can be broken down into two steps. First, in the presence of appropriate acceptors, but in the absence of either the sugar donor or cation, if necessary, the enzymes could function to bind acceptor molecules. In this way, the glycosyltransferases could function as do lectins. Second, the enzymes might bind carbohydrates and

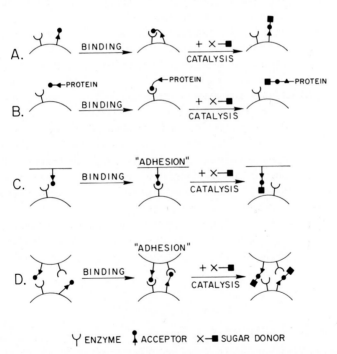

FIG. 2. Four classes of cell surface glycosyltransferase reactions. (A) Binding and transfer to endogenous acceptor (cis-glycosylation). (B) Binding and transfer to exogenous, soluble acceptor. (C) Binding (adhesion) and transfer to exogenous, insoluble acceptor (e.g., chondroitin sulfate derivatized to plastic surface). (D) Binding (adhesion) and transfer to acceptor on adjacent cell surfaces (transglycosylation).

also act catalytically to modify the bound molecules. The catalysis reaction would then allow a cell to release a bound carbohydrate. If these cell surface enzymes bound or modified specific carbohydrates on their own or other cell surfaces, or those sugars present on various extracellular macromolecules, they could play direct roles in cell interactions such as recognition, migration, and growth control.

III. Measurement of Glycosyltransferase Activities

A. ASSAY SYSTEMS

A glycosyltransferase activity is most often measured by incubating the enzyme with an acceptor, cation (if necessary), and the appropriate sugar nucleotide that is radioactively labeled in the sugar moiety. At the end of the incubation, the glycosylated product is separated from the unused radioactive substrate and any of its breakdown products. A common method of separation is high-voltage electrophoresis of samples in 1% sodium borate-impregnated paper (Roth *et al.*, 1972). In this assay system, the sugar nucleotides and most of their breakdown products migrate well away from the origin, while acceptors generally remain at or near the origin. Alternatively, the acceptors may often be precipitated by addition of trichloroacetic acid (TCA) or phosphotungstic acid and collected on filters that are permeable to the soluble sugar nucleotides.

Other assay systems have been devised that consist of acceptors derivatized to supports such as agarose or glass beads or modified polystyrene surfaces. At the end of incubations, the supports are washed repeatedly with solutions of concentrated detergents such as sodium dodecyl sulfate and denaturants such as 8 *M* urea to remove all noncovalently bound radioactivity. Remaining radioactivity can then be easily determined (Yogeeswaran *et al.*, 1974).

Glycosyltransferase activity toward endogenous acceptors can be assayed by autoradiography. This assay, however, can localize only the products of the enzyme reaction and not the enzyme molecules themselves (Porter and Bernacki, 1975).

B. ASSAY CONDITIONS

To quantitate the levels of enzyme activity from any source, the enzyme must be assayed under optimal conditions. Saturating levels of substrate, exogenous acceptor (if used), and necessary cations must be determined for each activity to be assayed, and the reaction performed such that these levels are present throughout the incubation period. Sparato *et al.* (1975) measured cell surface sialyltransferase activity on chick embryo cells and compared this activity to that found on fibroblasts transformed by Rous sarcoma virus. These authors showed that the relative activities of these cells varied depending on the incubation conditions under which the activities were measured. Clearly, no valid information can be gained unless the glycosyltransferase activities are measured under optimal conditions.

C. Obligatory Control Experiments

When intact cells are assayed for glycosyltransferase activity, control experiments must be performed to demonstrate that the observed activity is indeed a result of cell surface glycosyltransferases.

1. *Breakdown of Sugar Nucleotide Substrates*

Intact cells and plasma membrane preparations can contain high levels of phosphatase and/or pyrophosphatase activities that can degrade sugar nucleotides to free sugars and sugar phosphates as shown in Fig. 3. These enzymes can therefore reduce the amount of sugar nucleotide available to the glycosyltransferases and possibly invalidate the results of the assay. Breakdown of substrate becomes a critical problem if subsaturating levels of substrate are used in an incubation.

The simplest method to prevent breakdown of sugar nucleotides is to include molar excesses of competing substrates for the hydrolytic enzymes. One substrate of this type is 5'-adenosine monophosphate (AMP), which inhibits pyrophosphatase activities on BHK cells by over 95% when present at millimolar levels. Concentrations of 5'-AMP greater than 3 mM inhibit galactosyltransferase activity toward N-acetylglucosamine on intact BALB/c 3T12 cells (Webb, 1980). Consequently, for the assay of a particular glycosyltransferase activity, concentrations of a competing substrate must be found that effectively inhibit breakdown of sugar nucleotide, but have no other effects on the assay.

A report (Kirschbaum and Bosmann, 1973a) suggesting that millimolar concentrations of folic acid activated glycosyltransferase activity appears to be an artifact of sugar nucleotide degradation. Geren and Ebner (1974) demonstrated that folic acid inactivated phosphatase activity and, if the concentrations of the sugar nucleotide were subsaturating, more substrate was available in incubations with folic acid. Thus, folic acid can artifactually produce an increase in trans-

$$\text{UDP-Galactose} \xrightarrow[\text{phatase}]{\text{pyrophos-}} \text{Galactose-1-P} + \text{UMP} \xrightarrow{\text{phosphatase}} \text{Galactose} + \text{P}$$

Fig. 3. An example of sugar nucleotide breakdown to sugar phosphate and free sugar (Patt and Grimes, 1977; Shur, 1977b; Webb, 1980).

ferase activity when the assay is performed with suboptimal levels of sugar nucleotide substrate.

2. Utilization of Breakdown Products

Although sugar nucleotides have not been shown to be transported into any cell type, the free sugars that result from their breakdown can be internalized. Thus, if hydrolysis of radioactive sugar nucleotide does occur, the labeled, free sugars can be transported into the cells and the resulting cell-associated radioactivity erroneously interpreted as endogenous glycosyltransferase activity. If hydrolysis of sugar nucleotide cannot be prevented effectively, then uptake of labeled monosaccharides by intact cells can be inhibited in either of two ways. First, an excess of unlabeled, free sugar can be included in the incubation in order to decrease the specific activity of the labeled monosaccharide. Second, the sugar transport system in question can be blocked by an appropriate inhibitor such as cytochalasin B (Shur, 1977b; Webb, 1979).

3. Release of Internal Enzymes

A frequent concern when cell surface glycosyltransferases are assayed is that the observed activity does not arise from internal enzymes that have been released by lysed cells into the supernatant of the incubation. Since the level of any surface glycosyltransferase activity is often a small fraction of the intracellular levels, a leakage of internal enzyme activity could easily be interpreted as arising from cell surface enzymes. Partial lysis of cells and subsequent leakage of internal enzymes is a particular concern when the cells to be used in the assay have been dissociated, usually by trypsinization or treatment with EDTA (Hirschberg et al., 1976). To determine if activity is present in incubation supernatants, the cell preparation can first be incubated in a complete assay mixture minus only the radioactive sugar nucleotide substrate. After a typical incubation time, the cells are pelleted by centrifugation, the supernatant transferred to another assay tube containing the radioactive substrate, and the reaction allowed to proceed for another incubation period. Experiments using this procedure with BALB/c 3T12 cells showed that less than 20% of the intact cell galactosyltransferase activity toward GlcNAc was present in the supernatant, and free of the cells, after a typical incubation of 1 hour (Webb, 1980).

In addition, the total amount of activity present when a cell preparation is disrupted by sonication can be determined, and the activity measured on intact cells may be expressed as a percentage of the total activity. This value can then be compared to other, independent measures of cell integrity; for example, the presence of soluble enzymes such as lactate dehydrogenase in incubation supernatants, or the ability of cells to exclude vital dyes. Cell surface transferases may be present when intact cell activities are significantly higher than can be accounted for by any indicators of cell lysis.

IV. Cell Surface Glycosyltransferase Activity on Cultured Cells

A. DEMONSTRATION OF CELL SURFACE GLYCOSYLTRANSFERASE ACTIVITY

The studies that have examined glycosyltransferase activity on intact, cultured cells as well as plasma membrane fractions obtained from these cells are listed in Table I. Enzyme activities on intact cells have not always been fully characterized and many investigators have neglected some or all of the controls that are necessary to substantiate surface localization of transferase activity. As discussed in Section II, it is crucial to demonstrate that the exogenous sugar nucleotide, and not one of its breakdown products, is the authentic sugar donor. One piece of evidence is the demonstration that large, molar excesses of appropriate, unlabeled free sugar have no effect on the incorporation of radioactivity onto sugar acceptors. Those experiments that include this control are noted in Table I. Since many of the results listed in Table I have been reviewed by Shur and Roth (1975), only the more recent will be fully discussed here.

Cervén (1977) has recently reported endogenous sialyltransferase activity on the surfaces of intact Ehrlich ascites cells that were passed in Swiss albino mice. Several control experiments support the hypothesis that the radioactivity incorporated onto intact cells from CMP-[³H]NeuNAc was a measure of cell surface sialyltransferase activity. First, the acid-precipitable radioactivity incorporated onto cells was shown to be NeuNAc by chromatography in three solvent systems after hydrolysis from the acceptors. Second, only 1.5% of the total CMP-[³H]NeuNAc was degraded by the end of the experiment, and the addition of 100-fold excesses of unlabeled NeuNAc to the assay did not effect incorporation of radioactivity onto cells. Third, leakage of enzyme into the incubation supernatant was ruled out. Autoradiography of labeled cells suggested that, for short times of development of the autoradiogram, approximately 75% of the 580 cells examined were uniformly labeled, while 10% showed a patchy distribution of grains. Greater than 95% of the cells excluded Trypan blue when tested after 1 hour incubations. Estimation of the apparent K_m and V_{max} for the endogenous reaction with CMP-[³H]NeuNAc gave values of 10 μM and 70 pmole/10^8 cells · 30 minutes, respectively.

Bernacki (1974) measured the endogenous sialyltransferase activity of intact leukemic L-1210 cells and utilized pretreatment of the cells with neuraminidase to stimulate this activity 6-fold. Porter and Bernacki (1975) examined leukemic L-1210 cells autoradiographically in the electron microscope after pretreatment with neuraminidase and incubation with CMP-[³H]NeuNAc. Results showed that over 80% of the 780 grains counted were localized over the plasma membrane. A number of control experiments suggest that the observed labeling with CMP-[³H]NeuNAc was a measure of cell surface sialyltransferase activity and that the surface-localized grains in the autoradiograms reflected transferase:acceptor interaction.

TABLE I

REPORTS OF CELL SURFACE GLYCOSYLTRANSFERASE ACTIVITIES[a,b]

Cell or tissue type	Source of enzyme	Labeled substrate	Acceptor	Free sugar control	Reference
HeLa cells	Plasma membranes	UDP-Glc UDP-Gal	Collagen	NA	Hagopian et al. (1968) Bosmann (1969)
Ehrlich ascites cells	Plasma membranes	UDP-Gal	Collagen	NA	Molnar et al. (1969)
	Intact cells	CMP-NeuNac	Endogenous	Yes	Cervén (1977)
Chick embryo fibroblasts and Rous sarcoma transformants	Intact cells	UDP-Gal GDP-Man UDP-GalNAc UDP-GlcNAc UDP-Glc	Endogenous	No	Bosmann et al. (1973); Morgan and Bosmann (1974)
	Intact cells	CMP-NeuNAc	Endogenous	No	Sparato et al. (1975)
Hamster fibroblasts and polyoma transformants	Intact cells	CMP-NeuNAc	Endogenous	Yes	Sasaki and Robbins (1974)
Swiss 3T3, SV3T3, PY3T3, SVT2 mouse fibroblasts	Intact cells	CMP-NeuNAc UDP-Gal UDP-GalNAc UDP-GlcNAc GDP-Man	Endogenous	Yes	Patt and Grimes (1974)
		UDP-Gal CMP-NeuNAc	AS-AG-fetuin AS-fetuin	Yes	Patt and Grimes (1974)
3T3, PY3T3 mouse fibroblasts	Intact cells	CMP-NeuNAc	Endogenous	Yes	Datta (1974)
BALB/c 3T3, 3T12, SV3T3 mouse fibroblasts	Intact cells	UDP-Gal CMP-NeuNAc UDP-GlcNAc UDP-GalNAc	Endogenous	Yes	Webb (1980); Webb and Roth (1974); Roth et al. (1974, 1977); Roth and White (1972)
		UDP-Gal	AS-AG-fetuin AS-OSM, GlcNAc	Yes	

3T12 mouse fibroblasts	Plasma membranes	UDP-Gal	AS-AG-fetuin Glc-NAc	NA	Cummings *et al.* (1979a)
SV3T3 mouse fibroblasts	Intact cells	Prelabeled with Gal	Hyaluronic acid Chondroitin sulfate	NA	Turley and Roth (1979)
BALB/c 3T3, SVT2, PBC, KMSVT, 3T12, 3T12T mouse fibroblasts	Intact cells	UDP-Gal UDP-Glc UDP-GalNAc UDP-GlcNAc CMP-NeuNAc GDP-Man	Endogenous	Yes	Patt *et al.* (1975)
3T3, MSV3T3, RSV3T3, PY3T3 mouse fibroblasts	Intact cells	UDP-Glc GDP-Man CMP-NeuNAc UDP-Gal UDP-GalNAc UDP-GlcNAc CMP-NeuNAc UDP-GlcNAc UDP-Glc UDP-Gal	Endogenous As-fetuin AS-AG-minus GlcNAc-fetuin A-Glc-collagen A-Glc-AG-collagen	No	Bosmann (1972b)
C3H-2K, SVC3H-2K mouse cells	Intact cells	UDP-Gal UDP-Glu GDP-Fuc GDP-Man CMP-NeuNAc	Endogenous As-AG-fetuin Endogenous AS-fetuin Endogenous Endogenous Endogenous AS-fetuin	No	Sudo and Onodera (1975)
	Plasma membranes	CMP-NAN UDP-Gal	AS-fetuin AS-AG-fetuin	NA	Sudo and Onodera (1975)

11

(continued)

TABLE I (continued)

Cell or tissue type	Source of enzyme	Labeled substrate	Acceptor	Free sugar control	Reference
L-5178Y mouse lymphoma	Intact cells	UDP-Gal	Endogenous AS-AG-fetuin GlcNAc	No	Bosmann (1974)
Mouse melanoma lines 26 and 27	Intact cells	UDP-Glc UDP-Gal UDP-GlcNAc UDP-GalNAc GDP-Fuc GDP-Man CMP-NeuNAc	Endogenous	No	Bosmann et al. (1973)
RAJI lymphoma	Intact cells	CMP-NeuNAc	Endogenous	Yes	Kilton and Maca (1977)
Mouse leukemic L-1210 cells	Intact cells	CMP-NeuNAc	Endogenous	Yes	Porter and Bernacki (1975); Bernacki (1974)
BHK, PYBHK fibroblasts	Intact cells	CMP-NeuNAc	Endogenous	Yes	Datta (1974)
BHK fibroblasts	Intact cells	CMP-NeuNAc UDP-Gal GDP-Man	Endogenous AS-fetuin Endogenous Endogenous	Yes	Patt and Grimes (1975)
A31 fibroblasts	Intact cells	UDP-Gal	AS-AG-fetuin	Yes	LaMont et al. (1977)
BHK fibroblasts	Intact cells	UDP-Gal	GlcNAc, Glc	Yes	Yogeeswaran et al. (1974)
Hamster cells BHK, PYBHK	Intact cells	CMP-NeuNAc UDP-Gal	CDH CDH	No	Yogeeswaran et al. (1974)
NIL2E	Intact cells	CMP-NeuNAc UDP-Gal UDP-GalNAc	CDH CDH Globoside	No	Yogeeswaran et al. (1974)

(continued)

Cell type	Preparation	Sugar donor	Acceptor		Reference
NIL2K	Intact cells	UDP-Gal	CDH	No	Yogeeswaran et al. (1974)
		UDP-GalNAc	Globoside		
BHK, NIL2E	Intact cells	Prelabeled with Gal	CDH	NA	Yogeeswaran et al. (1974)
			Globoside		
BHKPY, NILPY	Intact cells	Prelabeled with Gal	CDH	NA	Yogeeswaran et al. (1974)
			Globoside		
Rat dermal fibroblasts	Intact cells	CMP-NeuNAc	AS-BSM	No	Lloyd and Cook (1974)
	Plasma membranes	CMP-NeuNAc	AS-BSM	NA	Lloyd and Cook (1974)
Rat crypt and villus intestinal cells	Intact cells	UDP-Gal	AS-AG-fetuin	No	Weiser (1973a,b)
		UDP-Gal	Endogenous	No	
		CMP-NeuNAc		No	
		GDP-Fuc		No	
		GDP-Man		No	
		UDP-Glc		Yes	
		UDP-GlcNAc		No	
		UDP-Gal	GlcNAc	No	
		CMP-NeuNAc	Lactose		
	Plasma membranes	UDP-Gal	AS-AG-fetuin	NA	Weiser et al. (1978)
		CMP-NeuNAc	AS-fetuin	NA	
Chick neural retinal cells	Intact cells	UDP-Gal	Endogenous	No	Roth et al. (1971a)
			AS-AG-orosomucoid		
			AS-OSM, GlcNAc		
Chick liver cells	Intact cells	UDP-Gal	GlcNAc	Yes	Porzig (1978)
		UDP-Gal	Endogenous and cell surface glycoprotein fragments	No	Arnold et al. (1973)
		UDP-GlcNAc			
		CMP-NeuNAc			
		GDP-Fuc			
		GDP-Man			
		UDP-GA	p-Nitrophenol-Sephadex	Yes	Levine and Roth (1980)

TABLE I (continued)

Cell or tissue type	Source of enzyme	Labeled substrate	Acceptor	Free sugar control	Reference
Rat liver cells	Plasma membranes	UDP-Gal UDP-Glc UDP-GlcNAc GDP-Man	Endogenous	NA	Merritt *et al.* (1977)
		UDP-Gal CMP-NeuNAc	AS-AG-fetuin AS-fetuin	NA NA	Aronson *et al.* (1973) Pricer and Ashwell (1971)
Rat kidney cells	Plasma membranes	UDP-Gal CMP-NeuNAc UDP-Glc	AS-AG-fetuin AS-fetuin Collagen Tamm-Horsfall glycoprotein	NA	Kirschbaum and Bosmann (1973b)
Rat lymphocytes	Intact cells	UDP-Gal	Endogenous Ovomucoid Ovomucoid- Sephadex	Yes	Verbert *et al.* (1976); Cacan *et al.* (1976)
		CMP-NeuNAc	Endogenous AS-glycophorin fragments	Yes	Verbert *et al.* (1977)
Thymus and spleen lymphocytes	Intact cells	UDP-Gal	Endogenous AS-AG-fetuin	No	LaMont *et al.* (1974)

14

Cell/Source	Preparation	Donor	Acceptor	Endogenous	Reference
Rabbit erythrocytes	Intact cells	UDP-Gal	Endogenous AS-AG-fetuin	No	Podolsky et al. (1974)
	Purified cell surface enzyme	UDP-Gal	AS-AG-fetuin	NA	Podolsky et al. (1974)
Human platelets	Intact cells	UDP-Glc	Denatured collagen	Yes	Jamieson et al. (1971); Barber and Jamieson (1971) Jamieson (1974)
	Plasma membranes	UDP-Glc		NA	Smith et al. (1977)
		UDP-Glc UDP-Gal UDP-Glc	Poly-Glc-lysine Collagen	NA	Bosmann (1971)
Mouse sperm	Intact cells	CMP-NeuNAc CMP-NeuNAc UDP-Gal UDP-GlcNAc	Egg cell surface AS-fetuin GlcNAc Zona pellucidae	Yes Yes	Durr et al. (1976) Shur and Bennett (1979)
Gastrulating chick embryos	Intact embryos	UDP-Gal UDP-GlcNAc GDP-Fuc CMP-NeuNAc	Endogenous	Yes	Shur (1977a,b)
Chlamydomonas	Intact cells and membrane vesicles	UDP-Gal UDP-Glc UDP-GlcNAc GDP-Man GDP-Fuc CMP-NeuNAc	Endogenous	No	McLean and Bosmann (1975); Bosmann and McLean (1975)
Dictyostelium discoideium	Plasma membranes	CMP-NeuNAc GDP-Man UDP-Glc UDP-GlcNAc	Endogenous Endogenous	Yes NA	Columbrino et al. (1978) Sievers et al. (1978)

(continued)

15

TABLE I (continued)

Cell or tissue type	Source of enzyme	Labeled substrate	Acceptor	Free sugar control	Reference
Geodia cydonium	Purified aggregation factor	UDP-Gal	AG-aggregation receptor AG-antiaggregation receptor	NA	Muller et al. (1979)
		UDP-GA	A-glucuronylated-aggregation receptor	NA	Muller et al. (1978)
		CMP-NeuNAc	AS-aggregation receptor	NA	Muller and Zahn (1977)
BALB/c fibroblasts	Intact cells	GDP-Man	Endogenous (lipid-intermediates)	Yes	Patt and Grimes (1977)
Chick liver cells	Intact cells	GDP-Man	Endogenous (lipid-intermediates)	Yes	Arnold et al. (1975, 1976)
Hen oviduct cells	Intact cells	GDP-Man	Endogenous (lipid-intermediates)	Yes	Struck and Lennarz (1976, 1977)
Rabbit erythrocytes and reticulocytes	Plasma membranes	UDP-Glc UDP-GlcNAc GDP-Man	Endogenous (lipid-intermediates)	NA	Parodi and Martin-Barrientos (1977)

[a] Reports of cell surface glycosyltransferase activities on yeast, bacteria, and plant cells have been omitted from this table.

[b] "Purified" aggregation factor from the sponge, *Geodia*, has been reported to contain three glycosyltransferase activities. Although these activities have not been localized on the surfaces of sponge cells, Muller et al. (1978) have suggested that as aggregation factors interact with sponge cell surfaces, the glycosyltransferases in the factors act on carbohydrate acceptors on the cell surfaces and thereby influence cell adhesion (Section VII). NA, Not applicable; AS-fetuin, asialo-fetuin; AS-AG-fetuin, asialo-agalacto-fetuin; AS-OSM, asialo-ovine submaxillary mucin; A-Glc-collagen, aglucosyl-collagen; A-Glc-AG-collagen, aglucosyl-agalacto-collagen; CDH, ceramide dihexoside; AS-BSM, asialo-bovine submaxillary mucin; AS-AG-orosomucoid, asialo-agalacto-orosomucoid; AS-glycophorin, asialo-glycophorin.

Webb (1980) has further characterized the endogenous galactosyltransferase activity on intact BALB/c 3T3 and 3T12 cells. When assayed under optimal conditions that included 3 mM 5'-AMP to inhibit sugar nucleotide breakdown, the K_m and V_{max} constants with respect to UDP-galactose concentration were 21 μM and 12 pmole/10^5 cells·hour, respectively, for 3T3 cells and 15 μM and 2 pmole/10^5 cells·hour, respectively, for 3T12 cells. Further controls have shown that UDP-galactose was indeed the sugar donor in the endogenous reactions with both cell types.

Recent experiments by LaMont et al. (1977) and Cummings et al. (1979b) suggest that, in the presence of α-lactalbumin, intact BHK and BALB/c 3T12 fibroblasts can synthesize lactose from glucose and UDP-galactose. Brew et al. (1968) showed initially that the soluble UDP-galactose:GlcNAc galactosyltransferase in bovine milk no longer utilized GlcNAc as an efficient acceptor when α-lactalbumin, a protein also found in bovine milk, was included in the incubation mixture. Instead, the enzyme transferred galactose to glucose and synthesized lactose rather than N-acetyllactosamine. The interaction of α-lactalbumin and the enzyme was later shown to cause a reduction in the K_m of the enzyme for glucose and inhibit transfer to GlcNAc (Schanbacher and Ebner, 1970). Interestingly, α-lactalbumin does not inhibit transfer to large molecular weight acceptors such as asialo-agalacto-fetuin. These effects have been observed with various UDP-galactose:GlcNAc galactosyltransferase activities, both soluble and membrane-bound, including the activity in mammary gland, rat liver microsomes, and onion root tip (Jentoft et al., 1976).

LaMont et al. (1977) demonstrated that intact BHK/C13 cells transferred galactose to glucose only when they were incubated with UDP-galactose in the presence of α-lactalbumin. No attempt was made to measure inhibition of N-acetyllactosamine synthesis under identical conditions. A control experiment showed that no enzyme activity was leaked into the supernatant during the incubation. Cummings et al. (1979b) have shown a similar effect by α-lactalbumin on a cell surface galactosyltransferase of 3T12 cells. In the presence of α-lactalbumin, glucose was utilized as an acceptor, while the transfer of galactose to GlcNAc was inhibited by over 95%. It appears, therefore, that a galactosyltransferase modifiable by α-lactalbumin is present on the surfaces of at least these two cell types.

Powell and Brew (1976) recently described the effects of a derivative of UDP, dialdehyde-UDP, on the activity of soluble, bovine colostrum galactosyltransferase. Dialdehyde-UDP, synthesized by oxidizing the 2',3' cis-hydroxyls of UDP with equimolar concentrations of meta-periodate (Fig. 4), rapidly and irreversibly inactivated galactosyltransferase activity toward GlcNAc. In these experiments, activity was assayed after 1 hour preincubation of the enzyme in the presence of acceptor, low concentrations of Mn^{2+}, and 3 mM dialdehyde-UDP. The authors suggested that this inhibitor is specific for the active site of the enzyme since, (1) incubating the enzyme with radioactive dialdehyde-UDP re-

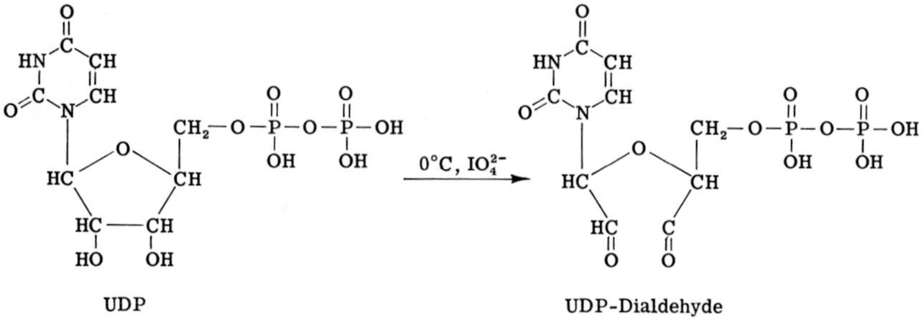

FIG. 4.   Synthesis of UDP-dialdehyde from UDP by periodate oxidation.

sults in a stoichiometric, 1:1 binding of the inhibitor to the enzyme, and (2) after reduction of $NaBH_4$ and fingerprinting of tryptic digests of the enzyme, one tripeptide can be isolated with a lysine residue that contains radioactive dialdehyde-UDP.

Cummings et al. (1979b) have shown that preincubation of intact 3T12 cells with 3 mM dialdehyde-UDP in the presence of low levels of Mn^{2+} inhibited transfer of galactose from UDP-galactose to both endogenous acceptors and to GlcNAc. No data as yet suggest that the inhibition is or can be made specific for a particular acceptor, or that inhibition of the enzyme affects its binding to acceptor. The preparation of dialdehyde derivatives of other mono- and diphosphate nucleosides would also seem possible. This battery of inhibitors may allow a number of the hypotheses for cell surface glycosyltransferase function to be tested, since the catalytic properties of populations of these enzymes might be selectively, and irreversibly, inactivated. In addition, autoradiography of radioactive dialdehyde derivatives bound by intact cells could eventually allow direct localization of cell surface enzymes.

Plasma membrane-enriched fractions have been prepared from Balb/c 3T12 cells by Cummings et al. (1979b) as a means to study cell surface glycosyltransferases. Characterization of these fractions revealed an 8-fold enrichment of Na^+,K^+-ATPase activity and a 13-fold enrichment of 5′-nucleotidase activity, both plasma membrane markers. Galactosyltransferase activity toward GlcNAc and asialo-agalacto-fetuin was present in the plasma membrane fractions and shown to be enriched only by 1.5–2-fold over homogenates. The K_m values of this enzyme toward UDP-galactose and GlcNAc were 12.7 μM and 4.6 mM, respectively. In addition, α-lactalbumin stimulated lactose production in the plasma membrane fractions by 60-fold while inhibiting transfer to GlcNAc by over 85%.

Dialdehyde-UDP inhibited the galactosyltransferase activity of these fractions by greater than 90%. When [³H]dialdehyde was incubated with intact cells in culture and their plasma membranes isolated, these fractions showed a 3.5-fold enrichment of bound radioactivity compared to other membrane fractions. If cells were first homogenized and then incubated with [^{3}H]dialdehyde-UDP, no membrane fraction showed enriched levels of radioactivity.

The inhibition by various nucleosides and nucleotides of endogenous sialyltransferase activity on intact RAJI lymphoma cells has been reported by Kilton and Maca (1977). Pretreatment of the cells with neuraminidase stimulated activity 6-fold, while posttreatment with neuraminidase removed 75% of the incorporated radioactivity. Addition of 1 mM sialic acid did not affect incorporation of radioactivity. These controls suggest that CMP-NeuNAc was the authentic sugar donor for the incorporated radioactivity. Insignificant amounts of TCA-insoluble, intracellular radioactivity were present at the end of the incubation. In addition, trypsin treatment of neuraminidase-treated cells before incubation with CMP-NeuNAc inhibited incorporation by over 90%.

When assayed at 2 mM levels, cytidine inhibited transfer onto neuraminidase-treated cells by 26%, whereas other nucleosides showed no effect. Also at 2 mM concentration, CMP inhibited the reaction by 75%, whereas the other nucleoside mono- and diphosphates had effects ranging from about 30 to 50% inhibition. Cyclic CMP and cyclic AMP at 2 mM inhibited activity by approximately 20%, whereas their dibutyryl derivatives inhibited activity by about 15%, although butyrate alone had the same effect. Whether the assay conditions used in this study were optimal for CMP-NeuNAc concentration was not reported. In addition, the effects of the various nucleosides and nucleotides on substrate breakdown were not tested.

Deppert and co-workers (1974; Deppert and Walter, 1978) have been unable to detect cell surface glycosyltransferase activities on a number of cultured cell types, although evidence for exogenous and endogenous cell surface activities has been presented by others for many of these cell types (see, for example, Sasaki and Robbins, 1972; Patt and Grimes, 1974, 1975; Webb and Roth, 1974; LaMont *et al.*, 1977). Assay conditions used in both studies by Deppert differed markedly from those previously reported. Most notably, concentrations of sugar nucleotide substrates were consistently much lower than those used by others. Also, Deppert and Walter and Deppert *et al.* failed to inhibit substrate degradation or use higher substrate concentrations, although complete degradation of sugar nucleotide was shown to occur during most of their assays for cell surface glycosyltransferase activities. Perhaps for these reasons the conclusions by these authors concerning the presence of glycosyltransferases on intact cells do not agree with those of others, who used different incubation conditions to assay for cell surface activities on intact cells from similar cell lines.

B. COMPARISONS OF GLYCOSYLTRANSFERASE ACTIVITIES ON NORMAL AND TRANSFORMED, INTACT CELLS

Several studies have compared glycosyltransferase activities on intact, cultured cells and their transformants to determine if transformation alters the levels of these cell surface enzymes. As reviewed by Shur and Roth (1975), there is no clear correlation between transformation and absolute levels of glycosyltransferase activity either on intact cells or, in separate studies, in cell homogenates. Cell surface glycosyltransferases have been shown to increase, show no change, or decrease upon transformation. Glycosyltransferase activities detected on intact cells used in these studies are listed in Table I.

Two studies that measured the activities of endogenous sialyltransferase on two cell types and their transformants concluded that transformation decreases this activity up to 50%. Sasaki and Robbins (1974) used normal and polyoma-transformed hamster cells and showed that the activity of the transformed cells was about half that of the normal cells, although the kinetics of the activity were similar on both cell types. Datta (1974) measured the endogenous sialyltransferase activity of BHK, temperature sensitive, polyoma-transformed BHK, 3T3, and temperature sensitive, polyoma-transformed 3T3 cells while the cells were in confluent monolayer cultures. Results showed that expression of the transformed phenotype accompanied a drop in endogenous cell surface sialyltransferase activity from 25 to 55%, depending on the cell type. Control experiments suggested that both these studies employed assay conditions that measured surface transferase activity.

A series of experiments by Patt and Grimes (1974) investigated the cell surface glycolipid and glycoprotein glycosyltransferases of Swiss 3T3, Py3T3, and SV3T3, as well as BALB/c SVT2 cells. These authors demonstrated transfer of sialic acid, galactose, N-acetylgalactosamine, N-acetylglucosamine, glucose, and mannose from nucleotide sugar donors to endogenous acceptors on these cells, but observed no systematic variation of activity between normal and transformed cells. Other studies by Patt et al. (1975) investigated the cell surface glycosyltransferase activities of normal BALB/c fibroblasts and compared them to activities of transformed cells from the same cell line. No correlation could be demonstrated between transformation and surface transferase activities, or between transformed cells that produced tumors in immunologically incompetent hosts and those that did not.

In a more recent study, Patt and Grimes (1975) compared the activities of three cell types assayed either in monolayer culture or after EDTA treatment followed by centrifugation to a pellet. BHK cells showed large incorporation with CMP-[^{14}C]NeuNAc, UDP-[^{14}C]galactose, and GDP-[^{14}C]mannose in pellet incubations; but when assayed in monolayer, these activities fell by at least 97%. Both assays included 5'-AMP to inhibit breakdown of substrate. In the absence of this inhibitor, after a 1 hour incubation, monolayer BHK cells had hydrolyzed 96% of UDP-galactose, whereas pellet incubations degraded only about 40%. Although no data were given, monolayers of both normal Balb/c A31 cells and SV40-transformed A31 cells were said to show no detectable endogenous surface activity with the four substrates mentioned earlier in the presence of 5'-AMP.

Different results were obtained, however, when exogenous sialyltransferase activity toward asialofetuin was measured on monolayers and in pellets of BHK cells in the presence of 5′-AMP. Compared to pelleted cells, cells in monolayers showed very low levels of endogenous activity with CMP-[^{14}C]NeuNAc but demonstrated slightly higher exogenous activity than cells in pellets. The endogenous and exogenous galactosyltransferase activities of monolayer A31 and SVA31 cells were also compared. Both cell types showed negligible endogenous activity, but both displayed exogenous transfer to asialo-agalacto-fetuin. These results suggested that monolayers of these cells catalyze poorly, if at all, endogenous galactosyltransferase reactions. They can, on the other hand, actively catalyze reactions when appropriate acceptors are added. In addition, monolayer cultures have a much higher level of enzymes that degrade sugar nucleotides compared to cells that are in pellet incubations. All cells in this study were taken from confluent cultures.

C. Dependence of Cell Surface Glycosyltransferase Activity on Culture Density and Cell Contact

Surface transferases have been shown to be localized on the surfaces of some cell types, and it is possible that these enzymes could function in the interactions between cells, playing roles in all recognition and, perhaps, growth control. To study this possibility it is useful to discriminate between a cell's ability to catalyze transfer of sugars to acceptors on its own cell surface (cis-glycosylation) and to acceptors on adjacent cells (trans-glycosylation). A simple assay for endogenous activity, whether on confluent monolayers or in pellet incubations, cannot make this distinction.

One possible way to discriminate between these two types of glycosylation is to assay endogenous activities of the cells as a function of their density in culture. Several, but not all studies that have measured glycosyltransferase activities on intact cells as a function of culture density suggest that normal cells show differences in some endogenous activities, whereas transformed cells that are not contact inhibited in culture show few differences.

Roth and White (1972) demonstrated that exogenous galactosyltransferase activity toward GlcNAc on intact BALB/c 3T3 cells did not vary between cells that were taken from confluent or sparse cultures. By contrast, endogenous galactosyltransferase activity of these cells from confluent cultures was only one-third of the activity of cells from sparse cultures when assayed in pellet incubations. Autoradiographs of 3T3 cells in either sparse or confluent cultures after incubation with UDP-[^{3}H]galactose showed very little incorporation of label. However, when sparsely growing cells were harvested and allowed to pellet before incubating with UDP-[^{3}H]galactose, autoradiography revealed that most of the cells were heavily labeled.

By contrast, the spontaneously transformed BALB/c cell type, 3T12, which is tumorigenic, showed high levels of cell surface endogenous galactosyltransferase activity regardless of culture cell density. Autoradiography revealed that when cells in culture were incubated with UDP-[^{3}H]galactose, they showed very high endogenous activity whether or not cells were in contact. These results suggested that 3T12 cells could carry out the cis-glycosylation reaction and that cell contact was not necessary for high endogenous activities.

Two other sets of experiments that utilized 3T3 cells are consistent with these observations. Bosmann (1972b) showed that 3T3 cells transformed by any of three viruses showed high endogenous activities when assayed in pellet incubations harvested from either sparse or confluent cultures. Untransformed 3T3 cells from sparse cultures showed activities as high as some transformed cells, whereas normal cells from confluent cultures showed much lower activities. Patt and Grimes (1974) assayed Swiss 3T3 cells for endogenous activity as a function of their culture density. All glycosyltransferase activities measured showed higher total activities (glycoprotein plus glycolipid) in pellet incubations prepared from confluent cultures compared to those prepared from sparse cultures.

LaMont et al. (1977) have shown cell culture density dependence of endogenous galactosyltransferase activity on intact BHK cells. Compared to confluent cells, cells from sparse cultures showed much less activity. Polyoma virus-transformed BHK cells, on the other hand, showed no decrease of activity upon confluency; this activity actually increased somewhat as a function of cell density. Actively growing BHK and PyBHK cells both showed 2- to 3-fold higher activity than did cells whose growth was arrested by low serum levels. Upon release from arrest, culture supernatants of PyBHK cells showed large amounts of soluble galactosyltransferase activity compared to BHK culture supernatants.

Other experiments by Bosmann et al. (1973) determined the endogenous surface transferase activities of two melanoma cell lines differing in their abilities to cause tumors. The cell line with a higher level of metastasizing ability, when assayed from sparse cultures, showed higher levels of some, but not all, endogenous glycosyltransferase activities compared to cells of the same line taken from confluent cultures or the less malignant cell line taken from either sparse or confluent cultures.

Yogeeswaran et al. (1974) measured the ability of BHK and NIL cells and their polyoma transformants to glycosylate exogenous acceptors that had been conjugated to glass beads or cover slips. Using either exogenous sugar nucleotides or cells prelabeled with radioactive galactose, transformed cells glycosylated the derivatized acceptors 20–50% less than did untransformed cells. These experiments suggest that transformed cells are much less able to glycosylate trans-acceptors than normal cells using both endogenous and exogenous sugar donors.

In order to discriminate between cis- and trans-glycosylation, cell contact was minimized by spinning the incubation tube. Assayed under these conditions, the endogenous galactosyltransferase activity of intact, untransformed 3T3 cells was less than one-third of the activity measured when the cells were in a pellet (Roth and White, 1972). By contrast, 3T12 cells showed no difference in endogenous activity when assayed under these two conditions. Control experiments showed that UDP-galactose was the sugar donor in these incubations and that the transferase activity present in supernatants of incubations could not account for the differences (Webb, 1980). These results suggest that untransformed cells can glycosylate one another upon contact and that transformed cells have higher levels of cis-glycosylation. Patt and Grimes (1974), however, could demonstrate no systematic difference between cell surface glycosyltransferase activities on Swiss 3T3 cells when these cells were assayed in incubations that were shaken and incubations in which the cells were allowed to pellet, although the means by which the cells in this study were kept suspended was quite different from those used by others (Roth and White, 1972; Webb and Roth, 1974).

The demonstration that endogenous activities of intact, untransformed cells vary with culture density coupled with the decreases in activity measurable when untransformed cells are assayed with minimal cell contact suggests that these cells can glycosylate other cells upon contact. Some transformed cells, on the other hand, do not show this density dependence, but show constant high levels of endogenous activity and do not show a drop in activity when assayed in suspension incubations. These results are consistent with the possibility that transformed cells are capable of cis-glycosylation. This potential inability to trans-glycosylate could relate to differences in the growth and other behaviors of normal and transformed cells.

D. MODULATION OF CELL SURFACE GLYCOSYLTRANSFERASE ACTIVITY

Since endogenous activities of cell surface glycosyltransferases depend on the interaction of enzymes and acceptors on the cell surface, high levels of endogenous activity on 3T12 and other transformed cells suggest that the surfaces of the transformed cell lines allow a higher degree of enzyme:acceptor interaction.

Setlow et al. (1979) measured the endogenous galactosyltransferase activity of mouse L-cells as a function of incubation temperature. The activity, instead of showing linear (monotonic) response as a function of temperature, showed a distinct transition in rate at 18°C. A similar result was observed with these cells when the rate of surface antigen spreading was measured (Petit and Edidin, 1974). These experiments argue that, indeed, endogenous activity does depend on the ability of enzyme and acceptor to interact on cell surfaces and suggest that this interaction is a direct function of those variables that control relative mobility of components embedded in the plasma membrane. It seems reasonable to suggest that the alteration of membrane composition seen in many transformed cells could allow many more enzyme:acceptor interactions.

As discussed in Shur and Roth (1975), experiments on the mobility of membrane antigens and ability of lectins to agglutinate cells suggest that a difference between normal and transformed cells is that the binding sites for these polyvalent molecules can be clustered to a greater extent on transformed cell surfaces. If surface transferases and acceptors do not easily interact on untransformed cell surfaces, as lectin binding sites do not, then these enzymes might be more available to participate in interactions with adjacent cell surfaces.

Studies on the agglutination of untransformed cells by lectins as a function of the stage of their cell cycle show that normal cells in mitosis are agglutinated at low lectin concentration just as malignant cells are, whereas normal cells in other stages of the cell cycle require higher concentrations of lectin for agglutination to occur (Sharon and Lis, 1972). If this change in agglutinability of mitotic, untransformed cells reflect changes in plasma membrane properties, then endogenous transferase activities of normal cells could also be expected to show similar changes.

Webb and Roth (1974) measured the endogenous galactosyltransferase activities of 3T3 and malignant 3T12 cells while these cells were in mitosis. Transformed (3T12) cells showed no difference in activity when assayed in stationary and spinning incubations, regardless of cell cycle stage. Randomly growing, nontransformed 3T3 cells require cell contact for maximum endogenous activity. While they are in mitosis, however, this requirement is lost. The level of endogenous activity of mitotic 3T3 cells is similar to that of 3T12 cells throughout their cell cycle. Autoradiography of cells in culture with UDP-[^{3}H]galactose confirmed these results. Control experiments showed that these results could not be explained by differential cell viability, breakdown of substrate, or release of enzyme into supernatants.

Bosmann (1974) measured the endogenous and exogenous activities of intact lymphoma suspensions as a function of their cell cycle. In pellet incubations, activities of galactosyl-, sialyl-, and N-acetylglucosaminyl transferases showed highest activity in the S phase and very low activity in mitosis. Many conditions for the assay of these enzymes were not shown to be optimal, and control experiments to rule out breakdown of substrates were not reported. Nevertheless, it is possible that transferase activities vary during the cell cycle and that, moreover, they vary differently in different cell types.

E. MODEL FOR THE INVOLVEMENT OF SURFACE GLYCOSYLTRANSFERASES IN GROWTH CONTROL

Since the surface, endogenous transferase activities of some nontransformed cells, notably 3T3, seem to vary as a function of cell contact, whereas the activities of some transformed cells, such as 3T12 and SV3T3 do not, Cebula and Roth (1976) have suggested that the ability of a cell to glycosylate adjacent cell surfaces is fundamentally related to a cell's ability to regulate its growth. That is, trans-glycosylation may play a direct role in the signaling of cell contact. This hypothesis (Roth *et al.*, 1977) attempts to explain the altered growth properties of two transformed cell types and how these alterations could relate to the social behaviors of cells observed *in vivo.*

The model assumes that at least two membrane properties are necessary for growth control: one property sends a signal to an opposing cell when contact has been made (trans-glycosylation) and another receives the signal from an adjacent cell and transduces it so that the growth rate of the recipient cell is then reduced. Nontransformed 3T3 cells, which show contact inhibition of growth, can therefore both send and receive contact signals. Transformed, malignant 3T12 cells do not show regulation of their growth rates when they contact one another in culture, but do show contact inhibition of growth when placed on a monolayer of 3T3 cells, although 3T3 cells are not growth-inhibited by 3T12 cells. A possible explanation of this behavior would be that 3T12 cells cannot send contact signals, but can receive them. Transformed SV3T3 cells, which are also malignant, show no contact inhibition of growth, nor are they growth-inhibited on 3T3 monolayers. These viral transformants may, therefore, be unable to send or to receive contact signals.

Cis-glycosylation by 3T3 cells may be restricted by low, membrane mobility of enzymes and acceptors on their surfaces. Surface glycosylation on 3T3 cells would then occur largely by trans-glycosylation. Cells would respond to this signal and, accordingly, slow their growth rates. An increase in membrane fluidity would allow cis-glycosylation to occur much more often, causing the altered cells to slow their rate of growth. *In vivo,* cells of this type might have the properties of a benign, slowly growing tumor because of constant cis-glycosylation.

Any subsequent change or mutation that allowed these altered cells to escape from signal reception would be highly advantageous and would lead to drastic increases in mitotic rates. One change that could increase their rate of growth would be a defect in their ability to carry out the glycosylation reaction, which could occur by a number of means. Cells could develop a defect in their ability to supply sugar donors for the glycosylation reaction. Cells with this phenotype could not undergo cis-glycosylation and, as a result, the constant self-inhibition would cease. Growth of these cells would not be inhibited by contact with similar cells but, since they can respond to contact signals, they would be sensitive to inhibition by cells that can send signals. Transformed, 3T12 cells seem to fall into just such a category.

If, alternatively, a cell with an increased level of cis-glycosylation should lose its ability to respond to contact signals, then this cell would be insensitive to contact inhibition by any cell type. SV3T3 cells may fall into this category, since they will grow on monolayers of untransformed 3T3 cells.

Needless to say, a great deal of experimentation must be done before even a few of the aspects of this model can be validated. However, it does provide a specific, biochemical framework that allows direct experimental testing.

V. Cell Surface Glycosyltransferases and Recognition

Recognition events occurring during morphogenesis appear to involve the selective adherence of cells to one another. This selective adherence has been attributed to a variety of mechanisms including physicochemical forces such as calcium bridging, hydrogen bonding, and long-range, attractive Van der Waals forces (Grinnell, 1978), as well as to the specific binding between cell surface molecules (Grinnell, 1978; Marchase *et al.*, 1976). In 1970, Roseman proposed that the binding between enzymes located on one cell surface and substrates located on an adjacent cell surface may represent one mechanism by which cells initially recognize and selectively adhere to one another. The first evidence that cell surface glycosyltransferases might be involved in adhesion came from *in vitro* studies of neural retina cells (Roth *et al.*, 1971a). Since that time, cell surface glycosyltransferase activity has been implicated in the adhesion of many cell types as well as in a variety of other recognition events including cell migration, gametic interaction, bacterial adhesion, cell differentiation, and immune recognition.

A. CELL ADHESION

Evidence for the role of glycosyltransferases in cell adhesion rests primarily upon studies suggesting the involvement of carbohydrates and proteins in cell adhesion (Marchase *et al.*, 1976), the demonstration of glycosyltransferases on intact cell surfaces, and the ability of specific sugar acceptors to influence cell adhesion (Shur and Roth, 1975). Since these enzymes must recognize and bind carbohydrates, several studies indicate that surface glycosyltransferases and their endogenous acceptors may be involved in the adhesion of neural retina cells (Roth *et al.*, 1971a), blood platelets (Bosmann, 1971; Jamieson *et al.*, 1971), and cultured cell lines (Roth and White, 1972; Bosman, 1972b). As these studies have been recently reviewed by Shur and Roth (1975), they will be mentioned only briefly here. Several studies (Turner and Burger, 1973; Yen and Ballou, 1974; Wiese and Wiese, 1975; Barondes and Rosen, 1975) have suggested that carbohydrates were involved in cellular adhesion in lower organisms. The studies of Moscona and co-workers (1975) and Oppenheimer *et al.* (1969) also indicated that neural retinal cell and mouse teratoma cell adhesion was dependent on the presence of carbohydrate. Later studies showed that the treatment of neural retina cells with purified β-galactosidase affected their rate of adhesion to each other (Roth *et al.*, 1971b) and to tectal halves (Marchase,

1977). Several cell lines (Chipowsky *et al.*, 1973) and cultured rat hepatocytes (Weigel *et al.*, 1978) attàched only to specific sugars immobilized onto polyacrylamide gels. Finally, the demonstration of cell surface-localized glycosyltransferases on neural retina cells, as well as the ability of exogenously added carbohydrate acceptors to perturb adhesion, provided evidence consistent with the possibility that early adhesive events might be mediated, in part, by transferase–acceptor complexes.

Lloyd and Cook (1974) observed that neuraminidase treatment of rat dermal fibroblasts stimulated their adhesion possibly by creating cell surface components that could act as acceptors for adjacent cell surface sialyltransferases. To test this possibility, these authors produced evidence for the localization of sialyltransferases on the cell surface by showing (1) the ability of cells to add sialic acid from CMP-$[^{14}C]$NeuNAc to large molecular weight, extracellular sugar acceptors, (2) electron microscopic evidence for the surface localization of transfer product from CMP-$[^{14}C]$NeuNAc, (3) a difference in the ability of intact cells to transfer sialic acid to exogenous acceptors compared to microsomal fragments, (4) the inability of free sialic acid to reduce incorporation of CMP-$[^{14}C]$NeuNAc, and (5) lack of hydrolysis of CMP-$[^{14}C]$NeuNAc to free sialic acid. Furthermore exogenous glycoprotein substrates with terminal β-GalNAc residues reversed the effects of neuraminidase.

Selective adhesion during embryogenesis has often been postulated as one mechanism for cellular recognition. Roth *et al.* (1971a) proposed that neural retinal cells may recognize and adhere to one another via an interaction between cell surface galactosyltransferases and their substrates on adjacent cells. Recently, Marchase (1977) proposed that an interaction between a UDP-Gal:ganglioside GM_2 galactosyltransferase and the ganglioside GM_2 might mediate recognition of retinal axons for the appropriate region of the optic tectum. During chick development, the axons from the dorsal retina innervate the ventral region of the optic tectum, whereas those from the ventral region of the retina innervate the dorsal tectum. A similar inversion apparently occurs in the anterior–posterior axis. Barbera *et al.* (1973) developed a collection assay in which single cells from either dorsal or ventral retinal halves adhered preferentially to the ventral and dorsal halves of the optic tecta, respectively. Marchase (1977) presented evidence that a trypsin-sensitive molecule located on the ventral retina and dorsal tectum was required for adhesion to dorsal regions of tectal halves, whereas terminal β-N-acetylgalactosamine residues on the dorsal retina and ventral tectum were necessary for adhesive recognition to the ventral optic tectum. Lecithin vesicles containing GM_2 bound preferentially to the ventral tecta. Also, a 30% increase in UDP-Gal:GM_2 transferase activity was demonstrated in the ventral half of the neural retina, although no difference was found in a number of other galactosyltransferase activities. An increase in the activity of this enzyme was shown to occur concurrently with the developmental age at which ventral retina first exhibited preferential adhesion for dorsal tecta. Marchase proposed that a double gradient consisting of a protein, possibly the glycosyl-transferase, and a galactosaminyl group occurring on a molecule such as the GM_2 ganglioside exists on both the retinal and tectal surfaces and could account for the observed adhesive specificities. The molecules containing galactosaminyl residues concentrated dorsally were proposed to interact in a lock-and-key manner with the protein molecules concentrated ventrally. However, no evidence of the surface localization of the glycosyltransferase was presented in this study and a gradient of GM_2 molecules could not be demonstrated using sonicates of retinal cells.

Porzig (1978) recently measured the galactosyltransferse activity on intact neural retina cells using N-acetylglucosamine as an exogenous acceptor. She demonstrated activity that appeared to be cell surface-localized as assessed by control of nucleotide sugar hydrolysis. Cells were viable by dye exclusion as well as transmission and scanning electron microscopy. Using saturating concentrations of UDP-galactose, she was able to quantitate and compare enzyme activity in different populations of retinal cells. Quantitative differences were not detected in cells from nasal, temporal, dorsal, or ventral quadrants or retinal halves, although enzyme activity was 2-fold higher in the mitotically active margin than in the postmitotic fundus. Transferase activity was noted to decline with developmental age. Porzig concluded, as did Marchase (1977), that this galactosyltransferase was probably not involved as a recognition molecule mediating anterior–posterior, ventral–dorsal recognition events.

B. CELL MIGRATION

Increasing evidence indicates that carbohydrates, particularly glycosamino-glycans, are efficient substrates during morphogenesis and gastrulation (Toole, 1976), although no single mechanism of glycosaminoglycan-mediated movement has been characterized. Additionally, several studies implicate the involvement of cell surface glycosyltransferases. Shur (1977a,b) demonstrated autoradiographically that migrating cells in the gastrulating chick embryo exhibited intense, transferase activity associated with the cell surface. Migrating primitive streak cells actively incorporated galactose from exogenously added UDP-Gal, fucose from GDP-Fuc, and NeuNAc from CMP-NeuNAc, while neural crest cells actively incorporated Gal, GlcNAc, and Fuc from the appropriate nucleotide derivatives. Glucosyl-, N-acetylgalactosaminyl-, and glucuronyl-transferase activities were relatively inactive under the assay conditions used. The enzyme–substrate complexes assayed during these sugar nucleotide incubations appeared to occur at the cell surface for the following reasons: (a) available levels of sugar phosphate and free sugar could not account for the autoradiographic patterns, (2) no evidence exists for the uptake of sugar nucleotides into these cells, and (3) neither supernatant nor homogenate glycosyltransferase activities could account for the autoradiographs. Shur suggested from these data and from recent evidence implicating glycosaminoglycans in cell movement (Toole, 1976) that migrating cells utilize cell surface glycosyltransferases as receptors for exposed oligosaccharides to promote adhesion between the cell surface and matrix, thus mediating movement. As cells move forward, surface–enzyme matrix interactions are broken via catalysis or by leaving bits of the cell membrane behind (Fig. 5). Additionally, the cell surface transferases may influence migration indirectly by altering adhesive interactions via their modification of the cell surface or substratum. All of these models predicted that the perturbation of extracellular glycosyltransferase activity should alter cell movement. The addition of 2.5–5.0 mM UDP-Gal and UDP-GlcNAc to gastrulating chick embryos caused abnormal development, whereas free Gal, free GlcNAc, and UDP-GA had no comparable effect. Also, UDP-Gal and UDP-GlcNAc had no effect when administered after gastrulation was completed. Although the direct cause of the abnormalities was not investigated further, the simplest explanation is that the sugar nucleotides interfered with cell movement and induction interactions by altering surface glycosyltransferase activity.

Recent studies (Karfunkel *et al.*, 1977) are consistent with this possibility. These authors have shown that the addition of slightly lower levels (0.1 mM) of sugar nucleotides influence both embryonic cell migration and axon elongation on cell monolayers *in vitro*. Both Shur's and Karfunkel's studies favor the hypothesis that glycosyltransferase activity on the cell surface can influence cell migration and axon elongation, but assume that cells can spontaneously glycosylate substrates as they move across them and that such activity is relevant to migration.

Yogeeswaran *et al.* (1974) demonstrated that BHK cells, when preloaded with [^{3}H]Gal, glycosylated glycolipids fixed to solid surfaces. This result indicated that cells attaching to high-molecular-weight acceptors can spontaneously glycosylate them. The relationship between spontaneous glycosylation and cell

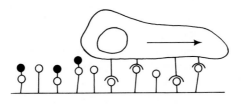

Carbohydrate Matrix

FIG. 5. Model of cell migration on a polysaccharide substratum. Cell migration is mediated by an adhesive interaction between cell surface glycosyltransferases and terminal sugars on the substratum. Catalysis and transfer of sugar leads to loss of adhesion (Shur, 1977a).

migration has also been investigated by Turley and Roth (1979). Virally transformed (SV40 3T3) cells were prelabeled with [^{3}H]Glc, [^{3}H]Gal, and [^{35}S]sulfate and then allowed to migrate, in serum-free medium, upon surfaces derivatized with glycosaminoglycans. Cells spontaneously glycosylated these glycosaminoglycans as they migrated. Labeled Gal-sulfate and GA were added to surfaces derivatized with chondroitin-6-sulfate, while GlcNAc and GA were added to hyaluronic acid. No identifiable radioactivity was detectable when cells were maintained on plates derivatized with polygalacturonic acid or when completely lysed cells were incubated on dishes derivatized with the glycosaminoglycans. Cell migration was most extensive on hyaluronic acid, yet chondroitin sulfate was the glycosaminoglycan that was most extensively glycosylated. Treatment of either glycosaminoglycan with testicular hyaluronidase reduced cell migration and increased the extent of glycosylation. Both results suggest that migration varies inversely with glycosylation and that the extent of migration may depend on the number of terminal sugars available for interaction with surface glycosyltransferases. Exposure to a large number of terminal sugars could increase adhesion and slow migration rates by increasing the number of sugar-enzyme complexes. Although SV40 3T3 cells appear to contain cell surface glycosyltransferases specific for chondroitin sulfate and hyaluronic acid, the experiments described above do not, of course, indicate whether the activity is incidental or instructive to migration.

C. GAMETE RECOGNITION

It has long been thought that complementary interaction between specific receptors on the egg and sperm cell surfaces may be responsible for specific attachment of gametes to one another. For instance, as early as 1910, Lillie suggested that the egg produced a ligand, fertilizin, that aided binding of the sperm to the egg.

Several studies have reported transferase activity at the surface of gametes, which raises the interesting possibility that an interaction between these enzymes and their substrates could be one of the mechanisms whereby gametes recognize and adhere to one another. Gametes of algae such as *Chlamydomonas moewussi* adhere to one another by flagellae. After a cytoplasmic bridge is formed between the two bodies of the gametes, the adhering flagellae are released from each other and are no longer agglutinable with other gametes. The mechanism of flagellar adhesion, which constitutes the species-specific gametic recognition event, is unknown, although several studies using proteases and glycosidases suggest an involvement of both proteins and carbohydrates. McLean and Bosmann (1975a) observed a greater, cell surface, glycosyltransferase activity toward endogenous acceptors on both intact gametic flagellae and flagellar vesicles than on the vegetative cells. Mixing of the (−) and (+) gametic flagella from *C. moewussi* stimulated glycosyltransferase activity. In a later study (Bosmann and McLean, 1975), the authors examined the glycosyltransferase activity of sexually incompetent (+) and (−) gametes of C. moewussi and also the (+) and (−) gametes of the sexually incompatible species, C. reinhardtii and C. moewussi. Unlike the result with the sexually competent, compatible gametes, mixing of sexually incompatible (−) gametes of C. moewussi with (+) gametes of C. reinhardtii or of the sexually incompetent vegetative gametes did not result in an enhancement of glycosyltransferase activity toward endogenous acceptors. In this study, sugar nucleotide breakdown was not assessed. However, it is unlikely that the uptake of free sugar could account for the difference in glycosyltransferase activity between vegetative and gametic cells of both C. moewussi and C. reinhardtii. Recently, Columbrino et al. (1978) examined whether the observed cell surface sialyltransferase activity of C. moewussi was in any way due to CMP-NeuNAc hydrolysis and uptake of NeuNAc. Free NeuNAc was taken up by the cells of both mating types, but mixing of the two types had no effect on this uptake. Addition of 100-fold molar excesses of free NeuNAc did not reduce the levels of incorporation observed when the two mating types were mixed and incubated with labeled CMP-NeuNAc. Furthermore, trypsin treatment of cells previously labeled with CMP-NeuNAc released nearly all of the radioactivity. These results suggested that sialyltransferase activity was present on the cell surface. The authors propose that an initial interaction between the transferases and their substrates represents one of the mechanisms of adhesive recognition, while catalysis may allow the release of the flagellae from each other. However, the possibility that enhanced glycosyltransferase activity is incidental to the mating reaction and merely represents increased membrane synthesis required for the elaboration and secretion of a hypothetical adhesive molecule, cannot be excluded by these studies.

Durr et al. (1976) demonstrated CMP-NeuNAc:asialofetuin sialyltransferase activity on sperm, although no endogenous sialyltransferase activity was detected. Autoradiographic analysis of intact mouse ova incubated with CMP-[³H]NeuNAc, on the other hand, indicated that endogenous acceptors for the enzyme existed on the ovum surface, although the ova showed no activity toward the asialofetuin. When sperm and ova as well as CMP-[³H]NeuNAc were incubated together, additional, dense grain patches were found on the egg surface, often associated with the penetrating sperm. Ova incubated identically, but in the absence of sperm, did not show these patches. Since the 500 g supernatant from the sperm preparation contained significant levels of sialyltransferase activity toward asialofetuin it was not clear whether activity associated with the sperm represents secondarily adsorbed enzyme or whether the enzyme was merely released by the sperm into seminal fluid. In these studies, hydrolysis of CMP-[³H]NeuNAc did occur, but the extent of hydrolysis did not exceed 17% of input sugar nucleotide. This level of free [³H]NeuNAc could not account for the observed activities since, when [³H]NeuNAc was substituted for CMP-[³H]NeuNAc at identical concentrations and specific activity, there was no sialic acid transfer to asialofetuin by sperm. The authors suggested that CMP-NeuNAc:asialofetuin sialyltransferase associated with intact sperm transfers sialic acid to the egg membranes and may represent one of the mechanisms by which sperm attach to eggs.

D. INTESTINAL CELL DIFFERENTIATION

Cells within the intestinal crypt are mitotically active and differ ultrastructurally and biochemically from the differentiated cells found toward the villus tip. Villus cells incorporate carbohydrate precursors into cell surface membrane glycoproteins, whereas crypt cells do not (Weiser, 1973a). Weiser (1973a,b) selectively isolated crypt and villus cells by sequential dissociation with EDTA. Intact crypt cells, as well as the isolated microvilli, showed high glycosyltransferase activity toward endogenous acceptors, but low sialyltransferase activity. Villus cells, as well as their isolated brush borders, showed low glycosyltransferase activity and high sialytransferase activity. Use of exogenous acceptors suggested that crypt cells were enriched in both glycosyltransferases and exogenous acceptors. Recently, Weiser *et al.*, (1978) purified plasma and Golgi membranes from both crypt and villus cells. Membrane preparations were isolated from both the lateral–basal and microvillus cell borders. These were enriched for Na^+,K^+-ATPase activity and, in the case of the microvillus membrane, glycosyltransferase activity. The Golgi membrane fraction was also enriched for glycosyltransferase, but was low in Na^+,K^+-ATPase and, unlike the plasma membrane fractions, was not labeled when purified from intact cells externally labeled with ^{125}I. Antibody prepared against the lateral–based membrane fractions reacted with the cell surface, but not the Golgi fractions. Sialyltransferase activity was lower and galactosyltransferase activity was higher in Golgi fractions compared to plasma membrane fractions from villus portions. Both the crypt and villus plasma membranes had high glucosyltransferase activities. The authors suggested that the inability to demonstrate glycosyltransferase activity in earlier studies of villus cells probably resulted from high levels of substrate degraded in microvillus incubations.

E. BACTERIAL ADHESION

The ability of bacterial strains or species to adhere specifically to mucosal surfaces appears to be a mechanism for their distribution in the mouth and gastrointestinal tract. Bacterial adhesive interactions, particularly those of *Streptococcus mutans,* are also important for plaque formation on teeth. These bacteria adhere to teeth via a pellicle composed of bacterially synthesized polymers and salivary constituents (Gibbons and Van Houte, 1973). It has recently been shown that the adhesion of *S. mutans* to such surfaces also requires the presence of dextran sucrase and levan sucrase adsorbed externally to the bacterial wall (Mukasa and Slade, 1973a,b,c). Mukasa and Slade (1973c) showed that the sucrase enzymes in a crude, cell-free enzyme preparation from the culture supernatant of *S. mutans* (Strain HS6, group a) were able to bind to the surface of heat-killed cells and subsequently synthesize both a water-insoluble and water-soluble dextran and levan from sucrose. These conditions resulted in the adherence of heat-killed *S. mutans* cells to a smooth glass surface. Control cultures of heat-killed cells exposed to heat-treated extract did not adhere to this surface.

The enzymes responsible for adhesion were partly purified by chromatography on agarose and hydroxylapatite gels and shown to have a molecular weight of 400,000 to 2,000,000. The purified complex contained equivalent amounts of dextran and levan sucrases, as well as 5–30% polysaccharide and negligible amounts of contaminating protein, as judged by polyacrylamide gel electrophoresis and gel diffusion. During purification, the ability of all fractions to produce adherence paralleled the enzyme activity responsible for the synthesis of the water-insoluble polysaccharide from sucrose.

Adsorption of the sucrases to the cell wall depended on the presence of dextran on the cell surface. Soluble dextran, as well as dextrase treatment of the bacterial surface, inhibited enzyme binding. Enzyme adsorption was also inhibited by antibody to α-d antigen polysaccharide on the bacterial wall and by antidextran globulin. Although the mechanism(s) by which this enzyme adsorbed to the cell surface and subsequently promoted adhesion to glass was not discussed, it is possible that enzyme–acceptor interactions between both the surface-associated dextran acceptor and the water-insoluble polymer could mediate one mode of adhesion to glass.

F. Immune Recognition

The presence of some glycosyltransferase activities on lymphocyte surfaces suggests that these enzymes could be involved in the recognition processes that characterize the immune system. Cell surface-associated galactosyl- and sialyltransferase activities have been shown to occur on the surfaces of rat lymphocytes as demonstrated by the ability of these cells to transfer galactose from UDP-Gal to ovomucoid derivatized to Sephadex beads or to endogenous acceptors on the cell surface (Cacan *et al.*, 1976; Verbert *et al.*, 1976), or to transfer NeuNAc from CMP-NeuNAc to intact cells and to fragments of asialoglycophorin (Verbert *et al.*, 1977). Evidence for the presence of cell surface galactosyltransferases included glycosylation of a nonphagocytozable acceptor (ovomucoid-Sephadex), increased agglutinability of cells by soybean agglutinin after incubation with UDP-Gal and glycosylation of endogenous, cell surface acceptors, and no transferase activity present in incubation supernatants. In these studies, uridine-5'-monophosphate was used to inhibit sugar nucleotide hydrolysis. The inability of a 100-fold molar excess of free NeuNAc to alter incorporation from CMP-NeuNAc onto endogenous acceptors, as well as the inability of broken cells to account for total amount of transfer to fragments of exogenous, asialo-glycophorin, suggested surface localization for these sialyltransferases (Verbert *et al.*, 1977). Intact cells did not add NeuNAc to asialo-orosomucoid or asialoglycophorin, whereas homogenates of cells did, implying that phagocytosis and release of sialyated products did not occur. These authors suggested that the cell surface sialyltransferase(s) is buried in the plasma membrane, since it did not utilize macromolecular acceptors, but was able to glycosylate glycopeptide fragments of asialoglycophorin.

LaMont *et al.* (1974) concluded in earlier studies that endogenous, cell surface galactosyltransferase activity of spleen lymphocytes greatly increased when these cells were stimulated with Concanavalin A to undergo blast transformation. However, Patt *et al.* (1976) showed that when 3 mM 5'-AMP was included in incubations to inhibit UDP-Gal breakdown, no glycosylation of endogenous, cell surface acceptors by intact, Concanavalin A-stimulated lymphocytes could be observed. Unfortunately, Patt *et al.* (1976) do not show results of incubations of native lymphocytes with UDP-Gal and 5'-AMP, so comparisons between their results and those of Verbert *et al.* (1976) cannot be made.

Parish (1977) has suggested that immune mechanisms that enable invertebrates to eliminate infectious agents may be mediated by complexes of glycosyltransferases attached to hemocyte cell surfaces. Recognition factors have been isolated from a variety of invertebrates that have opsonic properties as well as agglutinating and bactericidal activity. These factors exist as large molecules that appear to recognize carbohydrate structures and can be dissociated into smaller subunits. Parish proposed a model in which the glycosyltransferases used by the invertebrate to synthesize its own carbohydrate side chains could act as a mechanism for recognizing nonself. This model allows the maintenance of strict self:nonself discrimination, since glycosyltransferases have a very low affinity for their products. Thus, an oligosaccharide side chain would not be recognized and bound by the glycosyltransferases that made it. Such transferases would, however, bind to a comparatively wide range of "foreign" antigens or to cell surfaces of the invertebrate host that have been damaged due to the release of hydrolytic enzymes.

VI. Cell Surface Glycosyltransferases and Physiological Regulation

A. Hemostasis

Wounding of vascular endothelium exposes matrix components such as collagen fibrils that serve as adhesion sites for circulating platelets, thus initiating hemostasis. Certain platelet constituents such as ADP are then released causing the aggregation of additional platelets onto those already adhering to collagen (Jamieson, 1974).

Several studies suggest that carbohydrate moieties on collagen are necessary for platelet adhesion and aggregation. Brass and Bensusan (1976) treated polymeric, calf skin collagen with periodate using conditions that minimized the destruction of periodate-sensitive bonds other than those in the carbohydrate residues. Collagen treated in this manner showed reduced ability to interact with human platelets. Kang et al. (1973) isolated a cyanogen bromide fragment from the α_1 chain of chick skin collagen, which promoted platelet aggregation. Inhibition and periodate oxidation studies suggested an involvement of the fragment's carbohydrate moieties in promoting human platelet aggregation. However, fragments from other animal sources did not promote aggregation, suggesting that structural factors in addition to carbohydrate may also be necessary to stimulate platelet aggregation. Kang et al. (1973) were unable to demonstrate an effect of periodate oxidation of polymeric collagen on platelet aggregation. The discrepancy between these results and those of Brass and Bensusan could be due to differences in methods of assaying platelet adhesions and of periodate oxidation.

Two laboratories have reported the existence of plasma membrane glycosyltransferases on the surface of blood platelets. Collagen:glucosyltransferase has been implicated in the adhesion of these cells to endothelial collagen (Bosmann, 1971; Jamieson et al., 1977c; Jamieson, 1974; Barber and Jamieson, 1971). Surface sialyltransferases have been suggested for a role in the subsequent aggregation of platelets to one another (Bosmann, 1972a). These papers have been discussed in detail elsewhere (Shur and Roth, 1975).

Smith et al. (1977) have studied glucosyltransferase activities from human platelets. Purified (plasma) membranes contained 5–10% of glucosyltransferase activity using galactosylhydroxylysine as acceptor, whereas the remainder of this activity was soluble. Using 80-fold purified, soluble glucosyltransferase, the K_m for galactosylhydroxylysine was 4 mM, but when this acceptor was coupled to the α_1 chain of collagen, the K_m was lowered by three orders of magnitude to 2 μM. The authors suggested that the enzyme must interact with the protein portion of the collagen chain as well as the carbohydrate moieties.

B. Glycoprotein and Drug Clearance

One common function of some liver and kidney cells is to selectively recognize and remove various substances from the circulatory systems. Demonstration of cell surface glycosyltransferase activities on these cells suggests that these enzymes could possibly function in the recognition of circulating substances that are to be internalized by the cells and ultimately cleared.

Merritt et al. (1977) assayed well characterized, plasma membrane preparations from rat liver for various, endogenous glycosyltransferase activities.

After incubation with labeled sugar nucleotide, chloroform–methanol (2:1 v/v) was added, and after various extractions, incorporation of radioactivity was determined in the precipitated fraction (protein residue) and supernatant (lipid phase). No transfer of NeuNAc from CMP-NeuNAc was observed to either fraction. Significant incorporation into protein residue was seen with UDP-Glc and UDP-GlcNAc, however. Transfer of label from UDP-Gal and GDP-Man to both protein residue and lipid phase was observed. The authors state that, based on comparisons of activities of thiamine pyrophosphatase (often a reliable Golgi enzyme marker), the maximum amount of Golgi apparatus contamination of the plasma membrane fractions was 10%. Specific activities of endogenous galactosyltransferases in Golgi fractions was 7.5 pmole/min/mg protein, whereas specific activities in plasma membrane fractions were 8.8 pmole/min/mg protein for the protein residue and 16.4 pmole/min/mg protein for the lipid phase. This result strongly suggests that for at least the galactosyltransferase activities, contamination of the plasma membrane fractions by Golgi cannot explain the observed levels of incorporation.

Arnold *et al.* (1976a) demonstrated the presence of glycosyltransferases incorporating mannose, fucose, galactose, *N*-acetylglucosamine, and *N*-acetylneuraminic acid onto intact cells from dissociated chick embryo liver. Maximum activities were seen between the eighth and twelfth days of development. Glycopeptides trypsinized from the cell surface were purified by thin-layer chromatography and shown to compete with some endogenous activities by acting as exogenous acceptors.

Pricer and Ashwell (1971) have shown that the rate at which many serum glycoproteins are cleared from the blood by the liver is strongly dependent on the number of terminal galactose residues. If a small number of terminal sialic acid residues are removed, exposing terminal galactosides, the glycoproteins are rapidly removed from the blood. The demonstration of a sialyltransferase on liver plasma membranes (Pricer and Ashwell, 1971) raised the possibility that these asialoglycoproteins are initially bound to the liver cell membrane by interacting with the surface-associated glycosyltransferases. Others (Aronson *et al.*, 1973) have suggested that a galactosyltransferase on the liver cell membranes may be responsible for binding the asialoglycoproteins, since agalactofetuin inhibited the binding of asialofetuin to liver plasma membrane and since α-lactalbumin inhibited the binding of asialofetuin to plasma membranes. This protein is known to modify the acceptor specificity of some galactosyltransferases from *N*-acetylglucosamine to glucose. However, Hudgin and Ashwell (1974) purified the asialoglycoprotein-binding protein 200-fold, but could demonstrate no sialyl- or galactosyltransferase activity in this preparation. At present, no data links surface glycosyltransferases to the binding of asialoglycoproteins to liver plasma membranes.

Kidney tubules specifically adsorb soluble, low-molecular-weight glycoproteins that can pass through the glomerulus. The possibility that glycosyltransferases may be involved in this specific protein resorption, as well as in basement membrane synthesis, is being investigated, since both intact glomeruli and kidney plasma membrane fractions have considerable transferase activity (Kirschbaum and Bosmann, 1973b). Briefly, Kirschbaum and Bosmann (1973b) demonstrated the presence of a glucosyl:collagen-, galactosyl:agalactofetuin-, and sialyl:asialofetuin-transferase activity in kidney plasma membrane fractions and on intact cells. Although controls were not included in these studies to show that the sugar nucleotides were the authentic donors or that enzyme activity was surface-localized on intact glomeruli, the enrichment of sialyltransferase activity on isolated plasma membranes does suggest a cell surface locale for this particular enzyme. However, the role, if any, of these enzymes in renal physiology remains to be determined.

VII. Cell Surface Glycosyltransferases in Lower Organisms

A. CHITIN SYNTHETASES IN YEAST

Chitin is an analog of cellulose and is composed of N-acetylglucosamine residues assembled by N-acetylglucosaminyl transferases or chitin synthetases. These enzymes occur on purified yeast plasma membranes and are involved in the formation of primary septa of budding yeast (Duram *et al.*, 1979). These enzymes exist in particulate or zymogen form on the plasma membrane and are activated by unknown mechanisms. Chitin apparently remains associated with isolated zymogen particles that catalyze its formation. Therefore, Duram *et al.* (1970) were able to examine the distribution of chitin synthetases by localizing the product chitin with autoradiographic and fluorescent techniques. Using this approach, the authors demonstrated that the synthetases were uniformly distributed on the plasma membrane and were not concentrated at the site of bud formation, suggesting that the initiation of chitin synthetase activity must occur by the localized activation of a zymogen rather than the delivery of the enzyme to specific sites.

B. CELLULOSE SYNTHETASES

Cellulose is a polysaccharide composed of $\beta(1-4)$ linked glucose molecules and occurs extracellularly in plants and in some bacteria, as well as in many invertebrates. Growing myofibrils of cellulose lie parallel to the plasma membrane and seem to be synthesized bidirectionally by complexes of glycosyltransferases. The plasma membrane is the major site for the synthesis of cellulose microfibrils (Brown and Wilson, 1976), although in some plants, notably

Pleurochymis schiffelii, Golgi membranes also have the capacity to synthesize microfibrils and may contribute to their formation. Cellulose synthesis in the Gram negative bacterium *Acetobacter xylinum* also occurs extracellularly with the glucosyltransferases apparently embedded in the outer membrane, since the isolated outer envelope is required for cellulose synthesis.

C. GLYCOSYLTRANSFERASES IN *Dictyostelium discoideum*

The slime mold, *Dictyostelium discoideum,* undergoes a transition from a unicellular or vegetative phase to a multicellular or aggregated phase when cells are transferred from nutrient medium to a nonnutrient buffer. The competence to aggregate may involve changes in the cell surface, particularly in carbohydrate moieties. Sievers *et al.* (1978) examined isolated plasma membranes of *Dictyostelium* for transferase activity. Crude membranes were separated on a sucrose gradient, and the plasma membrane fraction was identified by marker enzymes including alkaline phosphatase, cAMP phosphodiesterase, the absence of enzymes characteristic of intracellular membranes, and by labeling with radioactive 1-fluoro-2,4-dinitrobenzene, as well as cAMP binding capacity. Mannosyl-, glucosyl-, and *N*-acetylglucosaminyltransferase activities were present in the plasma membrane fraction of both amebae and aggregating cells. However, the specific activities of mannosyl- and *N*-acetylglucosaminyltransferases were 1.4- and 5-fold higher, respectively, in aggregating cells than in vegetative cells. These authors postulate that the membrane-associated glycosyltransferases participate in the biosynthesis of membrane glycoproteins and/or glycolipids, which are then involved directly in the aggregation process.

D. GLYCOSYLTRANSFERASES IN THE SPONGE, *Geodia cydonium*

Muller and colleagues have studied aggregation of dissociated cells from *Geodia cydonium* and described an aggregation factor present in the intercellular space that binds to receptors on the cell surfaces of the sponge cells, causing them to aggregate (Muller and Zahn, 1973). Three glycosyltransferase activities are associated with the purified aggregation factor, but can be separated from it: sialyltransferase (Muller *et al.,* 1977), galactosyltransferase, and glucuronyltransferase (Muller *et al.,* 1978).

In a recent paper, Muller *et al.* (1979) described a molecular mechanism for the aggregation of sponge cells based on the interactions of glucuronic acid residues present on the cell surface aggregation receptor with the glucuronyltransferase activity of the aggregation factor and a glucuronidase also shown to be present on intact cells. Sponge cells lost their ability to aggregate when incubated under conditions optimal for the cell surface glucuronidase. No such loss of activity was observed if the cells were incubated after cell surface glucuronidase had been inactivated or detergent-extracted. Incubation of deglucuronylated cells with UDP-GA and glucuronyltransferase purified from the aggregation factor resulted in an increase in agglutination activity. However, the assay for cell aggregation was only the measurement of aggregate size after many hours of incubation. The authors suggested that cell agglutination and separation may be mediated by glucuronylation of aggregation receptors by a glucuronyltransferase at the cell surface. Unfortunately, the authors never attempted to show direct transfer of labeled glucuronic acid from UDP-GA to the cell surface aggregation receptor by incubating purified aggregation factor with its glucuronyltransferase activity with intact, aggregating sponge cells and labeled UDP-GA. Inhibition or activation of cell surface glucuronidase should influence the amount of label present on the receptor. In addition, the availability of endogenous sugar donors for the cell surface transferase reaction was not studied or discussed.

VIII. Genetic Control of Glycosyltransferase Activity

Although cell surface-associated glycosyltransferases have been implicated in a variety of recognition phenomena, the genetic control of their activity has not yet been investigated. However, mutant strains of *Escherichia coli, Melandrium,* and mammalian cell lines that are deficient in intracellular glycosyltransferase activity have been studied and may provide some information on the control of surface glycosyltransferases.

Abramsky and Tatum (1976) studied the control of the branching enzyme, 1,4-glucon:1,4-glucon-6-glycosyltransferase, which catalyzes the synthesis of a highly branched polysaccharide believed to be important in determining characteristics of the bacterial cell wall and culture morphology. When cultured on potato starch, the R2508 strain of *E. coli* was deficient in this activity. However, glycosyltransferase activity of both the wild-type and mutant strain was identical when sucrose was used as the carbon source. Although cell extracts of both the wild-type and mutant showed decreased enzyme activity in the presence of potato starch, the wild-type activity was reduced by 50-70%, whereas that of the mutant was completely inhibited. Addition of the mutant inhibitor to wild-type extracts inhibited transferase activity completely. The inhibitory activity in both mutant and wild-type extracts could be removed by fractionating the extracts on a DEAE column. The authors therefore suggested that glycosyltransferase activity is regulated by an inhibitor whose production is influenced by the nature of the carbon source. In the case of the mutant, when the carbon source is a highly branched potato starch, the inhibitor is produced in either greater amounts or in a more potent form than in the wild-type.

The glycosylation of the flavone, isovitose, in *Melandrium* has been genetically analyzed using mutant strains. These studies provide evidence for another mechanism of regulating glycosyltransferase activity in eukaryotes.

The flavone, isovitexin, occurs naturally in petals of *Melandrium* flowers and can be glycosylated at carbon position 6 or 7. An arabinosyltransferase adds arabinose to the C 6 position of isovitexin, while both a xylosyltransferase and a glucosyltransferase compete to add either xylose or glucose, respectively, to the C 7 position. When the flavone is not glycosylated, the petals are small, slender, and tend to curl more easily than wild-type petals. Transferase activities appear to be controlled by multiallelic dominant genes designated g^A, g^X, and g^G, respectively (Van Brederode and Van Nigtevecht, 1973, 1974). In plants containing both the alleles g^X and g^G, only isovitexin-glucoside can be detected, suggesting that g^G is dominant over g^X. However, the petal extracts of these plants also contain the xylosyltransferase activity toward isovitexin, which suggests that the dominance is not related to transcriptional or translational control. One hypothesis suggested by these authors to explain the lack of isovitexin-xyloside in g^X g^G plants was that the glucosyltransferase had a higher affinity for isovitexin than the xylosyltransferase. The apparent K_m of the xylosyltransferase for this acceptor, however, was determined to be lower than that of the glucosyltransferase. Inhibition of the xylosyltransferase by the isovitexin-xyloside was also shown not to occur. The V_{max} of xylosyltransferase for UDP-xylose using isovitexin as acceptor was 8-fold higher than that of the glucosyltransferase for UDP-glucose with isovitexin as acceptor. The authors suggested, therefore, that the lack of isovitexin-xyloside in the heterozygote results from the much faster rate of catalysis of the glucosyltransferase compared to the xylosyltransferase. This hypothesis was not directly tested, however, by quantitating the amounts of products formed when physiological levels of UDP-glucose, UDP-xylose, and isovitexin were added to extracts of heterozygotes or to a mixture of extracts from $g^G g^G$ and $g^X g^X$ plants.

Narasimham *et al.* (1977) have isolated a lectin-resistant Chinese hamster ovary cell mutant that, unlike wild-type cells, is unable to transfer *N*-acetylglucosamine to blood glycoproteins containing one terminal mannose and one terminal *N*-acetylglucosamine. However, mutants, like the wild-type cells, were able to transfer *N*-acetylglucosamine to structures containing two free mannose residues. Preliminary work suggests that the two glycosyltransferases involved may, as in *Melandrium,* compete for the same acceptor site. Use of these mutants will likely facilitate elucidation of the control mechanisms for glycosyltransferase activity in mammalian cells.

IX. Involvement of Lipid–Sugar Intermediates in Cell Surface Glycosyltransferase Activity

Lipid–sugar intermediates, as well as the enzymes that synthesize them, have been known to function in the synthesis of complex carbohydrates in bacteria and yeasts for many years. Only recently have vertebrates been shown to possess similar systems for glycoprotein synthesis. One of these systems is schematized in Fig. 6. To summarize a typical synthetic pathway, sugar–nucleotides react with an isoprenoid monophosphate to form a lipid mono- or diphosphate sugar. This molecule can then serve as the substrate for enzymes that either transfer additional sugars to the sugar moiety of the intermediate or, instead, transfer the sugar moiety itself to glycoprotein acceptors. Thus, lipid–sugars appear to be intermediates in the synthesis of carbohydrate side chains of some glycoproteins.

Since these intermediates are amphipathic molecules, they may "shuttle" sugars across membranes. Sugar nucleotides, which apparently cannot cross membranes, would react with an intermediate on one side of a membrane and then transfer the sugar to an acceptor on the other side. This shuttle system could function both in the Golgi apparatus and on the plasma membrane (Waechter and Lennarz, 1976).

A number of studies that utilized cells prelabeled with radioactive sugars and demonstrated transfer of sugar to acceptors external to the cells suggest that such a system of lipid–sugar intermediates may provide substrates for cell surface transferases (Yogeeswaran et al., 1974; Patt and Grimes, 1976; Turley and Roth, 1979). Other reports have demonstrated lipid–sugar intermediates in reactions of either intact cells or plasma membranes with exogenously added sugar nucleotides.

A series of reports by Arnold (Arnold et al., 1973, 1976b) investigated the involvement of lipid–sugar intermediates in incubations of intact, 8-day chick liver cells with GDP-mannose. The authors showed that these intact cells catalyzed the transfer of mannose from GDP-mannose to glycoproteins on the cell surface through mannosylphosphoryl–lipid intermediates in a two-step reaction. The first step involved the transfer of mannose to a lipid intermediate and the second the rate-limiting transfer from the intermediate to glycoprotein acceptors. At all times, both reaction products were associated with the cells. Mg^{2+} and Mn^{2+} stimulated both steps of the reaction, whereas EDTA inhibited both.

After intact liver cells were incubated with GDP-mannose for brief times, a mannosylphosphoryl-lipid was isolated that, when added back to intact cells in

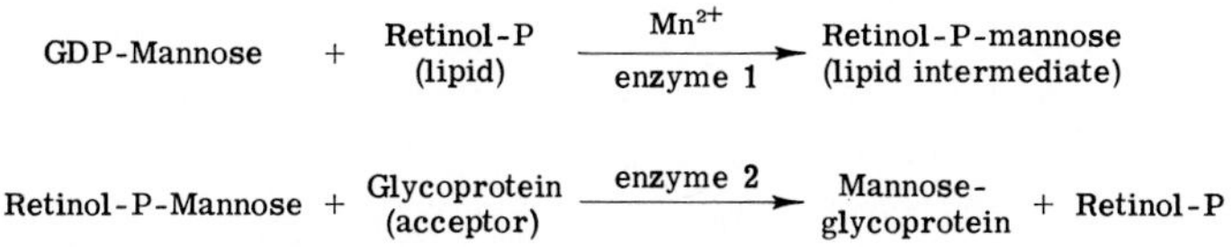

$$\text{GDP-Mannose} \quad + \quad \begin{array}{c}\text{Retinol-P}\\\text{(lipid)}\end{array} \quad \xrightarrow[\text{enzyme 1}]{Mn^{2+}} \quad \begin{array}{c}\text{Retinol-P-mannose}\\\text{(lipid intermediate)}\end{array}$$

$$\text{Retinol-P-Mannose} \quad + \quad \begin{array}{c}\text{Glycoprotein}\\\text{(acceptor)}\end{array} \quad \xrightarrow{\text{enzyme 2}} \quad \begin{array}{c}\text{Mannose-}\\\text{glycoprotein}\end{array} \quad + \quad \text{Retinol-P}$$

FIG. 6.   An example of glycosyltransferase activity that utilizes a lipid-sugar intermediate: transfer of mannose to glycoprotein via a retinol phosphate intermediate by rat liver microsomal fraction (DeLuca, 1978).

the absence of GDP-mannose, could function as substrate for the transfer of mannose to glycoprotein acceptors. These authors also state that a similar intermediate was observed when intact liver cells were incubated with UDP-GlcNAc instead of GDP-mannose. The lipid moiety of the intermediate was labeled by preincubating the cells with radioactive mevalonic acid, but was not characterized further. When cells were preincubated with [³H]retinol and then incubated with GDP-[¹⁴C]mannose, thin-layer chromatography showed that the endogenous lipid acceptor was not retinol, since no double-labeled material was found to cochomatograph with a mannosylphosphoryl-lipid standard. However, this experiment did demonstrate that exogenously added retinol can act as an acceptor for GDP-mannose, because a double-labeled intermediate was identified, but its rate of migration differed from the mannosylphosphoryl-lipid standard.

Patt and Grimes (1974, 1976) showed the synthesis of mannosyl-lipids by intact Balb/c fibroblasts. The possibility that the intermediates were formed from breakdown products of GDP-mannose was carefully excluded. In addition, a number of control experiments using autoradiography, Trypan blue exclusion, and titrations with broken cells suggest that the observed activities were present on surfaces of intact cells.

Two mannosylphosphoryl-lipids were observed in the transfer of mannose from GDP-mannose to glycoprotein acceptors. The first lipid cochromatographed with mannosylphosphoryldolichol in three solvent systems. Acid hydrolysis and gel-exclusion chromatography showed that the second lipid that contained mannose was composed of approximately 10–12 sugar residues, but these were not characterized further. When the glycoprotein fractions that had received the transfer of mannose were digested with Pronase, the oligosaccharides released were approximately the same size as those found in the second class of lipid intermediates. The kinetics of mannose transfer suggested that mannose was transferred from GDP-mannose to mannosylphosphoryldolichol and then transferred to an oligosaccharide-lipid that, at some point, can serve as substrate for the final transfer to glycoprotein acceptors.

Struck and Lennarz (1976, 1977) demonstrated that intact hen oviduct cells incorporate mannose from GDP-mannose into mannosylphosphoryldolichol, oligosaccharide-lipid, and finally glycoprotein acceptors. The oligosaccharide-lipid was identified as the same intermediate observed earlier in cell-free preparations from hen oviduct (Lucas *et al.,* 1975). By contrast, the glycoprotein to which mannose was finally transferred was considerably lower in molecular weight than those observed earlier *in vitro.*

Intact cells, and not broken cells or membrane fragments, were shown to synthesize these mannosyl-lipids by a number of control experiments. First, GDP-mannose was identified as the direct mannose donor for the initial reaction. Second, studies that utilized metabolic inhibitors such as 2-deoxyglucose and experiments that titrated broken cells into the incubation mixture suggested that the majority of the observed activities resulted from enzymes present on surfaces of intact cells rather than products that were soluble in TCA. In addition, experiments using GDP-mannose labeled in the guanosine moiety showed that insignificant amounts of labeled GDP were incorporated into nucleic acids when intact cells were incubated with this substrate. This result suggested that broken or "leaky" cells, which could presumably incorporate the breakdown product, radioactive GDP, were present in very low numbers.

Parodi and Martin-Barrientos (1977) purified membranes from rabbit reticulocytes and showed that these membrane fractions contain enzymatic activities and lipid–sugar intermediates similar to those involved in the glycosylation of protein by porcine liver preparations. Membrane preparations of reticulocytes consisted of both plasma membranes and mitochondrial membranes. In contrast, membrane preparations of erythrocytes, which lack mitochondria, consist almost entirely of plasma membranes. Because erythrocyte membrane preparations contained identical activities for the synthesis of sugar derivatives of dolichol compared to reticulocyte preparations, these authors therefore suggested that the reticulocyte activities were present in the plasma membranes of their preparations and not in the mitochondrial membranes.

The lipid–sugar intermediates synthesized by reticulocyte membrane preparations were analyzed by paper chromatography after mild acid hydrolysis and identified as mannosylphosphoryldolichol, glucosylphosphoryldolichol, N-acetylglucosaminyldiphosphoryldolichol, mannosylchitobiosyldiphosphoryldolichol, and other intermediates that lead to the synthesis of the most complex lipid whose polysaccharides were GlcNAc, Man, and Glc in the ratio of 2:12:4. The molecular weights of the glycosylated proteins ranged from 25,000 to 70,000 but none of these proteins corresponded to the major glycoproteins of reticulocyte membranes.

Erythrocyte plasma membranes contained all the activities and intermediates found in the reticulocyte membranes, but were unable to transfer the most complex polysaccharide–lipid to glycoprotein acceptors, suggesting that these membranes lacked either the final enzyme that transfers the polysaccharide, the glycoprotein acceptors, or both. No mixing experiments were performed between the two cell types to determine the deficiencies of the erythrocyte system.

These results are consistent with experiments of Harris and Johnson (1969) in which reticulocytes (but not erythrocytes), when preincubated with [^{14}C]GlcNAc, appeared to glycosylate proteins on the outside of the plasma membrane, since the majority of the transferred, labeled sugars were released by mild proteolytic treatment of the intact reticulocytes. Parodi and Martin-Barrientos suggested that reticulocytes utilize the lipid–sugar intermediate system to translocate sugars from the internal side of the plasma membrane to the outer (noncytoplasmic) side.

X. Conclusion

Many lines of evidence suggest that some glycosyltransferase activities are present at the cell surface of some cell types. Except for a few simple organisms, the function, if any, that glycosyltransferases perform on the surfaces of cells is unknown. In some cases, these enzymes may function synthetically and glycosylate acceptors at the cell surface. Perhaps several, spatially distinct, membrane-bound systems of glycosyltransferases exist in a cell, each with its own particular collection and order of enzymes. Each synthesizing system could serve as a separate "template" for carbohydrate structure, and cooperation between these systems would be necessary to complete the glycosylation of some proteins and lipids. The plasma membrane may be the site of one of these ensembles of enzymes, replete with lipid–sugar intermediates, as some studies suggest.

One of the most intriguing possibilities for cell surface glycosyltransferase function is that they are involved in intercellular recognition. Evidence that surface transferases do function in this manner is, however, largely correlative. Nearly all of the studies of cell surface glycosyltransferases have been descriptive ones, and few attempts have been made to perturb specific enzyme activities and observe the effects on cell interactions. A primary difficulty has been the inability to affect particular transferase activities. One means to test directly the functions of these enzymes may be through specific, active-site inhibitors that can mimic native sugar donors or acceptors and completely and rapidly inactivate particular classes of glycosyltransferases; for example, the dialdehyde derivatives of nucleotides (Section IV,B). By inhibiting a single class of enzymatic activity, its involvement in a recognition event could be evaluated. By utilizing biochemical, immunological, and genetic probes of enzyme activity, an insight into the functions of these enzymes, and ultimately the molecular bases of intercellular recognition, may be gained.

REFERENCES

Abramsky, T., and Tatum, E. L. (1976). *Biochim. Biophys. Acta* **421**, 106.
Arnold, P., Hommel, E., and Risse, H. (1973). *Biochem. Biophys. Res. Commun.* **54**, 100.
Arnold, P., Hommel, E., and Risse, H. J. (1976a). *Mol. Cell. Biochem.* **10**, 81.
Arnold, P., Hommel, E., and Risse, H. J. (1976b). *Mol. Cell. Biochem.* **11**, 137.
Aronson, N. N., Jr., Tan, L. Y., and Peters, B. P. (1973). *Biochem. Biophys. Res. Commun.* **53**, 112.
Barber, A. J., and Jamieson, G. A. (1971). *Biochim. Biophys. Acta* **252**, 533.
Barbera, A. J., Marchase, R. B., and Roth, S. (1973). *Proc. Natl. Acad. Sci. U.S.A.* **70**, 2482.
Barondes, S. H., and Rosen, S. P. (1975). *In* "Neuronal Recognition" (S. H. Barondes, ed.), p. 331. Plenum, New York.
Bernacki, R. J. (1974). *J. Cell. Physiol.* **83**, 457.
Bosmann, H. B. (1969). *Life Sci.* **8**, 737.

Bosmann, H. B. (1971). *Biochem. Biophys. Res. Commun.* **43**, 1118.

Bosmann, H. B. (1972a). *Biochim. Biophys. Acta* **279**, 456.

Bosmann, H. B. (1972b). *Biochem. Biophys. Res. Commun.* **48**, 523.

Bosmann, H. B. (1974). *Biochim. Biophys. Acta* **339**, 438.

Bosmann, H. B., and McLean, R. J. (1975). *Biochem. Biophys. Res. Commun.* **63**, 323.

Bosmann, H. B., Bieber, G. F., Brown, A. E., Case, K. R., Geistein, D. M., Klimmerer, T. W., and Lione, A. (1973). *Nature (London)* **246**, 487.

Brass, L. F., and Bensusan, H. B. (1976). *Biochim. Biophys. Acta* **444**, 43.

Brew, K., Vanaman, T., and Hill, R. L. (1968). *Proc. Natl. Acad. Sci. U.S.A.* **59**, 491.

Brown, R. M., and Wilson, T. H. M. (1976). *In* "International Cell Biology 1976–1977" (B. R. Brinkley and H. R. Porter, eds.), p. 267. Rockefeller Univ. Press, New York.

Cacan, R., Verbert, A., and Montreuil, J. (1976). FEBS *Lett.* **63**, 102.

Cebula, T. A., and Roth, S. (1976). *In* "Biogenesis and Turnover of Membrane Macromolecules" (T. S. Cook, ed.), p. 235. Raven Press, New York.

Cervén, E. (1977). *Biochim. Biophys. Acta* **467**, 72

Chipowsky, S., Lee, Y. C., and Roseman, S. (1973). *Proc. Natl. Acad. Sci. U.S.A.* **70**, 2309.

Columbrino, L. F., Bosmann, H. B., and McLean, R. J. (1978). *Exp. Cell Res.* **112**, 25.

Cummings, R., Cebula, T., and Roth, S. (1979). *J. Biol. Chem.* (in press).

Datta, P. (1974). *Biochemistry* **13**, 3987.

DeLuca, L. M. (1978). *In* "Handbook of Lipid Research" (H. F. Davis, ed.), Vol. 2, p. 1. Plenum, New York.

Deppert, W., Werchaw, H., and Walter, G. (1974). *Proc. Natl. Acad. Sci. U.S.A.* **71**, 3068.

Deppert, W., and Walter, G. (1978). *J. Supramol. Struct.* **8**, 19.

Durr, R., Shur, B., and Roth, S. (1976). *Nature (London)* **265**, 547.

Geren, L. M., and Ebner, K. E. (1974). *Biochem. Biophys. Res. Commun.* **59**, 14.

Gibbons, R. J., and Van Houte, J. (1973). *J. Periodontol.* **44**, 347.

Grinnell, F. (1978). *Int. Rev. Cytol.* **53**, 65.

Hagopian, A., Bosmann, H. B., and Eylar, E. H. (1968). *Arch. Biochem. Biophys.* **128**, 387.

Harris, E. D., and Johnson, G. A. (1969). *Biochemistry* **8**, 512.

Hirschberg, C. B., Goodman, S. R. and Green, C. (1976). *Biochemistry* **15**, 3591.

Hudgin, R. and Ashwell, G. (1974). *J. Biol. Chem.* **249**, 7369.

Jamieson, G. A. (1974). *In* "Biology and Chemistry of Eucaryotic Cell Surfaces" (E. Y. C. Lee and E. E. Smith, eds.), p. 67. Academic Press, New York.

Jamieson, G. A., Urban, C. L., and Barber, A. J. (1971). *Nature (London), New Biol.* **234**, 5.

Jentoft, N., Cheng, O. P., and Carlson, P. (1976). *In* "The Enzymes of Biological Membranes" (A. Martonosi, ed.), p. 343. Plenum, New York.

Kang, A. H., Beachey, E. W., and Katzman, R. L. (1973). *J. Biol. Chem.* **249**, 1054.

Karfunkel, P., Hoffman, M., Phillips, M., and Black, J. (1977). *In* "Formshaping Movements in Neurogenesis" (C. Jacobson and T. Eberdal, eds.), p. 23. Almqvist & Wiksell, Stockholm.

Kilton, L., and Maca, R. D. (1977). *J. Natl. Cancer Inst.* **58**, 1479.

Kirschbaum, B. B., and Bosmann, H. B. (1973a). *Biochim. Biophys. Acta* **320**, 416.

Kirschbaum, B. B., and Bosmann, H. B. (1973b). *Nephron* **12**, 211.

LaMont, J. T., Perrotto, J. L., Weiser, M. M., and Isselbacher, K. J. (1974). *Proc. Natl. Acad. Sci. U.S.A.* **71**, 3726.

LaMont, J. T., Gammon, M. T., and Isselbacher, K. J. (1977). *Proc. Natl. Acad. Sci. U.S.A.* **74**, 1086.

Levine, J., and Roth, S. (1980). *J. Cell Biol.* (Submitted for publication).

Lloyd, C. W., and Cook, G. M. W. (1974). *J. Cell Sci.* **15**, 575.

Lucas, J. J., Waechter, C. J., and Lennarz, W. J. (1975). *J. Biol. Chem.* **250**, 1992.

McLean, R. J., and Bosmann, H. B. (1975). *Proc. Natl. Acad. Sci. U.S.A.* **72**, 310.

Marchase, R. B. (1977). *J. Cell Biol.* **75**, 237.

Marchase, R. B., Vosbeck, K., and Roth, S. (1976). *Biochim. Biophys. Acta* **457**, 395.

Merritt, W. D., Morre, D. J., Franke, W. W., and Keenan, T. W. (1977). *Biochim. Biophys. Acta* **497**, 820.

Molnar, J., Chao, H., and Markovic, C. (1969). *Arch. Biochem. Biophys.* **134**, 533.

Morgan, H. R., and Bosmann, H. B. (1974). *Proc. Soc. Exp. Biol. Med.* **146**, 1146.

Moscona, A. A. (1975). *In* "Neuronal Recognition" (S. Barondes, ed.), p. 205. Plenum, New York.

Mukasa, H., and Slade, H. D. (1973a). *Infect. Immun.* **8**, 555.

Mukasa, H., and Slade, H. D. (1973b). *Infect. Immun.* **9**, 419.

Mukasa, H., and Slade, H. D. (1973c). *Infect. Immun.* **10**, 1135.

Muller, W. E. G., and Zahn, R. K. (1973). *Exp. Cell Res.* **80**, 95.

Muller, W. E. G., Arendes, I., Kurelec, B., Steffen, R., and Muller, I. (1977). *J. Biol. Chem.* **252**, 3836.

Muller, W. E. G., Zahn, R. K., Kurelec, B., Muller, I., Uhlenbruck, G., and Vaith, P. (1978). *Hoppe-Seyler's Z. Physiol. Chem.* **359**, 529.

Muller, W. E. G., Zahn, R. K., Kurelec, B., Muller, I., Uhlenbruck, G., and Vaith, P. (1979). *J. Biol. Chem.* **254**, 1280.

Narasimham, S., Stanley, P., and Schachter, H. (1977). *J. Biol. Chem.* **252**, 3926.

Oppenheimer, S. B., Edidin, M., Orr, C. W., and Roseman, S. (1969). *Proc. Natl. Acad. Sci. U.S.A.* **63**, 1395.

Parish, C. R. (1977). *Nature (London)* **267**, 711.

Parodi, A. J., and Martin-Barrientos, J. (1977). *Biochim. Biophys. Acta* **500**, 80.

Patt, L. M., and Grimes, W. J. (1974). *J. Biol. Chem.* **249**, 4157.

Patt, L. M., and Grimes, W. J. (1975). *Biochem. Biophys. Res. Commun.* **67**, 483.

Patt, L. M., and Grimes, W. J. (1976). *Biochim. Biophys. Acta* **444**, 97.

Patt, L. M., VanNest, G. A., and Grimes, W. J. (1975). *Cancer Res.* **35**, 438.

Patt, L. M., Endres, R. D., Lucas, D. O., and Grimes, W. J. (1976). *J. Cell Biol.* **68**, 799.

Podolsky, D. K., and Weiser, M. M. (1975). *Biochem. J.* **146**, 213.

Podolsky, D. K., Weiser, M. M., LaMont, J. T., and Isselbacher, K. J. (1974). *Proc. Natl. Acad. Sci. U.S.A.* **71**, 904.

Petit, V., and Edidin, M. (1974). *Science* **184**, 1183.

Porter, C. W., and Bernacki, R. J. (1975). *Nature (London)* **256**, 648.

Porzig, E. F. (1978). *Dev. Biol.* **67**, 114.

Powell, J., and Brew, K. (1976). *Biochemistry* **15**, 3499.

Pricer, W. E., Jr., and Ashwell, G. (1971). *J. Biol. Chem.* **246**, 4825.

Roseman, S. (1970). *Chem. Phys. Lipids* **5**, 270.

Rossler, K., Penckert, W., and Risse, H.-J. (1978). *Mol. Cell. Biochem.* **20**, 3.

Roth, S., McGuire, E. J., and Roseman, S. (1971a). *J. Cell Biol.* **51**, 525.

Roth, S., McGuire, E. J., and Roseman, S. (1971b). *J. Cell Biol.* **51**, 536.

Roth, S., Roelke, M., and Dorsey, J. (1977). *In* "Growth Kinetics and Biochemical Regulation of Normal and Malignant Cells" (B. Drewinko and R. M. Humphrey, eds.), p. 245. Williams & Wilkins, Baltimore, Maryland.

Sasaki, T., and Robbins, P. M. (1974). *In* "Biology and Chemistry of Eucaryotic Cell Surfaces" (E. Y. C. Lee and E. E. Smith, eds.), p. 125. Academic Press, New York.

Schanbacher, F. L., and Ebner, K. E. (1970). *J. Biol. Chem.* **245**, 5057.

Schwaezer, M., and Hill, R. (1977). *J. Biol. Chem.* **252**, 2346.

Setlow, V. P., Roth, S., and Edidin, M. (1979). *Exp. Cell Res.* (in press).

Sharon, N., and Lis, H. (1972). *Science* **177**, 949.

Shur, B. D. (1977a). *Dev. Biol.* **58**, 23.

Shur, B. D. (1977b). *Dev. Biol.* **58,** 40.

Shur, B. D., and Bennett, D. (1979). *Dev. Biol.* (in press).

Shur, B. D., and Roth, S. (1975). *Biochim. Biophys. Acta* **415,** 473.

Sievers, S., Risse, H.-J., and Sekeri-Pataryas, K. H. (1978). *Mol. Cell. Biochem.* **20,** 103.

Smith, D. F., Koscow, D. P., Wu, C., and Jamieson, G. A. (1977). *Biochim. Biophys. Acta* **483,** 263.

Sparato, A. C., Morgan, H. R., and Bosmann, H. B. (1975). *Proc. Soc. Exp. Biol. Med.* **149,** 486.

Struck, D. K., and Lennarz, W. J. (1976). *J. Biol. Chem.* **251,** 2511.

Struck, D. K., and Lennarz, W. J. (1977). *J. Biol. Chem.* **252,** 1007.

Sudo, T., and Onodera, K. (1975). *Exp. Cell Res.* **91,** 191.

Toda, G., Oka, T., Oda, T., and Ikeda, Y. (1975). *Biochim. Biophys. Acta* **413,** 52.

Toole, B. P. (1976). *In* "Neuronal Recognition" (S. Barondes, ed.), p. 275. Plenum, New York.

Turley, E. A., and Roth, S. (1979). *Cell* (in press).

Turner, R. S., and Burger, M. M. (1973). *Nature (London)* **244,** 509.

Van Brederode, J., and Van Nigtevecht, G. (1973). *Mol. Gen. Genet.* **122,** 215.

Van Brederode, J., and Van Nigtevecht, G. (1974). *Genetics* **77,** 507.

Verbert, A., Cacan, R., and Montreuil, J. (1976). *Eur. J. Biochem.* **70,** 49.

Verbert, A., Cacan, R., Deveire, P., and Montreuil, J. (1977). *FEBS Lett.* **74,** 267.

Waechter, C. J., and Lennarz, W. J. (1976). *Annu. Rev. Biochem.* **45,** 95.

Wailey, A., and Cook, G. M. W. (1973). *Biochim. Biophys. Acta* **323,** 55.

Webb, G. C., and Roth, S. (1974). *J. Cell Biol.* **63,** 796.

Webb, G. (1980). In preparation.

Weigel, P. H., Snell, E., Lee, Y. C., and Roseman, S. (1978). *J. Biol. Chem.* **253,** 330.

Weiser, M. M. (1973a). *J. Biol. Chem.* **248,** 2536.

Weiser, M. M. (1973b). *J. Biol. Chem.* **248,** 2542.

Weiser, M. M., Neumeier, M. M., Quaroni, A., and Kirsch, K. (1978). *J. Cell Biol.* **77,** 722.

Whaley, W. G., Dauwalder, M., and Kephart, J. E. (1972). *Science* **175,** 596.

Wiese, L., and Wiese, W. (1975). *Dev. Biol.* **43,** 264.

Yen, P. H., and Ballou, C. E. (1974). *Biochemistry* **13,** 2428.

Yogeeswaran, G., Laine, R. A., and Hakomori, S. (1974). *Biochem. Biophys. Res. Commun.* **59,** 591.

INTERNATIONAL REVIEW OF CYTOLOGY, VOL. 65

The Transport of Steroid Hormones into Animal Cells

ELEONORA P. GIORGI

Department of Biological Chemistry, The Hebrew University, Jerusalem, Israel

I. Introduction

The transport of steroid hormones from the extracellular fluid into animal cells has generally been assumed to occur by simple diffusion of the free steroid across the cell membrane. However, if by simple diffusion one means diffusion within the membrane following distribution of the permeant between aqueous external phase and the membrane according to the partition coefficient of the solute in the membrane, there have been very few clear-cut demonstrations of the correctness of this assumption (Gorski and Gannon, 1976). Recently, evidence has been presented suggesting the existence of facilitated diffusion systems for steroid hormones in some target tissues and in liver (see Section V). Cholesterol and vitamin A, compounds that are as liposoluble as steroids, are transported by carrier systems (Rask and Peterson, 1976; Goldstein and Brown, 1977). Interactions with membrane proteins by the anesthetics, another important group of liposoluble compounds, is also a strong possibility (Seeman, 1972; Lawrence and Gill, 1975; Richards and Hesketh, 1975; Franks and Lieb, 1978). These new findings make desirable a closer scrutiny of the mode of transport of steroid hormones into animal cells.

The existence of specialized transport systems for steroid hormones in a restricted number of cell types to the exclusion of all other cells is a tenable proposition. The plasma membrane maintains the integrity of the cell and at the same time regulates the exchanges between the cell interior and the environment. Specialized transport systems (facilitated diffusion and active transport) have developed mainly for transport of water-soluble compounds (nutrients, nucleosides, and electrolytes) to which the cell membrane would otherwise be impermeable. Active transport and facilitated diffusion systems of this type are universally distributed in the cells of animal organisms, probably because they provide to the cells compounds that are indispensable to the maintenance of cellular function. On the other hand, the requirement for steroid hormones differs from cell to cell and from tissues that are totally dependent on hormones for development and growth to tissues where hormones have apparently no action. Target tissues, during differentiation, acquire a complex machinery for the expression of hormone action: the "two-steps mechanism," as first proposed by Jensen and co-workers (1968). In the first step, the hormone binds to specific receptors in the cytoplasm; after activation of the steroid–receptor complex, which involves a conformational change, the second step transfers the complex to the nucleus, where it interacts with chromatin acceptor sites (for reviews, see King and Mainwaring, 1974; Liao, 1975; Gorski and Gannon, 1976; also two volumes edited by Pasqualini, 1977). It is conceivable that, at the time of differentiation, the cells also develop special transport systems in order to insure sufficient hormone supply. By the same token, cells processing large amounts of steroids (liver, kidney, and placenta) might require a greater rate of transport of steroid hormones than that afforded by simple diffusion of these compounds across the lipid bilayer of the cell membrane. Apart from these cases, the universal distribution of specialized modes of transport for nutrients, nucleosides, and electrolytes may not apply to transport of steroid hormones. Thus, the main questions to be answered are (1) whether the classical definition of simple diffusion applies to the transport of steroid hormones into animal cells; (2) whether target tissues have a selective permeability for some hormones; and (3) whether permeability to hormones changes in various physiological and pathological states and what agents, if any, modify the entry of hormones into target and nontarget tissues.

From the intensive investigations of the last decade into the mechanism of action of steroid hormones, it has become apparent [although this view has recently been challenged (Szego, 1978)] that these hormones do not act, at physiological concentrations, through a direct modification of the plasma membrane. A notable exception is the induction of meiosis in the amphibian oocyte by progesterone (Wasserman and Masui, 1975; Godeau et al., 1978). Steroid hormones, however, have an indirect action on the rate of transport of nutrients into cells of target tissues and on cell adhesiveness (Riggs et al., 1968; Mills and Spaziani, 1968; Ballard and Tomkins, 1970; Härkönen et al., 1975). The effect of glucocorticoids on cell proliferation might be mediated by membrane modifications (Sachs et al., 1974). The plasma membrane is also the site of action of anesthetic steroids (Lawrence and Gill, 1975; Richards and Hesketh, 1975). It would be impossible to deal adequately with these interesting aspects of hormonal action within the limits of this article. This article, therefore, will be concerned only with a discussion of the mode of transport of steroid molecules into animal cells and of its biological implications.

II. The Role of Plasma Proteins in the Transport of Steroid Hormones into Animal Cells

The plasma of all mammals contains one or more glycoproteins that bind steroids with high affinity. The most widely distributed of these is transcortin or corticosteroid-binding globulin (CBG), which, in species not possessing a specific globulin for sex steroids (sex hormone-binding globulin, SHBG), binds also this latter group of hormones. Albumin, present in the plasma of all species, binds steroids with low affinity, but practically infinite capacity. If association occurs only with albumin, the percentage of steroid bound in plasma is 60–80% of the total; binding to carrier proteins increases the percentage of bound steroid to 85–98%. [The subject has been extensively examined by Westphal (1971), and the reader is referred to this treatise for further details.] Interstitial fluid bathing the cells contains albumin and carrier proteins in proportions similar to those found in plasma, but the concentration is approximately half (Dewey, 1959). Because of the lower protein content, the capacity of interstitial fluid is less than that of plasma:thoracic duct lymph, whose composition resembles closely that of intercellular fluid, binds only 71% of the hydrocortisone present compared to over 80% bound in plasma (Sandberg *et al.*, 1960).

Within the cells, too, steroids are bound to proteins. In target tissues, binding to specific receptors accounts for approximately 10–20% of the total hormone present intracellularly; this bond is very strong, the association constant being of the order of 10^{13} M^{-1} for some hormones. The remainder of the intracellular hormone is associated loosely and nonspecifically with soluble proteins and with various subcellular structures (Kowarski *et al.*, 1969; Mainwaring and Peterken, 1971). This type of association might occur in nontarget tissues as well. Due to the low affinity of the binding of steroids to cellular components other than the specific receptors, a substantial fraction of hormone might be free inside the cells; as yet, no direct measurements have been reported. Thus, both at the external and at the internal face of the membrane, steroids are present in a bound and in an unbound form. According to the second law of thermodynamics, passage of solutes across the cell membrane is driven by a gradient between the intra- and extracellular concentration of chemically active solute, where chemical activity refers to those molecules of the solute that are in no way prevented from crossing the cell membrane (Stein, 1967; Davson and Danielli, 1970). The evidence that free rather than bound steroids cross the membrane is impressive, as we shall soon see. The role of the plasma proteins is therefore crucial, insofar as they determine the distribution of steroids between the bound and free forms and thus control the concentration gradient across the cell membrane. Similarly, intracellular binding of hormones to receptors in target tissues might control the gradient from the internal face of the membrane. We shall first review the evidence in favor of the hypothesis of passage of unbound steroid through the cell wall, and then we shall examine how the concentration gradient is regulated.

A. EVIDENCE FOR TRANSFER OF FREE STEROIDS ACROSS THE MEMBRANE

Since the classical studies of Slaunwhite and co-workers (1962) and Hoffmann and co-workers (1969), it has been conclusively demonstrated that, if the concentration of free steroids in the circulation plasma is reduced by raising the level of carrier globulins by some artificial means, less steroid enters the tissues. More recently, it has been shown that antibodies, produced in animals by active immunization with antigenic albumin–steroid derivatives, completely abolish the biological effects of the hormones. Regression of the sex accessory organs is observed in adult rats immunized against estradiol[1]; immunization of pregnant

---

[1]The following trivial names of steroids have been used: estradiol, estra-1,3,5,(10)-triene-3,17β-diol; estradiol-17α, estra-1,3,5,(10)-triene-3,17α-diol; estrone, 3-hydroxy-estra-1,3,5,(10)-triene-17-one; estriol, estra-1,3,5,(10)-triene-3,16α,17β-triol; pregnenolone, 3β-hydroxy-pregn-5-en-20-one; progesterone, pregn-4-en-3,20-dione; 20α-OH-progesterone, 20α-hydroxy-pregn-4-4en-3-one;

rats against testosterone causes feminization of the male fetuses (Hillier *et al.*, 1973; Bidlingmaier *et al.*, 1977). Antibodies produced by active immunization bind steroids very tightly; the association constants can be as high as 0.5×10^{11} M^{-1} (Williams and Underwood, 1974) as opposed to $K_A = 10^9$ M^{-1} for the plasma globulin with the highest affinity for steroids. The fraction of steroid free in plasma is practically nil (Hillier *et al.*, 1973). It is significant in this regard that, in the blood of pregnant rats immunized against testosterone, the level of the hormone was 100 times the normal, and yet no response was evoked in the fetal tissues. However, it is known that only free steroids can readily cross the capillary wall, while passage of steroid–protein complexes is very slow. This is the case for all complexes between small ligands and proteins (Simpson-Morgan and Sutherland, 1976). Therefore, when less free hormone is present in the circulation, the rate of transfer of steroids across the capillary wall into the interstitial spaces decreases. It is this reduction in transfer of steroids into the interstitial spaces that might be the factor limiting uptake by the tissues, rather than the impermeability of the cell membrane to steroid–protein complexes. In order to establish beyond doubt the existence of a permeability barrier to bound steroids at the membrane level, it is necessary to resort to *in vitro* systems. It has been reported that addition of plasma proteins to serum-free medium decreases the uptake of steroids by cultured tissues and at the same time prevents the expression of the biochemical and morphological effects of the hormones (Lasnitzki and Franklin, 1972, 1975; Peck *et al.*, 1973; Lasnitzki *et al.*, 1974; Mercier-Bodard *et al.*, 1976). Similarly, transport of corticosterone into isolated rat hepatocytes is inhibited by the addition of either 4% albumin or serum to the medium; the extent of inhibition of transport is inversely proportional to the percentage of free steroid in the medium (Rao *et al.*, 1977b). These *in vitro* studies thus show conclusively that the bulk of the steroid crossing the membrane is in the free form. Other interesting aspects of these studies will be discussed later.

17α-OH-progesterone, 17α-hydroxy-pregn-4-en-3-one; deoxycorticosterone, 21-hydroxy-pregn-4-en-3,20-dione; corticosterone, 11β,21-dihydroxy-pregn-4-en-3,20-dione; cortisone, 17α,21-dihydroxy-pregn-4-en-3,11,20-trione; hydrocortisone (cortisol), 11β,17α,21-trihydroxy-pregn-4-en-3,20-dione; aldosterone, 18,11-hemiacetal of 11β,21-dihydroxy-3,20-dioxo-pregn-4-en-18-al; androstanedione, 5α-androstan-3,17-dione; 5α-dihydrotestosterone, 17β-hydroxy-5α-androstan-3-one; etiocholanolone, 3β-hydroxy-5β-androstan-17-one; dehydroepiandrosterone, 3β-hydroxy-androst-5-en-17-one- testosterone, 17β-hydroxy-androst-4-en-3-one; androstenedione, androst-4-en-3,17-dione. Synthetic steroids: ethinyl estradiol, 17α-ethinyl-estra-1,3,5(10)-triene-3,17β-diol; mestranol, 17α-ethinyl-3-methoxy-estra-1,3,5,(10)-triene-17β-ol; norgestrel, 13-ethyl-17α-hydroxy-18,19-dinor-pregn-4-en-3,20-dione; cyproterone, 6α-chloro-17α-hydroxy-1β,2α-methylene-pregn-4,6-diene-3,20-dione; cyproterone acetate, 6α-chloro-17α-acetoxy-1β,2α-methylene-pregn-4,6-diene-3,20-dione; dexamethasone, 16α-methyl-9α-fluoro-11β,17α,21-trihydroxy-pregn-1,4-diene-3,20-dione; prednisolone, 11β,17α,21-trihydroxy-pregn-1,4-diene-3,20-dione; triamcinolone, acetonide, 16α,17α cyclic acetal with acetone of 9α-fluoro-11β,16α,17α,21-tetrahydroxy-pregn-1,4-diene-3,20-dione.

B. The Regulation of the Gradient across the Membrane

Having established that transfer across the cell membrane is limited to free steroids, it is possible to explore how the concentration gradient is regulated. Although binding of steroids both extra- and intracellularly is undoubtedly important, other factors are involved in the regulation of the gradient. These factors include, beside the rate of transport of the steroids from the capillaries into the intercellular spaces, the time of residence of free and bound steroids in these spaces, and irreversible removal of the hormone from the cells either by metabolism or by release back into the extracellular space. Very few data are available on these parameters, and a precise evaluation of their influence on the concentration gradient is not possible.

However, at least one dynamic aspect in the regulation of the concentration gradient can be considered in some detail, namely, the reversibility of the binding of steroids to intra- and extracellular proteins. High affinity or tight bonding means slow dissociation of the steroid–protein complex, and vice versa, because the association constant is defined as $K_A = k_1/k_{-1}$, where k_1 and k_{-1} are the rate constant of association and dissociation, respectively. It has been suggested by Westphal (1971) that dissociation of the steroid–hormone complex either in blood or in the interstitial spaces may allow transfer into the cells of an amount of hormone much greater than that present in the unbound form. This point has been investigated experimentally (Giorgi and Moses, 1976). Slices of prostate tissue were perfused across a semipermeable membrane with buffer or protein solutions containing radioactive testosterone. In this preparation, only unbound steroid could enter from the medium into the tissue (Fig. 1). Prostate tissue does not release testosterone back into the medium, since the steroid is almost completely metabolized in the cells. The total amount of testosterone passing into the tissue could therefore be measured by determining the radioactive hormone left in the perfused medium. The percentage of perfused steroid that diffused into the tissue during 60 minutes of perfusion with buffer was 15% (mean of six experiments); that diffusing from 4% albumin and from plasma was 7.5 and 2.7%, respectively. Free testosterone was measured in plasma by ultrafiltration and found to be 3.2% both before and after perfusion. Dissociation of testosterone from plasma protein had therefore occurred, but the amount of steroid diffusing into the tissue was 2.7%, that is, less than the amount free. On the other hand, diffusion from albumin was 7.5%, greater than the concentration of free testosterone in this solution, which was 5%. Dissociation of the albumin–steroid complex is very fast: in the liver, almost all the albumin-bound steroid is extracted during one passage through the portal circulation (Baird *et al.*, 1969). The steroid–globulin complex has a much slower rate of dissociation, in the region of 0.3 sec $^{-1}$ (Heyns and de Moor, 1971). However, the very slow rate of exchange of proteins in the interstitial spaces, and the reduced protein content

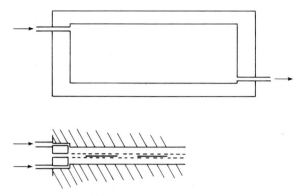

Fᴵɢ. 1. Apparatus for the determination of dissociation of steroids from plasma proteins. Slices of tissue (represented by the continuous lines in the bottom diagram) were disposed in one layer and covered on each side by a Cuprophane membrane (interrupted lines), which is impermeable to proteins but permeable to steroids. The membrane-slice preparation was sealed between two Perspex blocks (diagonally hatched sections), and a continuous flow of plasma or protein solution was forced through the interstices between membrane and the Perspex blocks. The upper diagram shows the shape of a block, the receptacle for the membrane-slice preparation, and the direction of flow. The receptacle was 0.3 cm deep and 10 cm² in area.

in the intercellular fluid (Dewey, 1959; Sandberg *et al.*, 1960) might favor some degree of dissociation of steroids from the proteins, especially from albumin.

The effect of addition of plasma proteins to tissues *in vitro* on uptake of steroids has been studied by Peck and co-workers (1973), Mercier-Bodard and co-workers (1976), and in great detail by Lasnitzki and collaborators (Lasnitzki and Franklin, 1972, 1975; Lasnitzki *et al.*, 1974). The results of some of the experiments of Lasnitzki and collaborators are summarized in Table I. Explants of rat ventral prostate were incubated for 22 hours in defined medium containing various amounts of serum and of testosterone and 5α-dihydrotestosterone. At very low, "physiological" concentrations of steroids, addition of serum reduced steroid uptake proportionally to the amount of serum. Furthermore, plasma from pregnant women, which contains 190–300 $M \times 10^{-9}$ SHBG, was more effective in diminishing steroid uptake than plasma from male subjects, which contains only 20–40 $M \times 10^{-9}$ of the globulin. However, when the concentration of testosterone and 5α-dihydrotestosterone was raised to 17 and 5 $M \times 10^{-9}$, respectively, sufficiently high to saturate all the specific binding sites in the medium, the inhibition of uptake was the same for the two types of sera. Further experiments (Lasnitzki *et al.*, 1974) investigated the effect of 5% male serum in the medium on the binding in prostate explants of 5α-dihydrotestosterone formed intracellularly (Table I, bottom two lines). This hormone binds to specific receptors in rat ventral prostate, whereas testosterone is bound to the receptors only to

TABLE I

Effect of Serum on Uptake of Androgens by Explants of Rat Ventral Prostrate[a,b]

Steroid ($M \times 10^{-9}$)	Serum in medium (%)	Uptake of testosterone		Uptake of 5α-dihydrotestosterone	
		Male serum (%)	Pregnancy serum (%)	Male serum (%)	Pregnancy serum (%)
Testosterone					
0.0017	5	68	18		
0.017	5	68	18		
0.017	10	37	7		
17.0	10	23	27		
5α-Dihydrotestosterone					
0.0005	5			45	16
0.005	5			45	16
0.005	10			32	8
0.005	20			15	1
5.0	5				28
Testosterone					
1.9	5	60		79[c]	
190000	5	90		44[c]	

[a] Values expressed as percentage of control (100).

[b] Explants of rat ventral prostate were grown for 3 days in medium 199. On the third day, fresh medium was added, with or without the amounts of serum and tritiated testosterone or 5α-dihydrotestosterone indicated in the table. The explants were further incubated for 20–24 hours, and total counts were determined in the tissue and in the medium (Lasnitzki and Franklin, 1972, 1975; Lasnitzki et al., 1974).

[c] Concentration in the tissue of [^{3}H]5α-dihydrotestosterone, measured after chromatographic separation.

a very small extent (see review in Mainwaring, 1977). When the concentration of testosterone in the medium was below that required to saturate the receptors, binding of testosterone was reduced by serum addition to 60% of the control, whereas binding of α-dihydrotestosterone was reduced only to 79%. Since there was a significant decrease in the biological effect of the hormone on tissue growth and on maintenance of cellular morphology, it might be inferred that binding to the receptors was reduced by the addition of serum to the medium. However, this high affinity binding was not affected as much as the less strong binding of testosterone. On the other hand, when the same amount of serum and a 10,000-fold amount of testosterone were added to the medium, the intracellular binding of 5α-dihydrotestosterone was reduced by more than half. This is probably because, at this high concentration of testosterone in the medium, intracellular formation of the hormone was sufficiently high to saturate the receptors, while binding to nonspecific sites, which have a very low affinity for the hormone, was somewhat suppressed. Thus, it could be postulated that the extent of intracellular binding of hormones, and therefore of free steroid concentration in the cells, depends on the relative affinity of extra- and intracellular proteins. In fact, inhibition by serum of hormone binding to the prostate receptors is possible because there is only a small difference in the affinity for 5α-dihydrotestosterone between SHBG and androgen receptors, $K_A = 0.5$ and 2.0×10^9 M^{-1}, respectively (Rosen et al., 1975). When the difference in affinity between protein at the outside of the cells and intracellular receptors is very great, a different equilibrium distribution of steroids is observed.

Peck and collaborators (1973) incubated uterine horns and diaphragm from immature rats in media containing labeled estradiol and progressively increasing amounts of albumin. The endometrium receptors have a very high affinity for the estrogen, $K_A = 10^{12}$ M^{-1}; the affinity of albumin is only 10^4 M^{-1}. The decrease in hormone uptake by albumin was slightly lower in endometrium than in diaphragm, a nontarget tissue that apparently contains no receptors; even more interesting, in endometrium, the nuclear retention of estradiol was not affected by the external albumin. This preferential binding of hormones to the receptors is also observed in vivo: intravenous administration of low doses of labeled estradiol to immature female rats leads to the accumulation of the hormone in the nuclei of target tissues (Jensen et al., 1968). Binding to the plasma proteins, which are depleted of endogenous hormone in immature animals, should be maximal under these conditions. On the other hand, in immature animals, the receptors in the target tissues are completely unsaturated; therefore, all the steroid entering the cells becomes bound and is retained for a long time, due to the slow rate of dissociation of the receptor–hormone complex. If the intracellular concentration of free steroid remains much lower than the concentration in the interstitial spaces, competition for the external steroid might take place and some steroid might be forced to dissociate from the proteins in the interstitial spaces. Dissociation of hormone from plasma proteins does indeed occur if the uptake of the hormone is irreversible (Giorgi and Moses, 1976).

In conclusion, when the receptors are saturated and the greater proportion of intracellular steroid is bound to nonspecific sites, the concentration gradient across the cell membrane is probably regulated by the proteins in the interstitial spaces and by the rate of transfer and removal of steroids from and into the blood (Fig. 2a). This mechanism of regulation obviously applies also to nontarget tissues. When, on the other hand, the receptors are not saturated, the concentration gradient may be regulated mainly by the extent of binding to the receptors (Fig. 2b). In order to verify the latter hypothesis, it would be necessary to devise a method for the measurement of free intracellular steroids.

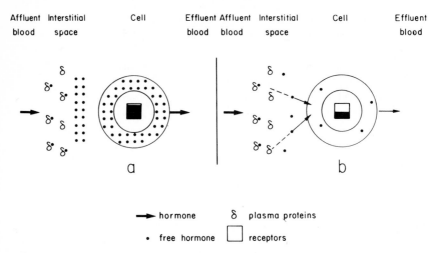

FIG. 2.   Hypothetical model of the regulation of the concentration gradient of free steroid in target tissues. (a) Under conditions of saturation of the receptors, the concentration of free steroid in the interstitial spaces is in equilibrium with that inside the cells, and the passage of steroid across the cell membrane is regulated by the rate of transfer from blood and of removal from the cell (arrows). (b) When the receptors are not saturated, the concentration of free steroid inside the cells is lower than that in the interstitial spaces and some steroid might dissociate from the plasma proteins.

C. OTHER EFFECTS OF PLASMA PROTEINS IN TARGET TISSUES

Accumulation of albumin and plasma proteins in the extracellular spaces is characteristic of some target tissues: in particular, the uterus and the prostate (Peterson and Spaziani, 1969; Peck *et al.*, 1973; Cowan *et al.*, 1975; Rosen *et al.*, 1975). It has been suggested that this accumulation may result in an increased concentration of hormones in close proximity to the cells so that the levels of hormones required to produce tissue stimulation are reached specifically in these tissues. According to the model illustrated in Fig. 2, whether such an accumulation of binding proteins will increase or decrease transfer of steroids into the cells depends on the balance between the dissociation of the steroid–protein complexes in the interstitial spaces and the degree of saturation of the receptors within the cells.

It is also possible that the carrier protein may determine which steroids are transferred into the cells by favoring transport of unbound steroids over that of steroids extensively bound, irrespective of the relative total concentration of the hormones in plasma. In a series of experiments on rat kidney, Funder and co-workers (1973) have shown the presence in the tissue of receptors specific for aldosterone, the active mineralocorticoid, and of receptors with equal affinity for aldosterone and for less active corticoids such as corticosterone. However, infusion *in vivo* of labeled aldosterone leads to preferential binding of the hormone to both types of receptors in the kidney, even when high does of unlabeled corticosterone are infused simultaneously. These authors conclude that, owing to the fact that corticosterone binds to CBG to a greater extent than does aldosterone, competition between the two hormones for the receptor is prevented.

Another example of preferential uptake of hormones due to the presence of carrier proteins in blood has been reported by Keller *et al.* (1969). A raised concentration of CBG in the blood of rats treated chronically with estrogen resulted in an enhanced induction of alanine aminotransferase in the liver; no effect was evident on pancreatic alanine aminotransferase. Since the free corticosterone in blood was decreased, the effect had to be attributed to the CBG–corticosterone complex itself. The response was limited to the liver, either because this organ is capable of detaching steroids from the complex with CBG (see Section VI,B) or because liver sinusoids are much more permeable to proteins than are other capillary systems.

The presence of small quantities of specific plasma globulins inside the cells of uterus and prostate has been reported (Peterson and Spaziani, 1969; Milgrom and Baulieu, 1970; Rosen *et al.*, 1975), but the evidence is indirect, and so far, no studies on isolated cells or on tissues perfused free of blood have been conducted to clarify this issue. Transfer into cells of carrier globulins or of their complex with steroids is perhaps possible, since passage of various macromolecules into cells has been demonstrated (Szego, 1978). Moreover, it has been suggested recently that plasma carrier proteins can become "complexed" with the plasma membrane of target cells (Koch *et al.*, 1978).

Finally, an exchange of hormones between plasma carrier proteins and some components, most likely proteins, on the membrane has been hypothesized (Westphal, 1971). This hypothesis has received some support from the recent discovery that proteins binding steroids with high affinity are present on the cell membrane of some target tissues (see Section VI,B).

III. Physicochemical Characteristics of Steroid Molecules

In order to understand the nature of the process of permeation of steroids across the cell membrane, it is necessary first to establish the behavior of the steroid molecules in simple systems such as solvents or lipid membranes that possess physicochemical characteristics resembling those of the membrane components. From the behavior of the steroids in these systems, and from a particular assumed model of molecular arrangement in the membrane, some predictions on the permeability of the cell wall to the hormones can be derived. The extent of agreement between predictions and results of permeability studies in cells will indicate whether the process of permeation in the cell membrane is accurately described by the proposed model. This is the biophysical approach, pioneered by Davson and Danielli (1970), Stein (1967), and Diamond and Wright (1969). The biochemical approach to the investigation of the nature of the process of permeation, that is the search for modifications of cell membrane composition that affect permeability to steroids, will be introduced in the next section, together with studies on the permeability of animal cells to hormones. In this section, the physicochemical characteristics of steroid molecules and of membrane components will be reviewed, together with old and new evidence on the behavior of steroids in solvents and in artificial or biological membranes.

A full description of the composition and physical properties of the plasma membrane of animal cells would be outside the scope of this article. [For a concise, but complete, treatment of this subject, the reader should consult the excellent paper by Bretscher and Raff (1975).] Information essential to a comprehension of steroid–membrane interactions must however be sketched briefly. Plasma membranes are composed of proteins and lipids in a weight to weight ratio varying from 0.5 in myelin to 1.5 in liver membranes; the water content has been estimated at approximately 20%. The lipids are represented by polar compounds (phospholipids, sphingolipids, glycolipids) and cholesterol; the ratio of cholesterol to polar lipids approaches unity in plasma membranes. The phospholipids are amphipathic molecules because they present both highly hydrophobic regions, the acyl chain of the fatty acids, and a polar head, the glycerol-phosphoric acid group. Cholesterol is also highly hydrophobic in its cyclopentanoperhydrophenantrene ring and lateral branched chain, with only a hydroxyl (OH) group at C 3 capable of forming hydrogen bonds. The molar ratio of cholesterol:polar lipids and the length and degree of saturation of the acyl fatty residues of the polar lipids are the main factors that influence membrane permeability (Brockerhoff, 1974). The structure of the acyl chains of the phospholipids could provide an almost infinite range of variations in the composition of membrane lipids. Membrane proteins are also capable of forming both hydrogen and hydrophobic bonds; their polarity is determined principally by their position relative to the lipid matrix and by the degree of order of the surrounding water molecules (Urry, 1972). Some of the membrane proteins have been identified as enzymes, immunoproteins, or complexes involved in transport of solutes, but the function of most of the proteins present in the membrane remains unknown. The exterior of the plasmalemma is coated by the glycocalyx, a thin layer formed by chains of glycoproteins, protoglycans, and lipids. It is not clear as yet whether the glycocalyx is an integral part of the membrane.

A unifying concept of the molecular arrangement of the membrane that explains all the known data on selective permeation of solutes has not been produced as yet. The model, introduced in 1972 by Singer and Nicolson to explain the lateral movements of membrane components, has received numerous experimental confirmations and is the most generally accepted. According to this model, the membrane lipids are arranged in a bilayer, with the polar heads of the phospholipids at the external and internal surfaces of the membrane; intercalated within the lipids, there are cholesterol molecules and proteins, some of which are loosely adsorbed at the surface, some partially embedded in the lipid matrix, and some spanning the whole thickness of the membrane. Intrinsic long-range ordering of membrane components is absent, and the membrane is in a liquid crystal state: hence the denomination of "fluid mosaic."

The alternative model of Solomon (1960) was proposed in the attempt to explain the anomalously high permeability of biological membranes to water. As we shall see shortly, the pore model is not relevant to transport of steroids across the membrane. The "fluid mosaic" concept of Singer and Nicolson (1972) will therefore be the model representing the membrane structure when the nature of the permeation process of steroids across membranes is discussed in Section IV.

A. Energetic Considerations

The methylene groups forming the backbone of the steroid molecule confer a strong hydrophobic character to the steroid nucleus, whereas the oxygenated groups contribute a certain degree of hydrophilicity. Like all amphipathic molecules, steroids can form both hydrogen and hydrophobic bonds with other molecules. The hydrogen bonds are formed between a donor (OH, O or, less probably, H) and an acceptor atom when the two atoms are at a distance of between 0.25 and 0.34 nm. According to Vandenheuvel (1963), because of the restricted mobility of the lipid molecules in the membrane bilayer and because of their bulkiness, the colinearity condition essential for hydrogen bond formation is impossible in many instances; the chances for indirect hydrogen bonding through water molecules appear to be greater, particularly with molecules such as steroids, which possess some affinity for water. Formation of hydrogen bonds is spontaneous and breaking of an OH—H bond requires an expenditure of energy approximately 16–42 kJ/mole: this energy requirement might affect, first of all, the rate of entry of the permeant into the membrane (Stein, 1967). It is probably significant in this regard that the activation energy for diffusion across acetyl acetate membranes has a value of 11.5 kJ/mole for progesterone, a compound with no hydrogen function, and a higher value, 16.1 kJ/mole, for testosterone, which bears an OH at C 17 (Barry and El Eini, 1976).

The energy of activation for the hydrophobic bond is approximately 5.8 kJ for each pair of methylene groups 0.3 nm apart. Since at least ten pairs of —CH_2— may be involved and the van der Waals forces are additive, a total of 58 kJ/mole steroid can be reached (Vandenheuvel, 1963). van der Waals forces cause interactions also between proteins and steroids; values of 80 kJ/mole have been reported for binding of hormones to specific receptors, but the values are much lower, 11.5 kJ/mole, in nonspecific adsorption (Williams and Gorski, 1973).

The hydrophobic forces are of sufficient magnitude to remove the steroid molecules from the aqueous phase into the membrane and form associations of steroids with the lipids in the nonpolar region of the membrane, or with proteins, that favor the diffusion of the steroid molecules across the cell wall. Permeability is directly proportional to the hydrophobicity of the permeant; lipid membranes are more permeable to steroids with extended lateral chains, as for instance $(CH_2)_3-(CH_2)_{10}$ saturated esters of testosterone (Stevens and Green, 1972).

Transfer of a methylene group from the aqueous phase to a completely nonpolar phase typical of lipids involves a reduction of free energy of 2.9–3.3 kJ/mole —CH_2—: since the change in free energy during adsorption by erythrocytes of hydrophobic compounds, namely anesthetic alcohols, is -2.8 kJ/mole, it can be argued that the site of interactions of this type of compounds is the lipid matrix (Seeman, 1972). However, transfer from water into the nonpolar region of proteins reduces free energy by 0.42–2.10 kJ/mole —CH_2—: as a consequence, in facilitated diffusion by carrier proteins, the energy of activation for membrane permeation is lower than expected from the solubility of the permeant in lipids (Stein, 1967). Thus, the activation energy for facilitated diffusion of hydrocortisone into liver cells is 75 kJ/mole, whereas the activation energy for simple diffusion of prednisolone into hepatoma cells is 81.4 kJ/mole (Rao *et al.*, 1976b; Graff *et al.*, 1977).

Another energetic parameter that affects permeability is the external temperature. Temperature dependence has been observed in most, if not all, studies on cell permeability to steroids (see Sections IV and V), although the coefficient of temperature Q_{10} (the increase in permeability for a difference of 10%C) has been measured only in a few instances. An increase in permeability with temperature depends on a greater fluidity of the membrane lipids and on a greater kinetic energy of the permeant molecules (Davson and Danielli, 1970; Sackmann and Träuble, 1972). In transport by facilitated diffusion, acceleration of the dissociation of the carrier–protein complex may accompany a raise in temperature; this effect is generally small, and it would be impossible to discriminate between simple and facilitated diffusion on the basis of the temperature coefficient. Very high coefficients, $Q_{10} > 5$, are observed in active transport, as this form of transport requires energy from external sources.

B. Mass, Configuration, and Conformation

The dimensions of a steroid molecule with 19 carbon atoms are $1.5 \times 0.7 \times 0.5$ nm, approximately. The estimated size of the hypothetical membrane pores is below 0.5 nm (Sha'afi *et al.*, 1973), which would seem to exclude molecules of the size of the steroids. Variations in mass from one steroid to another are very small: 270 MW for estrone, the compound with the smallest molecular weight among the natural steroid hormones, to 360 for hydrocortisone, the steroid hormone with the highest molecular weight. Some of the synthetic, biologically active steroids have larger mass, reaching 390 MW in dexamethasone. However, the minimum cross-sectional area is practically the same for all steroids, since it is independent of mass. The cross-sectional area seems to be all important for the insertion of permeant molecules between the spaces separating the acyl fatty chains of the phospholipids in the membrane (Fig. 3). It can be expected that changes in configuration and conformation of the steroid molecules may affect their progress through the membrane if the shape is not fitted closely to that of the lipids. Studies in artificial and biological membranes have indeed shown that conformational changes, such as those occurring in the structure and in the degree of unsaturation of the steroid ring, have quite distinct effects on the functional properties of lipid bilayer membranes. Only a few of the many configurational changes that are possible in steroid molecules bearing functional groups have been examined from this point of view; namely, the presence of a lateral chain at C 17 and the axial-equatorial configuration of the OH group at C 3. Isomers with respect to the 5α-5β ring structure or the C 3 OH α-β configuration have the same partition coefficient in lipids; the different effects on membrane permeability that are described below can therefore be attributed exclusively to molecular shape. It must be noted that these effects are apparent at concentrations of steroids in the millimolar range, that is, three orders of magnitude greater than the physiological concentration of hormones in the blood of mammals. Incorporation of 2-3 moles of steroids per 100 moles of membrane lipids is probably necessary to disrupt the ordering of the lipid bilayer and alter the permeability to solutes of lipid membranes (Butler et al., 1970; Heap et al., 1970; Weissmann et al., 1976). Physiological doses of testosterone, estradiol, and cortisone can, however, modify the permeability of lysosomes to enzymes in target tissues (Szego et al., 1971).

1. Conformational Changes

The 5α conformation of the steroid ring renders the molecule flat; in the 5β conformation, on the contrary, the A ring of the steroid is set at an angle to the other rings (Fig. 4). Unsaturation of the ring increases the rigidity of the molecule. It is conceivable that in an ordered array of lipids the flat, more rigid molecules are better accommodated in the interstices, whereas the angular molecule might tend to separate adjacent acyl chains, creating pockets of free volume. 5β steroids are in fact less mobile in artificial and biological membranes than the 5α isomers (Butler et al., 1970). They can also enhance the permeability of membranes to solutes. An increased permeability to cations has been observed in multilamellar liposomes of lecithin–cholesterol in the presence of etiocholanolone and 5β-pregnane derivatives (Bangham et al., 1965; Weissmann et al., 1965, 1976).

At concentrations of 5×10^{-4} M, etiocholanolone and 5β-pregnane deriva-

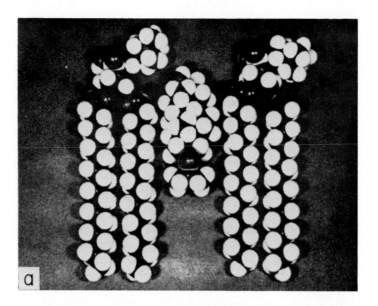

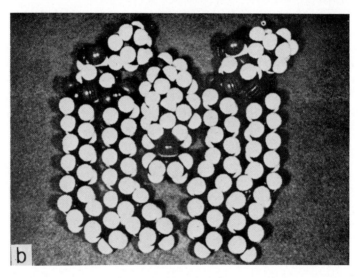

FIG. 3.   Steroid molecule sandwiched between two lipid molecules. The OH group at C 3 of the steroids is assumed to be anchored within the polar interface. (a) The hydrocarbon chains of the lipids are considered as rigid rods. (b) In order to obtain a better incorporation of the steroid nucleus between the lipid molecules, "kinks" (rotational isomers) of the hydrocarbon chains have been formed. (Reproduced from Sackmann and Träuble, 1972, with kind permission of The American Chemical Society.)

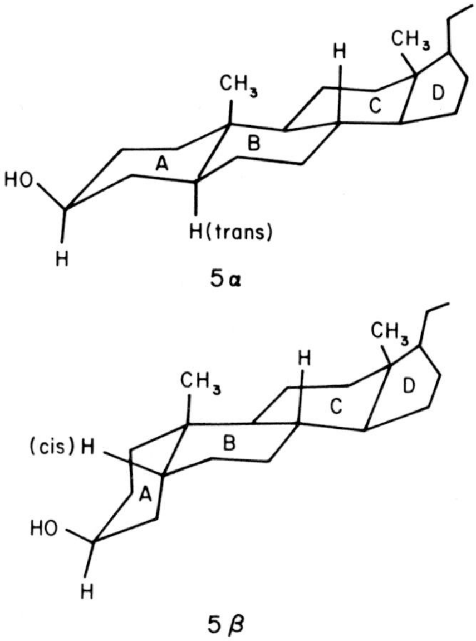

FIG. 4. The 5α and 5β conformation of the steroid nucleus. In both steroids, the CH_3 at C 13 and the side chain at C 17 are β(cis) with respect to the reference methyl group at C 10. The H atom attached to C 5 is, however, trans in the 5α steroid (rings A/B trans) and cis in the 5β isomer (rings A/B cis). In the "chair" form of the rings shown in the figure, the OH group at C 3 is axial (α), i.e., it is at right angles to the plane running through the rings of the steroid nucleus.

tives are active also on cell membranes: they enhance the permeability to enzymes of lysosomes from rabbit liver and cause hemolysis of erythrocytes. All the 5α compounds tested were, in contrast, inactive (Weissmann et al., 1965; Weissmann and Keiser, 1965).

Unsaturation of the steroid ring, e.g., the phenolic ring in estradiol, renders the molecule more effective in preventing release of CrO_4^{2-} from liposomes. However, hydrocortisone, a Δ_4 unsaturated steroid, is even more effective. Formation of hydrogen bonds between the three OH groups of the hormone and the phospholipids contributes to tightening of the membrane (Sessa and Weissmann, 1968; Weissmann et al., 1976).

The ring conformation is important also in interactions of steroids with membrane proteins. Thus, at concentrations of 2–10×10^{-4} M 5α-androstanedione inhibits glucose transport in erythrocytes noncompetitively, but, at the same concentrations, the 5β isomer acts as a competitive inhibitor (Lacko et al., 1975). It is possible that, because of the flat shape of the molecule, the 5α isomer binds both to the glucose carrier and to other sites, giving rise to noncompetitive inhibition. The nonflat 5β molecules would instead bind only to the glucose carrier.

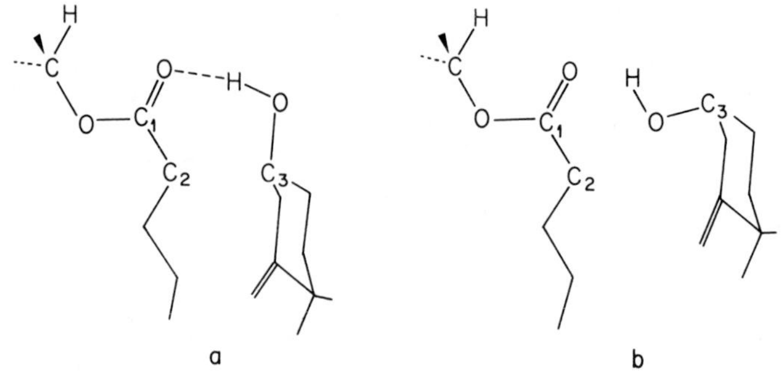

FIG. 5. Plausible spatial orientation of a 3β-hydroxy-pregn-5-en steroid (a) and of a 3α-hydroxy-pregn-5-en steroid (b) in the polar interface. Hydrogen bond formation between the phospholipid carbonyl C 1 and the 3β-hydroxyl group of the steroid is given in (a) as a dashed line. (Reproduced from Huang, 1976, with kind permission of Macmillan Journals).

2. Configurational Changes

According to studies with cholesterol derivatives in liposomes and erythrocytes, permeability of the membrane is reduced more by compounds with an equatorial OH at C 3 than by compounds with an axial OH (Masiak and LeFevre, 1974; Demel *et al.*, 1972). Huang (1976) attributes this difference to the fact that the sterol molecule is oriented at the interface with the aqueous phase in such a way that an equatorial OH at C 3, but not an axial OH, can form a hydrogen bond with the carbonyl oxygen at C 1 of the acyl chain of the phospholipids (Fig. 5). The hydrogen bond between the 3β-OH and the carbonyl at C 1 would somehow anchor the cholesterol molecule at the interface, thus increasing the tightness of the membrane surface. This hypothesis is supported by the findings of Lawrence and Gill (1975), who have observed that pregnane derivatives bearing a 3β-OH group do not affect the fluidity of liposomes composed of lecithin–cholesterol mixtures, while, on the contrary, compounds with the 3α-OH configuration increase the fluidity significantly. Studies with spin labeled steroid molecules (Hubbell *et al.*, 1970) suggest that an equatorial OH group at C 17 might also interact at the polar region of the membrane surface, since motion of C17β-OH molecules is restricted in biological membranes (see Section III,E).

C. PRESENCE OF POLAR GROUPS

The presence of polyfunctional oxygenated groups constitutes the main difference between steroid hormones and sterols, and no doubt contributes to the biological potency of the steroid molecules. Recognition of the position and configuration of functional groups on the steroid molecules by the intracellular receptors is in fact a prerequisite of hormone action (King and Mainwaring,

1974; King, 1976). The presence of polar groups determines also the solubility of the steroids in the membrane. This is mainly because of an increase in the activation energy for permeation: the more hydrogen bonds that have to be broken to transfer the molecules from water into the membrane, the more energy is expended (Stein, 1967). Therefore, while the solubility of steroids in water increases with the number of polar groups attached to the ring, the solubility in lipid membranes, in organic solvents, and in protein solutions decreases. This behavior of steroid in solution has become known as the *polarity rule*.

The residual hydrophobicity of steroids bearing OH groups is generally assessed from the distribution of the compounds between an aqueous and an organic phase: the partition coefficient. Partition coefficients of steroids in solvent systems of increasing polarity and in liposomes are collected in Table II, together with data on the solubility of hormones in water and in a solution of 5% albumin. The solvent systems have been chosen because of some properties similar to the physicochemical characteristics of the lipid bilayer or of regions of the bilayer. For instance, the least polar solvent hexane (a hydrocarbon) may have solubility properties similar to those of the interior region of the membrane formed by the hydrocarbon chains of the polar lipids; ether is also composed of methylene and methyl groups, but the —0— in the molecule may function as acceptor of hydrogen bonds; the most polar solvent octanol may reproduce some of the properties of the water–membrane interface, where the polar head groups of the lipids are situated. The solvent with both polar and nonpolar character, amyl caproate, may give some idea of the amphipathicity of the whole lipid bilayer (Wolosin *et al.*, 1978). From an inspection of Table II, it is evident that the polarity rule applies to steroids in all the systems: the solubility in water increases from the least to the most polar steroid, whereas it decreases in all other solvents. However, the ability to discriminate between polar and nonpolar steroids is different for each solvent. The greatest variations in solubility with increasing number of OH groups are observed in lecithin membranes (Column 9), the smallest in octanol (Column 7). Incorporation of cholesterol in liposomes enhances the solubility of the more polar steroids (Column 8).

Addition of more than one polar function has, however, a very different effect on the solubility of the steroids in the various solvents. From the data of Scheuplein and collaborators (1969) on a very carefully chosen series of steroids (some of which are not reported in Table II), an analysis of these effects in hexane and in amyl caproate can be made (Table III). In the apolar hexane, the first hydroxyl group added to C 18, C 19, or C 21 compounds has a much smaller effect than addition of a second group, whereas a third group has little effect. A similar trend can be observed in two other hydrocarbon solvents, heptane and decane (Beckett and Pickup, 1975; Giorgi, unpublished). The introduction of a carboxyl function together with a hydroxyl function (as in cortisone) diminishes the effect on solubility of the hydroxyl group, whereas insertion of a fluorine atom at C 9 (as

TABLE II. Steroid Solubility in Water and 5% Albumin and Partition Coefficients in Organic Solvents

Steroid	Solubility ($M \times 10^{-3}$)[a]		Partition Coefficients					
	Buffer	5% Albumin	Hexane/ water[b]	Ether/ water[c]	Amyl caproate/ water[b]	Octanol/ water[c]	Liposomes/ water[d]	Liposomes/ water[e]
C 21								
OH = O								
Progesterone	48.0		17.0	116.0	56.0	67.0	694	2601
OH = 1								
Deoxycorticosterone	107	1.03			46.0		93	513
OH > 1								
Corticosterone			0.024	6.0	6.8	46.0		
Cortisone	162	1.40	0.28	4.5	1.5	5.2	73	121
Hydroxycortisone			0.009	1.5	1.3	12.6		
Aldosterone							29	45
C 19								
OH = 0								
Androstenedione	140	2.50						315
OH = 1								
Dehydroepiandrosterone	104	2.85						1128
Dehydroepiandrosterone sulfate	172			0.04				194
Testosterone	101	2.15	2.6	30.0	16.0	65		514
5α-Dihydrotestosterone				82.0		87		
C 18								
OH = 1								
Estrone	8.2		3.0		80.0		477	4428
OH = 2								
Estradiol	15	0.31	0.63	49.0	66.0	98		2314
OH = 3								
Estriol			0.23					

[a] Equilibrated for 72 hours with Locke-Ringer's buffer (Eik-Nes et al., 1954).
[b] Radioactive steroids (Scheuplein et al., 1969).
[c] Radioactive steroids (E. P. Giorgi and W. D. Stein, unpublished data).
[d] Liposomes: 90% lecithin, 10% cholesterol. Equilibrated overnight (Snart and Wilson, 1967).
[e] Liposomes of egg lecithin. Equilibrated for 24 hours at 37°C (Heap et al., 1970).

TABLE III

EFFECT OF HYDROXYL GROUPS ON STEROID SOLUBILITY IN HEXANE AND AMYL CAPROATE[a]

Steroid ring	Order of addition	In hexane 1	In hexane 2	In hexane 3	In amyl caprate 1	In amyl caprate 2	In amyl caprate 3
Androst-4-en-3-one	17β-OH	6.5			3.4		
Pregn-4-en-3,20-dione	17α-OH	6.8	30.0	2.7	1.2	4.1	5.2
	17α-OH+CO		10.8			25.0	
	21-OH	5.7	25.0		1.9	2.6	
	11β-OH		125.0	11.0		4.4	8.6
	16αCH$_3$-9α-F			+56.0[b]			
Pregn-5-en-3-ol	17α-OH	2.7			1.1		
Estratriene	17β-OH	4.7			1.2		
	16α-OH		2.7		50.0		

[a] Data from Scheuplein *et al.*, 1969.
[b] In decane (E. P. Giorgi and W. D. Stein, unpublished data).

in dexamethasone) actually increases the solubility by over 50% because of the electron withdrawing property of the fluorine atom. Whereas 17α-OH and 21-OH give equivalent changes in solubility of the pregn-4-en-3-one compounds, addition of an 11β-OH decreases dramatically the solubility of the steroid; this is evidently a steric effect, due to the modification of the β, nonpolar side of the steroid molecule. On the other hand, all these trends seem to be reversed in the more polar solvent, amyl caproate, apart from the equivalence of 17α-OH and 21-OH groups. As can be seen from Table III, the greatest reduction of solubility in amyl caproate occurs on addition of the third OH group. A CO group decreases solubility further, and an 11β-OH is not more effective than groups at other positions of the ring. While the solubility of steroids in octanol and ether seems to conform to that in amyl caproate, in liposomes of lecithin, with or without added cholesterol, the greatest reduction in solubility is seen with the first OH attached as in the hydrophobic solvent hexane (see, for instance, progesterone versus deoxycorticosterone or versus testosterone in Table II). A similar finding in aqueous sols of lecithin has been reported by Hayes and Saunders (1966). This might indicate that the presence of polar groups at the end of the steroid molecule, which are involved mainly in interface phenomena, determine the solubility of the compounds in completely hydrophobic lipid phases.

It was implied earlier that solubilization of steroids by albumin is a function of the hydrophobicity of the protein. This calls for further clarification. If hydrophobic forces are at play in this interaction, a positive correlation should be demonstrable between binding affinity of the protein for the steroids and degree

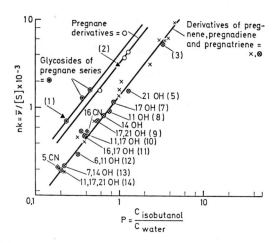

FIG. 6. Coefficient of correlation between the apparent association constant of steroids for 5% albumin (on the ordinate, see text for explanation of symbols), and the partition coefficient for the isobutanol/water system (P, on the abscissa). The numbers in parentheses indicate: 1,2, C 18 steroids; 3, progesterone; 5, deoxycorticosterone; 4,7–13,15, pregn-4-en-3,20-diones bearing one or more OH groups. (Reproduced from Scholtan *et al.*, 1968, with kind permission of *Artzeinmittel Forschung.*)

of hydrophobicity of the hormones as expressed by the partition coefficient. In Fig. 6, the partition coefficient in isobutanol/water of a number of pregnane derivatives bearing from 0 to 3 hydroxyl groups is plotted against the affinity for 4% albumin, $nk = \bar{v}/[S]$, where n is the number of binding sites on each albumin molecule, k is the association constant, $\bar{v}$ is the average number of steroid molecules bound per molecule of albumin, and [S] is the total concentration of steroid in the solution. The correlation between affinity for albumin and hydrophobicity of the steroids is excellent, which tends to prove that binding is due to hydrophobic forces. This type of binding can be best described as nonspecific adsorption (Westphal, 1971). Nonspecific adsorption of steroids to hydrophobic membrane proteins should therefore follow the polarity rule. On the other hand, binding to carrier proteins or to other membrane proteins specific for steroids will deviate from the polarity rule, since the steroid molecule might attach by a "lock and key" fit to special binding sites in the protein.

D. ELECTRICAL CHARGES

Functional groups of steroids, with the exception of the phenolic hydroxyl group of estrogens, are not dissociated at neutral pH, with the result that unconjugated steroids do not carry any electrical charge. The degree of dissociation of the phenolic hydroxyl group is, however, very small, viz. $pK > 10$ (Gower, 1975).

Steroid conjugated with sulfuric or glucuronic acid, in contrast, may carry an electrical charge, since their salts dissociate in almost one-third of the molecules at neutral pH. Sulfonated steroids behave as weak bases and steroid glucuronides behave as weak acids. Hydrolysis of the ester link does not seem to precede entry of the conjugates into endometrial, placental, or intestinal cells (Kapstein *et al.*, 1967; Tseng *et al.*, 1972a,b), and it is probable that the free, unionized form of the compounds penetrates the cells. Weak acids or bases of low molecular weight have been observed to cross the membrane in this form (Osterhout, 1935). It is therefore unlikely that permeation of steroid conjugates creates an electrical gradient across the cell membrane.

Few studies have been reported on the solubility of steroid conjugates. The solubility in water and the partition coefficients of dehydroepiandrosterone sulfate (DHAS) in lecithin vesicles, octanol, and ether are shown in Table II. The partition coefficient in lecithin, although reduced with respect to the corresponding free form, is comparable to that of the most polar steroids such as aldosterone or hydrocortisone. On the other hand, of the compounds listed in Table II, DHAS is the most soluble in water. Because of their hydrophilicity and because they bind only to albumin in plasma, the rate of removal of conjugated steroids from the circulation is much faster than removal of the unconjugated compounds (Gurpide *et al.*, 1963; Giorgi and Crosignani, 1969).

E. ORIENTATION AND MOVEMENT OF STEROIDS IN LIPID MEMBRANES

Physical techniques such as X-ray and neutron diffraction, nuclear magnetic resonance, and electron spin labeling have afforded the most direct evidence on the orientation and movement of molecules in artificial and biological membranes. Analyses of electron spin resonance spectra of spin labeled steroid molecules have been performed by Hubbell and co-workers (Hubbell and McConnell, 1969; Hubbel *et al.*, 1970) and Sackmann and Träuble (1972). Androstane derivatives labeled at C 3 with a nitroxide possessing an unpaired electron in a 2 π orbital at the nitrogen atom have been used in these studies. In sonicated dispersion of phosphatidylserine, the molecules of 5α-dihydrotestosterone and of the 5β isomer show no restriction in mobility in any direction around the long axis. This means that the different positioning of the nitroxide group of the 5α and 5β compound in the hydrophobic and hydrophilic regions of the membrane, respectively, has no effect on the motion of the molecules. The absence of anisotropic motion of steroid molecules in lipid bilayers has been confirmed by Sackman and Träuble (1972). These authors investigated in particular the effect of the concentration of labeled molecules and of temperature on the movement of spin labeled steroids in the bilayer. Up to a molar ratio of 1 mole of label to 100 moles of lipids, and above the transition temperature, no paramagnetic or spin exchange interactions among the spin labeled molecules are observed. Above the transition temperature, the bilayer is in a liquid–crystal state and the steroid molecules are randomly dispersed among the lipid chains. In these conditions, the rate of tumbling in all directions of the steroid molecules is very high, 0.5–0.3×10^{-8} seconds, i.e., on the order of tumbling found in organic solvents; the average travel distance is 1000 nm/second, one-thousandth the average distance traveled in self-diffusion in liquids. Analysis of the electron spin resonance spectra seems to exclude the possibility that steroid molecules are dissolved in the aqueous phase of the bilayer; on the contrary, it suggests that the steroid molecules move through the monolayer by free diffusion, along small pockets of free volume created by structural defects that develop along the acyl fatty chains. As shown by Hubbell and McConnell (1971) in membranes into which spin labeled lecithin had been incorporated, these structural defects, or "kinks", are rapidly interconvertible isomeric states of cis-trans conformation around single C—C bonds. These "kinks" produce some degree of static and dynamic disorder, especially below the C 8 of the acyl chain. The space between the polar head group and the gauche formation at C 8 corresponds roughly to the length of the steroid molecule (see Fig. 3). This might be the region of the membrane where the shape of the steroid molecule influences the interaction with the phospholipids, since these are arranged in parallel arrays. Moving toward the terminal methyl group of the lipid chains, the array becomes less and less ordered (Hubbell and McConnell, 1971).

The orientation and movement of spin labeled steroid molecules are, however, different in biological membranes (Hubbell and McConnell, 1969; Hubbell *et al.*, 1970). Both in unmyelinated nerve fibers from the walking leg of the lobster *Homarus americanus* and in human erythrocyte membranes, the steroid molecules show highly anisotropic motion. They rotate preferentially along the long axis, which is oriented perpendicular to the membrane surface; rotation about the other two axes is restricted. A hydroxyl group at C 17 seems to be bonded to the polar head groups of the phospholipids, and this might cause the restriction in movements of the steroid molecules around the horizontal axis.

The effect of various steroids and sterols on the degree of order in membranes composed of lipids extracted from bovine and human tissues has been investigated by Butler and co-workers (1970). This analysis was made possible by incorporating spin labeled cholestane molecules in the lipid films. In concentrations of 1–40 mole per 100 moles of lipids, neither saturated nor unsaturated androstane derivatives seem to have any ordering effect on the lipid membrane. Furthermore, it is suggested that the presence of an hydroxy group either at C 17 or at C 21 renders a steroid molecule completely ineffective in producing ordering of the lipid layer. This latter conclusion, however, seems at variance with the finding, previously mentioned, that hydrocortisone, a compound with an hydroxyl group at C 21, stabilizes lipid membranes (Weissmann *et al.*, 1976).

Thus, all these studies with spin labeled molecules indicate that androstane derivatives permeate artificial and biological membranes very fast through diffusion across the lipid matrix. The steroid molecules do not seem to react with proteins in intact membranes. In both unmyelinated nerve fibers and erythrocyte membranes, however, some interaction is observed after the membrane proteins are exposed by treatment with lytic doses of benzyl alcohol (Hubbell *et al.*, 1970).

IV. The Permeability of Cell Membranes to Steroid Hormones

Permeability of cells to solutes can be defined as the velocity of passage of solutes through the plasma membrane; therefore, it is assessed by measurements of velocity: the permeability coefficient P and the rate constant of transfer. Direct measurement of the permeability coefficient is based on the assumption that the nature of the permeation process is simple diffusion (Stein, 1967), whereas measurement of rate constants of transfer does not presuppose any particular model of the membrane and can be applied equally to simple and facilitated diffusion, or even to active transport. The unit of reference in P is the surface area of the cells. It would be practically impossible to determine this area accurately in cells of target tissues, such as endometrium or prostate, since the cells discharge secretory products into the glandular lumen and thus present many irregularities of the luminal surface. In these tissues, permeability is best assessed from the rates of transfer, where the relevant unit is weight. From the determination in one type of cells or tissue of either the permeability coefficient or of the rates of transfer for various steroids, it is possible to recognize the pattern of membrane permeability proper to each type of cells or tissue. A comparison of this pattern with the behavior predicted [on the basis of solubility studies (Section III,C)] for steroids in a membrane conforming to the "fluid-mosaic" model can give, as mentioned earlier, important information on the nature of the process of permeation of steroids across the cell wall. Further information on the nature of this process, and in particular on whether or not proteins are involved, can be obtained by biochemical studies. Biochemical studies on the effect of changes in membrane composition on steroid transport into animal cells will be described at the end of this section.

A. METHODS OF STUDY

The classical method for studying transport of solutes across the cell membrane is the zero trans procedure. This consists in measuring the change of concentration of the transported solute as a function of time, when the solute is initially present only at one face of the membrane. There are two zero trans experiments: influx and efflux. In influx experiments, the steroid, generally labeled with tritium, is added to the medium and the radioactivity is determined in the cells after set intervals. In the efflux experiments, the cells are first equilibrated with the steroid and then transferred to medium containing no steroid: the radioactivity released into the medium is measured at various times. The efflux method, however, is not often used because some metabolism of the steroids might occur during equilibration and this would invalidate the procedure. In order to obtain measurements meaningful in terms of cell membrane permeability, the transport of steroids must be calculated from the initial rate, the linear portion of the curve, transported steroid versus time. Transport of steroids, as we shall soon see, is very fast, and the initial portion of the curve remains linear, at 28°C, only for a few seconds in the case of some steroids, but never for more than 50 seconds, even in the case of the slowest compounds (Rao *et al.*, 1976b; 1977a,b; E. P. Giorgi and W. P. Stein, unpublished, 1978). Therefore, a critical feature in the experiments is the control of the time of exposure of the cells to the steroid, i.e., from the time of mixing of the solutions to the moment of cell separation.

Another well-known procedure for transport studies is that of equilibrium exchange. The concentration of labeled substrate is kept the same at the two faces of the membrane and the unidirectional flow of isotopically labeled substrate is followed in either direction across the membrane. The isotope tracer method of Gurpide and Welch (1969) may be considered an equilibrium exchange procedure. In this method, tissue is perfused with a continuous flow of medium containing labeled steroids, and reaches, within 60 minutes, a steady state with respect to steroid concentration in both tissue and perfusate [Eq. (1); Eq. (1a) in Fig. 7]. At this point, all the binding sites with high affinity in the tissue must be saturated and the free steroid outside must equal the free steroid inside the cell (Fig. 2a). The difference between this method and the usual equilibrium exchange procedures is that here both the steroid at equilibrium and the steroid transported out of the tissue are labeled, but the released steroid can be distinguished because of intracellular labeling from a precursor (Fig. 7). The unidirectional flow out, (β), is measured directly (see legend to Fig. 7), while the unidirectional flow in, (α), is calculated from:

Unidirectional flow in = unidirectional flow out + irreversible metabolism (1)

where the symbols of Eq. (1a) in Fig. 7 have been substituted by words. It is self evident that this method can be applied only to tissues that convert and/or metabolize steroid hormones.

In tissues in which steroid transport is relatively slow, but reaches a constant flux after a finite time, mathematical methods are used to derive a constant k with dimensions of velocity, sec $^{-1}$. If we assume that the concentration of steroid in the external medium remains unchanged (as is the case in experiments in which the volume of medium greatly exceeds that of the tissue) and that the rate constants of entry and exit are equal, the curve of net efflux versus time is

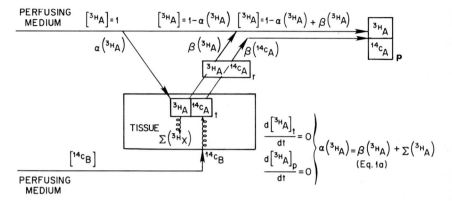

FIG. 7. The isotope tracer method for equilibrium exchange (Gurpide and Welch, 1969). The straight arrows represent movements of steroids and the coiled arrows represent conversions. α, Entry; β, release; $\Sigma(^{3H}X)$, sum of the concentrations of the irreversible metabolites of ^{3H}A; [ ], concentration of steroid in the medium. Subscripts: t, tissue; p, perfusate; r, released from tissue. Slices of tissue are perfused with a continuous flow of medium containing two steroids one the precursor of the other, and each labeled with a different isotope, ^{3H}A and ^{14C}B, until the steady state is reached (Eq. 1a). The steroids are then isolated from the tissue and the perfusate, identified by chromatographic techniques and the ratio $(^{3H}A/^{14C}A)$ measured. At the steady state, as can be demonstrated experimentally, the ratio $(^{3H}A/^{14C}A)_t$ equals the ratio $(^{3H}A/^{14C}A)_r$, and $\beta(^{3H}A) = (^{3H}A/^{14C}A)_t \times (^{14C}A)_p$. Also, the difference between $[^{3H}A]$ and $(^{3H}A)_p$ equals $\Sigma (^{3H}X)$, and $\alpha(^{3H}A) = \beta(^{3H}A) + \Sigma(^{3H}X)$.

expressed by:

$$[S^i]_t = [S^o] (1 - e^{-kt}) \qquad (2)$$

where $[S^i]$ and $[S^o]$ are the concentration of (free) steroid in the medium and in tissue, respectively, t is time, and k is the first-order rate constant. The first derivative of Eq. (2) is:

$$[S^i] = k[S^o] \qquad (3)$$

and k can be measured from the intercept on the $[S^i]$ axis at $t=0$ (Graff et al., 1977).

B. PERMEABILITY COEFFICIENTS

The permeability coefficient P is defined as the number of molecules of the penetrating species crossing, in unit time, a unit area of the membrane when a unit concentration difference is applied across the membrane. Therefore P has the dimensions of a velocity, and is expressed in cm sec $^{-1}$ (Stein, 1967). The permeability coefficient can be measured directly by the zero trans procedure, if

Fick's law applies, from:

$$\frac{dn}{dt} = P \, A \, dc \tag{4}$$

where n is the number of molecules of permeant crossing area A and c is the concentration gradient. In the case of steroids, since it is impossible to measure the concentration of free steroid in the cells, it is necessary to work with hormone-depleted cells. This method has been applied to cultured cells, hamster fibroblasts (NIL 8), and hepatoma cells (hypertetraploid cells, HTC) grown in monolayer (E. P. Giorgi and W. D. Stein, unpublished, 1978). The permeability of oral mucosa and skin has instead been calculated from the constant flux [Eq. (3)], but in oral mucosa, the derived rate constant k has not been normalized for area. However, the values measured in skin and oral mucosa do not depend exclusively on the permeability of the cell membrane to steroids, but reflect also penetration of the compounds through the intercellular spaces or specialized cellular structures, e.g., hair follicles and sweat ducts in skin (Scheuplein *et al.*, 1969; Beckett and Pickup, 1975).

The coefficients of permeability of human skin, human oral mucosa, and of cultured cells are collected in Table IV. It must be noted that the rate constant of transport in these cells or tissues does not vary with steroid concentration in the medium over a nanomolar to micromolar range, and therefore Fick's law applies. It can be seen that, in general, the pattern of permeability to steroids of the cells and tissues in Table IV bears a closer resemblance to the pattern of solubility of the compounds in organic solvents than to the pattern of solubility in either water or 5% albumin (Table II, Section III,C). In particular, the ability of the cell membrane to discriminate between polar and nonpolar steroids is very great and much greater than that of protein solutions. Since compounds with no polar functions, those most soluble in organic solvents, cross the membrane fastest and the polar steroids, the least soluble, are the slowest, it can be concluded that the steroid molecules become dissolved in the lipid matrix of the cell membrane. Thus, while the molecular interactions governing steroid permeation across the cell membrane will be the same as the interactions governing steroid partition between water and a bulk lipid phase, interactions of the compound with either membrane proteins or the glycocalyx are most probably of minor importance. In other words, the nature of the process of permeation of steroids into fibroblasts, hepatoma cells, and epithelial cells of the skin and oral mucosa is simple diffusion through the lipid bilayer of the membrane.

The selective permeability of the cell membrane to polar and nonpolar steroids deserves a closer examination, first of all because, as mentioned earlier, it may provide further insight into the process of permeation; second, because of its possible implications in hormone–tissue relationships. It would appear that the pattern of permeability to steroids in hamster fibroblasts and hepatoma cells

TABLE IV

PERMEABILITY COEFFICIENTS OF STEROIDS IN BIOLOGICAL AND ARTIFICIAL MEMBRANES

| | Permeability coefficient (cm sec^{-1}) | | | | |
| | Cultured cells[a] | | | | |
Steroid	NIL 8 ($\times 10^{-4}$)	HTC ($\times 10^{-4}$)	Skin[b] ($\times 10^{-7}$)	Oral mucosa[c,e] ($\times 10^{-3}$)	Lipid membrane[d] ($\times 10^{-4}$)
Progesterone	3.41	3.50	4.00	9.0	1.9
20α-OH-progesterone			1.78		
Deoxycorticosterone			1.28	7.5	
Cortexolone			0.20	0.9	
Corticosterone	0.78	0.99	0.17		
Cortisone	0.32	0.35	0.0026		
Hydrocortisone	0.38	0.11	0.50		1.0
Dexamethasone	0.51	0.54			
Aldosterone			0.20		
Dehydroepianodrosterone				12.0	
5α-Dihydrotestosterone	1.61	2.70		17.0	
Testosterone	0.61	0.66	2.0	5.4	
Estrone			0.007	7.5	
Estradiol	5.00	3.50	0.007	17.5	

[a] Cells grown in monolayer; area measured by microphotography (E. P. Giorgi and W. D. Stein, unpublished 1978).

[b] Data from Scheuplein et al., 1969.

[c] Data from Beckett and Pickup, 1975.

[d] Bilayer of phosphatidylethanolamine (70%), phosphatidylserine (30%) in 2% decane (M. Razin, R. Simons, and H. Ginsburg, unpublished data, 1978).

[e] Values not normalized for area (see text).

reproduces almost completely the relative solubility of steroids in octanol, a polar solvent. For instance, the permeability of NIL 8 cells to progesterone is 10 times that to hydrocortisone and less than that to estradiol (Column 2, Table IV); the solubility in octanol decreases by a factor of 10 from progesterone to hydrocortisone and by a factor of 1.26 from progesterone to estradiol (Table II, Section III,C). There are some similarities, too, between permeability of cultured cells to steroids and relative solubility of the compounds in ether, a hydrophobic solvent which however can accept hydrogen bonds, and in amyl caproate, a solvent that has both hydrophobic and hydrophilic properties; in particular, there is a great reduction in cell permeability to hydrocortisone with respect to corticosterone, that is, a marked effect of addition of a third OH group as in amyl caproate (Table III, Section III,C). On the other hand, there is little resemblance between the permeability to polar steroids in cultured cells and the very low solubility of this type of compounds in completely hydrophobic media like hexane or lecithin liposomes.

The pattern of permeability of skin and oral mucosa, however, agrees best with the pattern of steroid solubility in lecithin liposomes, with or without cholesterol (Table II, Section III,C), but in both tissues, permeability decreases progressively more for compounds bearing two or three OH groups, as does solubility in amyl caproate (Table III, Section III,C). Therefore, amyl caproate and ether are the solvents that represent at least some of the properties of the membrane of all the cells in which the coefficient P has been determined. Since the two solvents can form hydrophobic as well as hydrogen bonds, this may indicate that all the regions of the cell membrane interact with the steroids. Thus, within the framework of the "fluid mosaic" model of cellular membranes, the selectivity of cell permeability to nonpolar and polar steroids can be explained as follows:

1. Penetration of the polar steroids into the region of the membrane at the interface with water is retarded because hydrogen bonds are formed with the carbonyl at C 1 on the acyl chain of the phospholipids (Huang, 1976), or possibly, the ester linkage of the glycerol backbone (Giorgi and Stein, unpublished, 1978), whereas there is no impediment to the penetration of the nonpolar steroids, which do not form any hydrogen bonds.

2. Although the nonpolar steroids interact with phospholipids more than polar compounds, the mobility of the two types of compounds in the membrane may be the same (Diamond and Wright, 1969).

3. Mobility of molecules is favored at the interior of the membrane by the low degree of order of the terminal parts of the lipid chains (Butler *et al.,* 1970; Hubbell and McConnell, 1971), so that this region of the membrane does not delay the passage of either polar or nonpolar compounds.

Before we consider the biological significance of the findings described above, it must be pointed out how the selectivity toward the various steroids is much greater in cell membranes than in phospholipid membranes (Column 6, Table IV), whereas permeability to the hormones is not lower in the cells. These characteristics of the cell membrane may be of importance in determining hormone–tissue relationships. For instance, the permeability of fibroblasts to hydrocortisone and dexamethasone is high compared to that of HTC cells (Columns 2 and 3, Table IV). This difference in membrane permeability may account for the greater sensitivity of fibroblasts to inhibition of cellular growth by glucocorticoids. It is known that hydrocortisone at a concentration of 60×10^{-9} M prevents division of mouse fibroblasts, whereas this concentration of the hormone is optimal for induction of tyrosine aminotransferase in HTC cells and does not inhibit proliferation of these cells (Baxter and Tomkins, 1970; Jimenez de Asua *et al.*, 1977). Receptors for glucocorticoids have been identified in HTC cells and in skin (Overell *et al.*, 1960; Baxter and Tomkins, 1971). These cells have therefore become highly differentiated in order to respond to the hormonal stimulus. It is possible that the low permeability of HTC cells to the active hormone is another expression of this process of differentiation. However, a more specialized mode of transport has not developed in HTC cells, nor in skin, and steroid transport, as pointed out earlier, occurs by simple diffusion.

C. RATES OF TRANSFER OF STEROID HORMONES INTO CELLS AND TISSUES

Rates of transfer of steroids determined in cells and tissues by means of the zero trans procedure or by means of the isotope tracer method (Gurpide and Welch, 1969) are illustrated in Table V. All the experiments were performed in the absence of serum in the external medium but, in the perfusion experiments with human endometrium, human placenta, and human prostate, some residual 30% of the total initial proteins might still have been present in the interstitial spaces (see Cowan *et al.*, 1975). Uterus, prostate, placenta, and liver are target tissues for steroid hormones and contain specific receptors (for reviews, see King and Mainwaring, 1974; Mainwaring, 1977; Pasqualini, 1977), whereas the hepatoma cell line Novikoff is unresponsive to hormones and does not possess receptors for corticoids as other hepatoma lines do.

As can be seen from Table V, in liver, the rates for glucocorticoids are much higher at low concentrations of the steroids in the medium, because at these concentrations, a facilitated diffusion system is probably operative (Rao *et al.*, 1976b, 1977b). By definition, in facilitated diffusion, the membrane permeability is higher than would be expected from the partition coefficient of the steroid in the membrane (Stein, 1967). However, at concentrations above $2 \times 10^{-6} M$, when diffusion across the lipid matrix of the membrane is the prevailing mode of transport, the rate for corticosterone becomes 7 times that for hydrocortisone. The same ratio is found between the solubility of the two compounds in amyl caproate and ether (Table II, Section III,C). The permeability of liver cells to estrogens, too, is as high as the solubility of these hormones in amyl caproate or ether. This pattern of permeability is therefore similar to that observed in HTC cells, fibroblasts, and oral mucosa (Table IV).

It will be noticed that transfer of steroids into rat liver and Novikoff hepatoma cells is much faster than into any of the tissues in Table V. However, a precise comparison of transfer rates in cells and tissues cannot be made owing to the difference in temperature in the various experiments.

In human endometrium, the transfer rates of progestogens, androgens, and estrogens do not vary greatly. Addition of an OH group to progesterone or to androstenedione, however, reduces significantly the rate of transfer (see 20α-OH-progesterone and testosterone, respectively, in Table V, Column 6); addi-

TABLE V

RATES OF TRANSFER OF STEROIDS INTO CELLS AND TISSUES[a]

Steroid ($M \times 10^{-9}$)	Temperature (°C)	Rat liver cells[b]	Novikoff cells[c]	Rat uterus[d]	Human endometrium[c] (fmole/sec/mg protein/10^{-9} M)	Human placenta[c]	Human prostate[c]
Progesterone							
< 700	37				0.023	0.031	
20α-OH-progesterone							
< 700	37				0.013	0.030	
Corticosterone							
< 60	25	3.36					
>2000	25	2.00					
Hydrocortisone							
< 100	28	0.58					
>2000	28	0.30					
Prednisolone							
>2000	18		0.24				
Androstenedione							
< 110	37				0.026		0.008
5α-Dihydrotestosterone							
0.8	37						0.015
Testosterone							
5.0	37				0.020		0.011
3α,17β-Androstanediol							
< 700	37						0.011
Estrone							
>1200	15	10.00					
< 700	37				0.022	0.034	
< 40	37			0.70			
< 700	37				0.022	0.033	
>2000	15	8.00					
Estrone sulfate							
< 700	37				0.011	0.033	
Estradiol sulfate							
< 700	37				0.011	0.032	

[a] Calculated by assuming that 1 mg protein = 10 mg wet weight of tissue or 3×10^6 cells/ml.

[b] Cells isolated by collagenase digestion of liver; zero trans procedure (Rao et al., 1976b, 1977a,b).

[c] Cells grown in suspension; zero trans procedure (Plagemann and Erbe, 1976).

[d] Whole uterine horns; zero trans procedure (Peck et al., 1973).

[e] Perfusion of slices of tissue according to the equilibrium exchange method (Tseng et al., 1972a,b; Giorgi et al., 1971, 1974; Malathi and Gurpide, 1977).

tion of a second OH to estrone to form estradiol, in contrast, has no effect. This deviation from the general pattern of the permeability to estrogens thus renders the endometrial cells more permeable to estradiol, the biologically active hormone. Conjugation of estrogens with sulfuric acid halves the rate of transfer. In human placenta, on the other hand, the rates of transfer of free and conjugated estrogens are very similar (Table V, Column 7). This is another interesting deviation from the general pattern of cell permeability and has possible significance in physiological terms, as will be discussed later (Section V,E).

In the prostate, the rate of transfer of 5α-dihydrotestosterone and testosterone is 0.015 and 0.011 fmole/sec/mg/10^{-9} M, respectively, a much smaller difference than between the partition coefficient of the two compounds in ether: 82 for 5α-dihydrotestosterone and 30 for testosterone. In contrast, the faster transport of testosterone than of androstenedione, a compound without an OH group, is in opposition to the polarity rule.

Other aspects of transport of steroid hormones into uterus, prostate, and other target tissues are discussed in greater detail in the next section.

D. EFFECTS OF CHANGES IN MEMBRANE COMPOSITION ON PERMEABILITY OF CELLS TO STEROID HORMONES

1. *Changes in Lipid Composition*

Naturally occurring variations in the lipid composition of cell membranes, which might affect permeability to steroid hormones, have not been studied extensively. The only example clearly established so far is a high cholesterol content in the membrane of human erythrocytes of the genetic type Hb-SS, which present the sickling phenomenon, i.e., precipitation of deoxyhemoglobin gels in the affected erythrocytes (Erickson *et al.*, 1937). An interesting difference has been observed in uptake of steroids between erythrocytes of sickle cell anemia and normal erythrocytes with hemoglobin genotype Hb-AS and Hb-AA (Sogbesan *et al.*, 1974). The uptake of progesterone, testosterone, and estradiol is higher by 15–20% in Hb-SS cells than in those of the two other genotypes, while no difference is noted in the uptake of estrone, androstenedione, and cortisol. The changes are therefore quite selective for the individual steroids because the uptake of both a nonpolar compound, progesterone, and of more polar steroids, testosterone and estradiol, is affected. The possibility that these changes are the consequence of the altered cholesterol content in the membrane of Hb-SS erythrocytes is suggested by the results of the experiments on rat cell membranes described next. It is important to mention that testosterone and progesterone in micromolar concentrations can reverse the sickling phenomenon *in vitro;* this effect has been attributed to stabilization of the erythrocyte membrane by preferential incorporation of the steroids (Isaacs and Hayhoe, 1967).

An effect of the cholesterol concentration in cell membranes on permeability

TABLE VI

BINDING OF STEROIDS BY RAT LIVER PLASMA MEMBRANE AND MITOCHONDRIA[a]

| | | Steroid uptake (μmoles/μmole original membrane cholesterol) | | | |
| | | Plasma membranes | | Mitochondrial membranes | |
Steroid	Steroid concentration ($M \times 10^{-6}$)	Untreated membranes (Mean ± SEM)	Depleted membranes (Mean ± SEM)	Untreated mitochondria (Mean ± SEM)	Enriched mitochondria (Mean ± SEM)
Dexamethasone	38–49	0.28 ± 0.01	0.53 ± 0.03		
Cortisol	36–45	0.30 ± 0.02	0.32 ± 0.01		
Corticosterone	30–59	<0.02	0.22 ± 0.02	1.11 ± 0.03	0.78 ± 0.01
Deoxycorticosterone	16–27	0.12 ± 0.01	0.09 ± 0.01		
Progesterone	29–35	0.42 ± 0.01	0.25 ± 0.01		
Testosterone	23–60	0.35 ± 0.02	0.24 ± 0.01	0.31 ± 0.02	0.53 ± 0.03
Diethylstilbestrol	33	0.49 ± 0.02	0.34 ± 0.01		

[a] In control experiments, untreated plasma membranes were added to give a concentration of 59–91 μg of membrane cholesterol per 5 ml of medium. The same quantity of plasma membranes were added in the test experiments, but the cholesterol level in these membranes had been lowered to between 46 and 53% of that of the corresponding control. The two mitochondrial preparations contained 32 and 21 μg of cholesterol per ml and the cholesterol-enriched preparations 68 and 43 μg/ml, respectively. Incubations with steroids were carried out at 37°C for 75 minutes, after which the membranes were sedimented at 80,000 g for 3 minutes. Binding of radioactive compounds was determined by direct measurement of radioactivity in the membrane pellet after washing it twice by resuspending in saline and resedimenting. Nonradioactive compounds were determined in the washed pellet by adsorption in UV. Each result is the mean of four determinations (± SEM) using two different membrane preparations. (Reproduced from Graham and Green, 1969, with kind permission of the publishers, Pergamon Press.)

to steroids has been reported by Graham and Green (1969). Membranes from rat liver cells were incubated either with cholesterol-depleted or with cholesterol-loaded plasmalipoproteins in order to alter the membrane cholesterol content. This procedure resulted in an exchange of large amounts of the sterol between membrane and plasma lipoprotein, but did not seem to affect any other membrane components. Uptake of corticoids, progesterone, testosterone, and diethylstilbestrol was measured after exposure of the membranes to the hormones for 75 minutes. The results are illustrated in Table VI. In the cholesterol-depleted plasma membranes, uptake of the more polar steroids (corticosterone, dexamethasone, and cortisol) was enhanced with respect to untreated membranes, whereas, on the contrary, uptake of less polar compounds (deoxycorticosterone, testosterone, and progesterone) decreased. Cholesterol-enriched mitochondrial membranes showed a change in permeability in the opposite direction: uptake of the less polar compound, testosterone, was enhanced. These findings are at variance with the observation that the solubility of polar steroids in lecithin–cholesterol liposomes increases with the content of cholesterol (Snart and Wilson, 1967).

In human red blood cells, steroid uptake is not altered by prolonged incubation in buffer containing steroids in high concentrations (Brinkmann et al., 1972). Thus, even if there is any incorporation of steroids in the membrane, which seems doubtful under these experimental conditions, it has no effect on permeability. However, a reduction (or, less frequently, an increase) in cell permeability to polar steroids is observed in cultured fibroblasts in the presence of millimolar concentrations of steroids (E. P. Giorgi and W. D. Stein, unpublished, 1978). The mechanism of these changes is obscure.

2. Effects of Agents Perturbing the Membrane Structure

The importance of the interaction of steroids with the lipid matrix of the membrane is confirmed by experiments in which the integrity of the membrane is altered by treatment with phospholipase A_2, an enzyme that removes the fatty acid from C 2 of glycerol in glycerides. Thus treatment with phospholipase A_2 (10–20 μg/ml) reduces the uptake of triamcinolone acetonide by mouse pituitary adenoma cells or of estrogen, testosterone, and corticosterone by rat liver cells by 90 and 35%, respectively (Harrison et al., 1974; Rao et al., 1977a,b). Phospholipase C and D, however, are ineffective, even at concentrations of 200 μg/ml, probably because they have a less drastic effect on the membrane (Plagemann and Erbe, 1976; Rao et al., 1977a,b). Two reagents, ethanol and dimethyl sulfoxide, which probably decrease the water content of the membrane and the ordering of the lipid matrix, also inhibit uptake of triamcinolone acetonide in mouse pituitary adenoma cells (Harrison et al., 1977).

Proteolytic enzymes, with the exception of Pronase, have little effect on steroid transport (Harrison et al., 1974, 1977). A more specific enzyme, neuraminidase, which attacks the sialic groups in glycoproteins, in contrast, reduces uptake of triamcinolone acetonide by mouse pituitary adenoma cells by 50% (Harrison *et al.*, 1977). However, neuraminidase does not alter transport of steroids into rat liver cells, whereas β-glucosidase and β-galactosidase, enzymes that remove carbohydrate groups from the proteins, reduce steroid transport by 13–27% (Rao *et al.*, 1977a,b). Neuraminidase is ineffective, also, in Novikoff hepatoma cells (Plagemann and Erbe, 1976).

Thus, it would appear that specific alterations of membrane proteins affect steroid transport in cells (liver and mouse pituitary adenoma cells) in which facilitated diffusion of steroids has been claimed (Harrison *et al.*, 1974, 1975, 1976, 1977; Rao *et al.*, 1976a,b, 1977a,b), but not in cells (Novikoff hepatoma) in which transport occurs by simple diffusion (Plagemann and Erbe, 1976; Graff *et al.*, 1977). Whether enzyme treatment modifies specifically some protein carriers cannot, however, be stated without reservation. In some of the experiments, the treatment is reported to have considerably reduced cell viability, and this might have had an indirect effect on transport. It may also be significant that the greatest inhibition of transport, especially of androgens and estrogens in rat liver cells, was seen after treatment with lipases, rather than after treatment with specific protein enzymes.

Reagents blocking sulfhydryl (SH) groups have been widely employed. Compounds that do not readily penetrate the membrane, as for instance *p*-chloromercuriphenyl sulfonate (PCMPS) or *p*-chloromercuribenzoate (PCMB), have no effect on the uptake of corticoids by rat hepatoma cell or of androgens and progesterone by human erythrocytes, but inhibit transport of glucocorticoids in rat liver cells and estradiol in uterus (Brinkmann and van der Molen, 1972; Levinson *et al.*, 1972; Milgrom *et al.*, 1973; Rao *et al.*, 1976b). In the case of other SH reagents capable of penetrating the membrane (as, for instance, $HgCl_2$, fluoro-2,4-dinitrobenzene, and iodoacetate), it is difficult to distinguish between an alteration of the cell surface proteins that may affect transport of steroids and an effect on the intracellular binding of the hormones (Levinson *et al.*, 1972). Furthermore, the vast spectrum of action of SH blocking agents must also be taken into consideration before any effect of these agents on steroid transport is attributed to specific interactions with membrane proteins.

It is also worth mentioning that the polycyclic compounds, Persantin and Cytochalasin, which inhibit a variety of mediated diffusion systems, do not seem to interfere with transport of prednisolone in Novikoff cells (Graff *et al.*, 1977).

V. The Transport of Steroid Hormones into Target Tissues

A. Transport of Estrogens into the Uterus

1. *Rat Uterus*

The growth of the uterus is strictly dependent on estrogens, which also prime the tissue for the action of progesterone. The presence of uterine specific receptors for estrogens has been demonstrated in many species, and the physicochemical characteristics of these receptors have been thoroughly investigated (for review, see Gorski and Gannon, 1976). Transport of steroid hormones into the uterus of the rat has received much attention in the past few years, but the results are at variance with one another.

Williams and Gorski (1971, 1973) have reported that transport of estradiol into rat uteri and isolated uterine cells is temperature dependent. The uterine horns or the isolated cells were exposed for 5 minutes, to a solution containing either labeled estradiol alone or labeled estradiol plus an excess of unlabeled compound. This is a well established procedure used to distinguish between specific and nonspecific binding of hormones. The radioactivity present in the tissues or cells incubated with labeled estradiol represents the total binding or "uptake," i.e., nonspecific adsorption to various cellular components plus specific binding to the receptors. Binding to the receptors, however, is abolished if unlabeled hormone is added in amounts large enough to occupy all the receptors; the difference between radioactivity in the absence and in the presence of excess hormone yields the radioactivity specifically bound. In the experiments of Williams and Gorski (1971, 1973) on whole uteri, unlabeled hormone was again added to all samples just prior to homogenization in the cold. Omitting this precaution may cause, during homogenization, binding of free hormone trapped in the interstitial spaces to nonoccupied receptors and lead to an overestimate of the rate of binding. Both in whole uteri and in isolated cells, the uptake and the specific binding have been shown to vary with temperature, but the variations of uptake are much smaller than those of specific binding in the range up to 37°C and can be explained by changes in the fluidity of the membrane with the increase in temperature. The activation energy of uptake in isolated cells was calculated from an Arrhenius plot (log uptake versus reciprocal of absolute temperature) as 10.5 kJ/mole, a value compatible with a process of simple diffusion. The activation energy for binding to the specific receptors was calculated at 87.0 kJ/mole, which is similar to that found for protein–ligand associations leading to conformational changes.

A more elaborate method to investigate the characteristics of estradiol uptake by rat uteri has been devised by Milgrom and co-workers (1973). This method consists in measuring the time course of uptake and of binding of the hormones to the receptors, as in the method delineated earlier. In addition, the total binding capacity in cell-free extract is determined by charcoal adsorption at low and high temperature. After incubation with estradiol ($0.5-40 \times 10^{-9}$ M), the uterine horns were homogenized in the presence of excess labeled compound to prevent binding of free extracellular hormone to the receptors. Thus, the radioactivity measured in the tissue at the end of the incubations presumably represented only hormone located intracellularly. It was observed that the rate of uptake of estradiol by the uterine horns, which was measured after 5 minutes of exposure to the hormone, decreases at concentrations above 10×10^{-9} M and is temperature dependent. Preincubation of the uterine horns with iodoacetate, α-iodoacetamide, or PCMSP inhibits uptake of estrone and estriol, as well as of estradiol, whereas pretreatment with energy inhibitors or ouabain has no effect. Addition of micromolar amounts of estradiol or diethylstilbestrol, but not of corticoids, progesterone, or testosterone, reduces uptake of estradiol. A Scatchard plot (ratio bound to unbound hormone[2] versus bound) of the data of uptake at high estradiol concentrations was analyzed by computer (Baulieu *et al.*, 1970) and revealed a nonsaturable process, with rate constant 0.059×10^{-3} sec $^{-1}$/ uterus, and a saturable process with $K_A = 3 \times 10^{10} M^{-1}$. The nonsaturable process probably represents simple diffusion of the hormone across the cell membrane. On the other hand, the characteristics of saturability, temperature dependence, and sensitivity to SH blocking agents of the second process seem to suggest that this might be transport by facilitated diffusion. Since a comparison of estradiol uptake by whole tissue and binding to cell-free extracts showed that, after a 30 minute incubation, not all the receptors available in the homogenate were occupied, the authors postulate that saturation of the facilitated diffusion system, at high concentrations of estradiol in the medium, might set a barrier to binding of the hormone to the receptors (Milgrom *et al.*, 1973). With regard to this point, however, some of the pitfalls of comparisons between binding in intact tissues and in cell-free extracts should be mentioned. Slow penetration of the external solution in the tissue might prevent equilibrium with the receptors from being achieved as fast as in cell-free extracts; these extracts, furthermore, usually have a lower protein concentration and a lower osmolarity than is found in the cells, and this might increase binding of the hormones to the receptors.

A much more serious objection can be raised to the conclusions of Milgrom and co-workers (1973) on the ground that, because of the very long sampling times adopted in this study, it is doubtful that transport across the cell membrane has been specifically measured. These authors report that uptake of estrogen is linear up to 15 minutes, whereas others have observed linear rates up to approximately 10 seconds (Rao *et al.*, 1977a). How crucial the time of sampling is has been shown by the fact that experiments in which uptake of estradiol in rat uterus was measured at 15 second intervals gave completely different results (Peck *et al.*, 1973).

The results of these experiments, illustrated in Fig. 8, demonstrate that the uptake measured from initial rates at 15 seconds is linear up to estradiol concentrations of $40 \times 10^{-9} M$. Furthermore, there is no inhibition after treatment with SH reagents nor after addition of excess diethylstilbestrol. It is also evident that the rate of uptake in the uterus is the same as in the diaphragm, a nontarget tissue, although, as already mentioned, there are important differences between these two tissues with regard to hormone transfer to the nucleus.

---

[2]The unbound hormone was calculated from the experimentally determined values of bound hormones and K_A as follows: unbound hormone = bound $\times K_A$/(binding sites − bound).

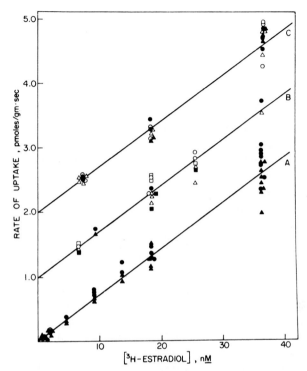

FIG. 8. Initial rates of uptake of [³H]estradiol by rat uterine and diaphragmic tissue as a function of substrate concentration. Tissues were incubated with shaking for 15 and 30 seconds and washed, and [³H]estradiol in the tissue determined after homogenization. A: (●), Normal uterine tissue; (▲), normal diaphragmic tissue. B: Initial rates were determined for control uteri, (●); uteri plus 1 × 10⁻³ M N-ethylmaleimide, (○); uteri plus 1 × 10⁻³ M, 2,4-dinitrophenol (□); control diaphragm strips, (▲); diaphragm strips plus 1 × 10⁻³ M, N-ethylmaleimide, (△); and diaphragm strips plus 1 × 10⁻³ M 2,4-dinitrophenol, (■). C: Initial rates were determined for uteri (circles) and strips of diaphragm (triangles) in the presence (open symbols) and absence (closed symbols) of diethylstilbestrol. Note that in the figure the rate for B and C have been displaced along the vertical axis by 1 and 2 pmole/gm/sec, respectively. (Reproduced from Peck et al., 1973 with kind permission of The American Chemical Society.)

It can be concluded that the evidence for simple diffusion of estrogen into uterine cells is incontrovertible; the evidence for a facilitated diffusion system supplementing the process of simple diffusion is much weaker. Another important point clearly demonstrated by all the studies mentioned above is that the intracellular receptors are not involved in the transport process (Williams and Gorski, 1971, 1973; Milgrom et al., 1973; Peck et all, 1973).

2. Human Endometrium

Transport of estrogens into human endometrium has been investigated by means of the tracer isotope technique (Gurpide and Welch, 1969; Tseng et al., 1972a).

As already mentioned, the entry of the two estrogens is almost the same, notwithstanding the difference in polarity between the two compounds. Entry remains constant over a range of concentrations of steroids in the medium from 0.3×10^{-9} M (lower than the physiological levels in female blood) to 17×10^{-6} M.

The effect of synthetic steroids on the entry of the estrogens into endometrium was also studied. Ethynyl estradiol and mestranol, which are estrogenic compounds, reduce estrogen uptake when present at a concentration of 15×10^{-6} M; norgestrel, a progestogen, is ineffective at this concentration (Tseng et al., 1972a). It might be interesting to note that both mestranol and norgestrel possess an OH group at C 17, but in norgestrel this is in the α configuration. This might explain the lack of effect of this compound on estrogen transport into the endometrium, in analogy with the observation that C 3α-OH compounds do not alter permeability to solutes in artificial membranes (Demel et al., 1972). On the other hand, the inhibition of estrogen transport by mestranol might depend on the interaction of the C17β-OH group with the membrane lipids and consequent tightening of the polar interface region of the membrane. The hypothesis that the reduction of estrogen transport is related to modification of the permeability of the lipid matrix is strengthened by the fact that the effect of mestranol is evident at a relatively high concentration, 15×10^{-6} M. On the basis of these results, it is suggested that transport of estrogen into human endometrium occurs either by simple diffusion or by facilitated diffusion through a system that cannot be saturated at physiological concentrations of the hormones (Tseng et al., 1972a). These authors also report that transport of estrogens into human endometrium does not vary with the phases of the menstrual cycle, nor are any changes of the parameters evident in abnormal hyperplastic endometrium.

B. Transport of Androgens into Human and Canine Prostate

The prostate gland is a typical target tissue, dependent on androgens for its differentiation, growth, and maintenance. The hormone active at the cellular level is 5α-dihydrotestosterone, a reduced metabolite of testosterone and the main androgen circulating in blood. Studies in vivo in male subjects have shown that the greatest proportion of 5α-dihydrotestosterone present in the gland originates from intracellular conversion of testosterone; other C 19 androgens contribute very little to the formation of the hormone (Becker et al., 1972; Harper et al., 1973).

The transport of testosterone, 5α-dihydrotestosterone, and androstenedione into human and canine prostate, the variations with steroid concentrations and the effect of antiandrogens and estrogens have been investigated by Giorgi and co-workers (1972, 1972a,b, 1973, 1974). The isotope tracer method was employed (Gurpide and Welch, 1969), but in some experiments it was slightly modified in order to measure entry of androgens at near physiological steroid concentrations. In these experiments, the perfusing medium contained, instead of [^{14}C]testosterone as precursor, the compound labeled with ^{3}H at C 17α, i.e., [17α-^{3}H]testosterone, and [1,2-^{3}H]5α-dihydrotestosterone, with a specific activity of 3.5 and 50 Ci/mmole, respectively. [17α-^{3}H]5α-Dihydrotestosterone, formed in the tissue by conversion from [17α-^{3}H]testosterone, was measured from the difference in tritium counts after removal of the tritium label at C 17 by oxidation to androstanedione. As can be seen from Table VII, in normal human prostate, entry of testosterone, 5α-hydrotestosterone, and androstenedione does not vary with increasing steroid supply. The pattern of permeability is: 5α-dihydrotestosterone > testosterone > androstenedione. The same pattern of permeability is evident in canine prostate (Giorgi et al., 1972b, 1974). Thus, it is probable that the bulk of androgens are transported into the prostate cells by simple diffusion. In human hypertrophic prostate, however, transport is decreased at high levels of androgens in the medium. This might be due to an alteration of the cell membrane, or otherwise might depend on the presence of large amounts of connective tissue with low permeability.

Notwithstanding the clear evidence for simple diffusion of androgens into prostate cells both in the human and canine gland, this transport is modified by the presence of other steroids. Addition to the medium of androstenedione (110 $\times$ 10^{-9} M) decreases the transfer of testosterone and, to a lower extent, of 5α-dihydrotestosterone (Table VII). The effect of estradiol is concentration dependent: the estrogen enhances entry of the androgens when added in concentrations below 110 $\times$ 10^{-9} M, but has an inhibitory effect at 330 $\times$ 10^{-9} M. Cyproterone acetate, an antiandrogen, also enhances entry of testosterone and 5α-dihydrotestosterone by 20%, whereas, on the contrary, at similar concentrations, the free alcohol, cyproterone (a less potent antiandrogen) reduces transport. In order to explore the action of the two antiandrogens, further experiments were performed with canine prostate. The concentrations of [17α-^{3}H]testosterone and [1,2-^{3}H]5α-dihydrotestosterone in the medium were maintained constant at 0.6 and 0.2 $\times$ 10^{-9} M, concentrations that are one-fifth those present in male dog plasma (Tremblay et al., 1972). Increasing amounts of cyproterone and cyproterone acetate were added to some of the media. In the presence of cyproterone or cyproterone acetate up to 130 $\times$ 10^{-9} M (i.e., 43 pmole/minute in Fig. 9), entry of testosterone and 5α-dihydrotestosterone into the prostate slices increases proportionally to the antiandrogen concentration in the medium. From the slopes of the lines in Fig. 9, it can be calculated that there is an increase of 1 pmole/minute in the entry of testosterone and 5α-dihydrotestosterone for every 500 and 700 pmole increase of antiandrogens in the medium, respectively. Since cyproterone and its acetate are highly hydrophobic compounds owing to the presence of an extra methyl group linking C 1 and C 2 and of a chlorine atom at C 6, it can be assumed that their rate of transfer is as fast, if not faster, than that of 5α-dihydrotestosterone, namely 0.015 fmole/second/mg/10^{-9} M (Table V, Section

TABLE VII

ENTRY OF ANDROGENS INTO HUMAN PROSTATE in Vitro[a,b]

Steroid added to medium	Entry of testosterone (concentration in medium)			Entry of 5α-dihydrotestosterone (concentration in medium)			Entry of androstenedione (concentration in medium)	
	$4.9 \times 10^{-9}\ M$[c]	$0.1 \times 10^{-6}\ M$[d]	$1.1 \times 10^{-6}\ M$[d]	$0.8 \times 10^{-9}\ M$[c]	$0.1 \times 10^{-6}\ M$[d]	$1.1 \times 10^{-6}\ M$[d]	$0.1 \times 10^{-6}\ M$[e]	$1.1 \times 10^{-6}\ M$[e]
Normal prostate								
Control (no steroid)		0.16 (N = 4)	0.19 (N = 4)					
Androstenedione ($0.1 \times 10^{-6}\ M$)		0.12 (N = 2)			0.22 (N = 3)	0.21 (N = 3)	0.13 (N = 3)	0.21 (N = 4)
Estradiol ($0.1 \times 10^{-6}\ M$)		0.18 (N = 1)			0.26 (N = 3)			
Hypertrophic prostate								
Control (no steroid)	0.23 ± 0.04 (N = 14)	0.20 ± 0.08 (N = 9)	0.12 (N = 5)	0.32 ± 0.07 (N = 14)	0.24 ± 0.04 (N = 7)	0.12 (N = 5)	0.15 ± 0.03 (N = 10)	0.14 ± 0.08 (N = 10)
Androstenedione ($0.1 \times 10^{-6}\ M$)		0.14 ± 0.06 (N = 12)	0.16 ± 0.07 (N = 12)		0.22 ± 0.09 (N = 12)	0.19 ± 0.12 (N = 11)		
Estradiol ($0.3 \times 10^{-6}\ M$)	0.29 (N = 5)	0.10 (N = 3)	0.16 (N = 5)	0.57 (N = 5)	0.36 (N = 3)	0.22 ± 0.07 (N = 7)	0.27 (N = 3)	0.16 (N = 5)
Cyproterone ($2.6 \times 10^{-6}\ M$)		0.07 (N = 4)			0.20 (N = 3)		0.11 (N = 1)	
Cyproterone acetate ($0.2 \times 10^{-6}\ M$)	0.27 (N = 5)			0.34 (N = 5)	0.14 (N = 3)		0.14 (N = 3)	
Cyproterone acetate ($2.4 \times 10^{-6}\ M$)		0.22 (N = 5)			0.25 (N = 3)		0.16 (N = 3)	

[a] Expressed as fraction of the steroid in the perfusing medium that enters the tissue. For calculation of entry, see legend to Fig. 7 (Section IV,A).

[b] Slices of prostate weighing 502 ± 6 mg were perfused at 37°C for 90 minutes with a continuous flow (18 ml/hour) of medium containing radioactive steroids and, in some cases, unlabeled steroids, to reach the required concentration. The values shown represent the mean ± S.D. of results of experiments in different specimens of tissue; N, number of experiments (Giorgi et al., 1971, 1972a, 1973, 1974).

[c] Experiments with [17α-^{3}H]testosterone and [$1,2$-^{3}H]5α-dihydrotestosterone.

[d] Experiments with [^{14}C]testosterone and [^{3}H]5α-dihydrotestosterone.

[e] Experiments with [^{3}H]testosterone and [^{14}C]androstenedione.

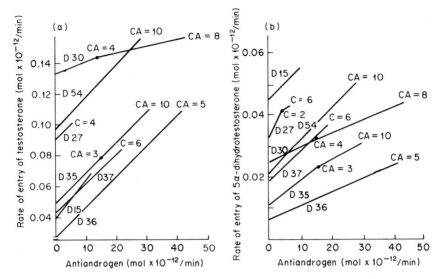

FIG. 9. Rates of entry of testosterone (*a*) and 5α- dihydrotestosterone (*b*) into canine prostate as a function of the concentration in the medium of antiandrogens. Slices of prostate were perfused for 90 minutes at 37°C with a continuous flow of [17α-³H]testosterone, 1.9 pmole/min, and [1,2-³H] 5α-dihydrotestosterone, 0.37 pmole/min. The weight of the tissue was 503 ± 5 mg. The entry was calculated as explained in the legend to Fig. 7 and in the text. The intercept on the ordinate is the rate of entry of testosterone and 5α-dihydrotestosterone in the absence of antiandrogens in the medium. Uptake by the tissue of [³H]cyproterone (C) and [³H]cyproterone acetate (CA) is given at each experimental point. D15, D36, etc., are the experiment numbers. The scale of the ordinate in (b) is two times that in (a). (Reproduced from Giorgi, 1976, with kind permission of *The Journal of Endocrinology*.)

IV,C). The molar ratio of transported compounds might therefore approximate 1 mole of testosterone per 80 mole of antiandrogens and 1 mole of 5α-dihydrotestosterone for 100 moles of antiandrogens. Cyproterone acetate enhances transport of 5α-dihydrotestosterone also at concentrations as high as 700 × 10^{-9} *M* (i.e., 180 pmole/minute in Fig. 10), while cyproterone above 220 × 10^{-9} *M* (i.e., 80 pmole/minute in Fig. 10) inhibits the entry of both androgens. Thus, the concentrations of cyproterone and of estradiol that inhibit androgen transport are both in the region of 220–330 × 10^{-9} *M;* on the other hand, in the absence of these compounds, an increase in the concentration of the androgens themselves up to 1.1 × 10^{-6} *M* does not reduce their transport (Table VII).

The effect of cyproterone and estradiol on androgen entry into the prostate is difficult to explain in terms of interaction between steroids and membrane lipids. A reversal of steroid effect on permeability (from enhancement to reduction) has not been described in either artificial or biological membranes. Furthermore, in these membranes, modifications of permeability to solutes have been obtained at

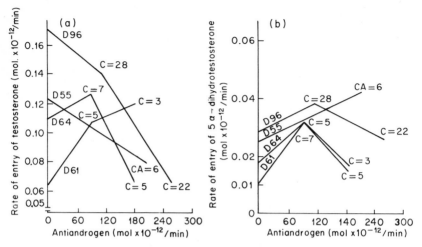

FIG. 10. Rates of entry of testosterone (*a*) and 5α-dihydrotestosterone (*b*) into canine prostate as a function of the concentration in the medium of antiandrogens, cyproterone, and cyproterone acetate. Experimental conditions were as in Fig. 9, but the concentration of the antiandrogens in the medium was higher. The weight of the tissue was 501 ± 5 mg. The scale of the abscissa is six times lower than that in Fig. 9. (Reproduced from Giorgi, 1976, with kind permission of *The Journal of Endocrinology*.)

millimolar concentrations of steroids (Bangham *et al.*, 1965; Weissmann and Kaiser, 1965; Weissmann *et al.*, 1965, 1976). However, it is known that leakage of enzymes from target tissue lysosomes is specifically increased by physiological doses of estradiol or testosterone (Szego *et al.*, 1971). Another discrepancy between the results obtained in artificial membranes and those obtained in the prostate is that in this gland both the polar estradiol and the nonpolar androstenedione reduce permeability of the membrane to steroids, whereas the nonpolar cyproterone acetate has the opposite action. A hypothesis that could account for these findings in human and canine prostate is that simple diffusion of 5α-dihydrotestosterone, and especially of testosterone, is supplemented by transport operated by polyvalent carriers present on the membrane. Transfer of the two androgens would be potentiated by the simultaneous transfer of estradiol and antiandrogens, until the polyvalent carriers became saturated. Since saturation appears to occur at lower levels of estradiol, androstenedione, and cyproterone than of testosterone and 5α-dihydrotestosterone, the affinity of the polyvalent carriers must be lower for the two androgens than for the other steroids. This order of affinity would indicate that the carriers are different entities from the intracellular receptors, which bind steroids in the order: 5α-dihydrotestosterone > testosterone > cyproterone acetate > cyproterone > estradiol > androstenedione (Mainwaring, 1977). Further studies would be necessary to establish the presence on the membrane of these, so far hypothetical, carriers.

C. Transport of Glucocorticoids into Thymocytes

Thymocytes contain specific receptors for glucocorticoid hormones and respond to these hormones with many metabolic changes, the most noticeable being a reduction in glucose uptake; at high concentrations, glucocorticoids can cause lysis of thymocytes (Munck and Leung, 1977). The kinetics of influx and efflux of glucocorticoids in thymocytes from adrenalectomized rats has been investigated by Munck and Brinck-Johnsen (1968) by means of the zero trans procedure. The results of the efflux experiments are illustrated in Fig. 11. The efflux curve of hydrocortisone, corticosterone, and dexamethasone shows two components: a fast efflux of steroid, which is almost completed in one minute, and a slower efflux, which does not become constant within 10 minutes. The latter component of the curve is due to release from the cells of steroids dissociat-

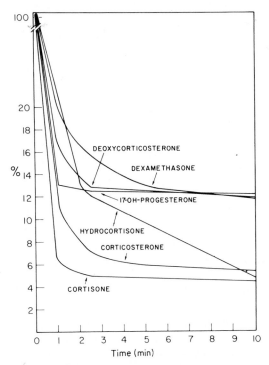

Fig. 11. Efflux of glucocorticoids from thymus cells. Thymus cell suspensions (3.2 ml; cells: buffer, 10:100 v/v) were incubated with tritiated steroids at concentrations of $30–110 \times 10^{-9}M$, and then diluted in 50 volumes of medium containing no steroids. The cells were centrifuged at set intervals after incubation in this medium, and radioactivity determined in the cell pellet and in the supernatant. On the ordinate: radioactivity in the cells as percentage of the total radioactivity present at the end of the first incubation. On the abscissa: time of sampling. Modified from Munck and Brinck-Johnsen, 1968.

ing from the receptors. The time constant of the rapid efflux is under 15 seconds for all steroids, but there are evident differences in the rapidity with which the various compounds cross the cell membrane: thus, less than 80% of the most polar compounds, hydrocortisone and dexamethasone, is released within one minute from the cells, while more than 90% of the intracellular cortisone leaves the cells within this time interval. However, there seem to be some deviations from the polarity rule: for instance, release of cortisone and corticosterone is faster than that of the less polar 17α-OH-progesterone (Fig. 11). The rates of efflux measured within 1 minute are linear for all compounds up to concentrations of 10^{-4} M and are temperature independent (Munck and Brinck-Johnsen, 1968). In influx experiments, it was found that near maximum uptake of the steroids is reached within 1 minute, that is, influx is as fast as efflux.

It was shown in further experiments (Bell and Munck, 1973), that in cell-free extracts from thymocytes the hydrocortisone–receptor complex dissociates in 3 minutes, which is equal to the time constant of the slow component of the efflux curve of the hormone (Fig. 11). This observation provides further evidence that the membrane of thymocytes is freely permeable to hydrocortisone. Therefore, the authors conclude that transport of glucocorticoids into thymocytes occurs by a process of simple diffusion. However, as mentioned above, the possibility that the membrane is selectively permeable to corticosterone, the natural, biologically active hormone, cannot be excluded.

D. Steroid Transport into Rat Liver Cells

The liver is the main organ metabolizing hormones, which are transformed into inactive reduced forms or conjugated with sulfuric and glucuronic acid. As discussed by Tait and Burstein (1964), the hepatic clearance is, in the case of many hormones, equal to the total clearance; whereas for other hormones, e.g., testosterone, extrahepatic clearance is minimal. The total turnover of steroid hormones in the liver is therefore in the region of 50 mg/day (Gower, 1975). The liver is also a target tissue for glucocorticoids, testosterone, and estrogens: these hormones induce the synthesis of specific liver enzymes involved in amino acid and carbohydrate metabolism (Keller et al., 1969). Specific receptors for glucocorticoids, but not for androgens or estrogens, have been isolated from the tissue (Litwack et al., 1973). The investigation of the mode of transport of steroids into the liver is therefore of outstanding interest.

A series of papers has been recently published on this topic (Rao et al., 1976a,b, 1977a,b). For the experiments, liver cells were isolated by perfusion of the intact organ with a Ca^{2+}-free solution containing collagenase; viability of the cells was monitored by the Trypan blue exclusion test. Transport of steroids into the cells was measured at intervals of seconds: it was established that the rates of transfer are linear up to 50 seconds at 27°C for hydrocortisone, 17 seconds at

25°C for corticosterone, and 15 seconds at 15°C for testosterone and the estrogens. In order to obtain precise times of sampling, a portion of the cell suspensions was pipetted directly onto a glass fiber disc fitted to a filter maintained under suction. It is also reported that no metabolism of the hormones occurred during the brief cell incubation. The rate of transport of hydrocortisone and corticosterone into rat liver cells was found to deviate from linearity at high steroid concentrations in the medium (Fig. 12a,b). Transport of hydrocortisone, but not of corticosterone, is inhibited by approximately 40% after incubation for 10 minutes with KCN; the unsaturable, KCN-insensitive influx is probably due to simple diffusion of the hormone through the lipid matrix of the cell membrane. Extrapolation to zero of the linear portion of the curve, rate of transport versus concentration of corticosterone (Fig. 12b), also reveals a nonsaturable component, which accounts for approximately 12% of the total rate of transport (Rao *et al.*, 1977b). Kinetic analysis of the curves in Fig. 12 suggests the existence of two saturable components for each steroid; one of the components has very high affinity for the hormones (Table VIII). It is noteworthy that the K_m of this

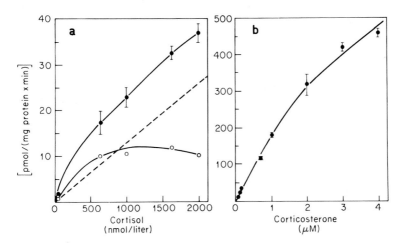

FIG. 12. Initial rates of uptake by isolated rat liver cells of cortisol (a) and corticosterone (b) as a function of substrate concentration. Cell suspensions, 0.1 ml corresponding to 0.5 mg protein, were added to 1 ml of Krebs-Ringer buffer containing tritiated steroids, and incubated for 30 seconds at 28°C (a) or 10 seconds at 25°C (b). Incubation was terminated by filtration of 1 ml suspension through Millipore filters. In (a): uptake of cortisol not corrected for diffusion (●————●); uptake due to diffusion (-------) was corrected by drawing a straight line through zero parallel to the linear part of the uncorrected uptake curve. The corrected uptake curve (○————○), due to the saturable process, was obtained by subtracting the uptake values due to diffusion from the uncorrected uptake curve. Each point represents the mean of triplicates; the vertical bars indicate standard deviation. (Reproduced: (a) from Rao *et al.*, 1976b, with kind permission of the publisher, Walter de Gruyter; (b) from Rao *et al.*, 1977b, with kind permission of North Medical, Amsterdam.)

TABLE VIII

KINETIC PARAMETERS OF STEROID TRANSPORT INTO RAT LIVER CELLS[a]

Steroid	Temperature (°C)	K_m ($M \times 10^{-9}$)	V_{max} (pmole/min/mg protein)
Hydrocortisone	28	91[b]	4[b]
	28	600[b]	17[b]
Corticosterone	25	56 ± 32[b]	20 ± 8[b]
	25	743 ± 183[b]	216 ± 90[b]
Testosterone	15	1600	910
Estrone	15	2200	2500
Estradiol	15	510	550

[a] Cells isolated by digestion with collagenase of liver *in situ*. Transport measured by the zero trans procedure at 30 seconds for hydrocortisone and 10 seconds for the other steroids (Rao *et al.*, 1976b, 1977a,b). The mean (± SD) of experiments with different cell preparations is shown.

[b] These values were corrected for simple diffusion, as explained in the text.

component for corticosterone is within the range of the physiological concentrations of free hormone in rat plasma. Transport of the two glucocorticoids by either the low or the high affinity system is competitively inhibited by hydrocortisone, corticosterone, and cortisone, and noncompetitively by dexamethasone, testosterone, estradiol, and estrone. The transport is temperature dependent, Q_{10} = 8, with an activation energy of 25 kJ/mole above 20°C. Pretreatment of the liver cells with phospholipase A_2 reduces the transport of corticosterone by 35%, whereas phospholipase D has no effect. β-Glucosidase and β-galactosidase decrease transport by 13%. These observations seem to indicate that integrity of both lipids and glycoproteins in the membrane is necessary for the transport of glucocorticoids into the cells. Functionality of SH groups might also be essential, since preincubation of the cells with 1-fluoro-2,4-dinitrobenzene and N-ethylmaleimide reduces uptake of corticosterone by approximately 60% (Rao *et al.*, 1977b). In addition, the rate of binding of cortisol to cell-free extracts of hepatocytes was measured and compared with the rate of transport in intact cells (Rao *et al.*, 1976a). The V_{max} is higher in intact cells than in cytosol (2.1 and 0.21 pmole/minute/mg protein, respectively), which might suggest that the cell membrane is not a barrier to entry of the hormone into the cells.

Taking into consideration the saturability, the temperature dependence, the sensitivity to treatment with various agents perturbing the membrane structure, and finally the high specificity of the transport of hydrocortisone and corticosterone, the presence of a mediated diffusion transport of the two glucocorticoids appears to be a strong possibility. Because of the difference in the kinetics parameters for the two compounds (Table VIII) and the low extent of reciprocal inhibition [50% at 160-fold excess of competitor (Rao *et al.*, 1976a, 1977b)], it seems probable that transport of hydrocortisone and corticosterone is operated by distinct systems. It is also evident that the transport systems have a different specificity from that of intracellular receptors; the latter bind dexamethasone more strongly than hydrocortisone (Litwack *et al.*, 1973), whereas dexamethasone is not a potent inhibitor of transport of the hormone.

On the other hand, the claim by this group of investigators that corticosterone and hydrocortisone are transported actively into rat liver cells does not seem justified. Metabolic inhibitors other than KCN have been shown not to prevent transport (Rao *et al.*, 1977b), and, even more important, a positive gradient of *free* glucocorticoid from the interior to the exterior of the membrane has not been demonstrated. Demonstration of such a gradient is an essential criterion for recognizing active transport systems of solutes (Davson and Danielli, 1970).

The transfer of testosterone, estradiol, and estrone into rat liver cells has also been studied by the same group of investigators (Rao *et al.*, 1977a). The transport rates of the three steroids at concentrations below 5×10^{-9} *M* are linear, but exhibit cooperativity (Fig. 13). A saturable component is observed in the micromolar range (Table VIII), and this is sensitive to antimycin A, but not to other metabolic inhibitors such as KCN and 2,4-dinitrophenol. Testosterone transport is not inhibited by treatment of the cells with SH blocking agents, whereas, on the other hand, the transport of estrogen is decreased considerably. This observation might imply involvement of membrane proteins in transport of the hormones. Furthermore, the fact that cooperativity is noted only up to a very low concentrations in the medium of estrogens, or of testosterone, might suggest that cooperativity is due to binding of the steroids to proteins that possess two or more dependent sites (Sips, 1948) and that have a very low capacity for the steroids. However, the specificity of the transport systems for estrogens and androgens has

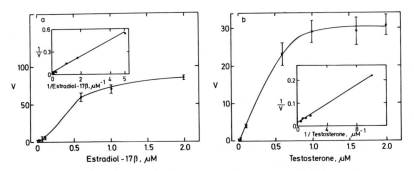

FIG. 13. Initial rates of uptake by isolated rat liver cells of estradiol (a) and testosterone (b) as a function of substrate concentration. On the ordinate: $V = V_{max}$ in pmole/minute/mg protein. The experimental conditions were as in Fig. 12, but cells were incubated with steroids at 15°C for 10 seconds. Each point in this figure represents the mean of triplicates; vertical bars indicate standard deviation. The insets show double reciprocal plots of the data. (Reproduced from Rao *et al.*, 1977a, with kind permission of Academic Press.)

not been investigated so far, and it is premature to state at this stage whether these hormones are transported into liver cells by facilitated diffusion.

E. Transport of Estrogens into Human Placenta

The placenta secretes much larger amounts of hormones that any other endocrine organ. The production of estrogens, derived from conversion in the placenta of precursors present in either the maternal or the fetal circulation, is in the region of 100 mg/day. Transport of estrogens into human placenta has some interesting aspects, as shown by the studies of Tseng and co-workers (1972b). For instance, it was observed that there is very little difference between entry of unconjugated and conjugated estrogens, particularly in term placenta. It would appear that the sulfates enter the cells without previous hydrolysis, although they are completely hydrolyzed before they are released from the cells (Tseng *et al.*, 1972b). This high permeability to hydrophilic compounds such as estrogen sulfates might be a characteristic feature of the placenta, since it was not observed in endometrium (Tseng *et al.*, 1972a). Furthermore, in term placenta, at the time when production of estrogens reaches its maximum, permeability to both free and conjugated estrogens significantly increases: whereas in early placenta, entry (expressed as fraction of the perfused steroid) is from 0.43 to 0.56; in late term placenta, it is 0.82 (Tseng *et al.*, 1972b).

Entry of unconjugated steroids remains unchanged over a wide range of concentrations, from 0.1×10^{-9} M to 17×10^{-6} M, but in one experiment, it was observed that transport of estrone and estradiol decreased progressively with concentration. Kinetic parameters calculated from this experiment were: $V_{max} = 0.4$ pmole/minute/mg protein and $K_m = 1.5 \times 10^{-6}$ M. This finding might indicate that labile carriers with large capacity for the estrogens are present in the cell membrane. This hypothesis finds some confirmation in preliminary results of M. E. Fant, R. D. Harbison, and R. W. Harrison (personal communication 1978), which show the presence of specific proteins binding glucocorticoids in the membrane fraction from human placenta.

F. Transport of Glucocorticoids into Cultured Cells

1. *Hepatoma Cells*

Plagemann and his associates have conducted a thorough study into the transport of the synthetic glucocorticoid prednisolone in the Reuber line of rat hepatoma cells, which contains receptors for glucocorticoids and is sensitive to hormonal action, and in the Novikoff line, which is hormone unresponsive (Plagemann and Erbe, 1976; Graff *et al.*, 1977). The zero trans procedure was used for the experiments. For determination of influx of prednisolone, the Reuber cells, which grow on monolayer, were washed free of medium; the solution of the radioactive steroid was then added to the dish and drained off after set intervals; Novikoff cells were suspended for brief periods in medium containing the radioactive steroid and centrifuged out of the solution. The minimum sampling time was 1 minute. As can be seen from Fig. 14a, the uptake of prednisolone by Novikoff cells at 10°C is very rapid and is not influenced by addition of a 2000-fold excess of dexamethasone or deoxycorticosterone. The efflux at 37°C is equally rapid (Fig. 14b). The rate of transport of the glucocorticoid is linear over a range of concentrations from 10×10^{-9} M to 5×10^{-3} M (Plagemann and Erbe, 1976). The coefficient of temperature is between 1.2 and 1.6; although, in a later paper, it is reported as 2.9 (Graff et $al.$, 1977). Similar results to those

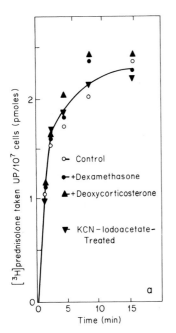
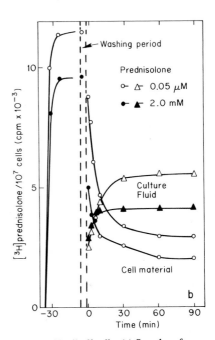

Fig. 14. Uptake and release of prednisolone by hepatoma Novikoff cells. (a) Samples of a suspension of 1×10^7 cells/ml in medium BM42A at 18°C were supplemented at zero time with [³H]-prednisolone, 50×10^{-9} M, and, in some experiments, with 100×10^{-6} M dexamethasone or deoxycorticosterone. Another sample was supplemented with 5×10^{-3} M KCN and 5×10^{-3} M iodoacetate, and incubated at 37°C for 10 minutes prior to addition of [³H]prednisolone. (b) Two samples of a suspension of 1×10^7 cells/ml were incubated for 30 minutes at 37°C with [³H]pred-nisolone, 50×10^{-9} $M;$ to one of the samples, unlabeled prednisolone, 2×10^{-3} M, was added. At the end of the incubation, the cells were collected by centrifugation, washed once in BM42B medium, resuspended to the same density in fresh medium, and incubated further at 37°C. Duplicate 1-ml samples were centrifuged at various intervals and the supernatant fluid (culture fluid) and the cell pellet (cell material) were analyzed for radioactivity. All points represent the mean of duplicates. (Reproduced from Plagemann and Erbe, 1976, with kind permission of Pergamon Press.)

shown in Fig. 14 have been obtained in Reuber cells (Plagemann and Erbe, 1976). In neither type of cells is prednisolone transport inhibited by pretreatment with neuraminidase or phospholipase C. Depletion of ATP from the cells or addition of metabolic inhibitors are also without any effect on transport.

These findings clearly point to a mode of transport of prednisolone into hepatoma cells by simple diffusion. This conclusion is reinforced by the observation that there is no countertransport of the steroid in the presence of corticosterone, whereas countertransport by analog compounds is characteristic of facilitated diffusion systems.

Another interesting observation concerns the interaction of prednisolone with the D-glucose transport systems in Novikoff cells. Although prednisolone inhibits transport of glucose in these cells and exhibits an affinity similar to that of the natural substrate, glucose does not prevent the transport of prednisolone into the cells. Thus, it is possible that the glucocorticoid blocks the D-glucose transport system without being itself transported (Graff et al., 1977).

The glucocorticoid responsive line HTC (hypertetraploid hepatoma cells) has become one of the best tools for the investigation of hormonal action, mainly owing to the pioneering work of the late Gordon Tomkins and his collaborators. The kinetics of association and dissociation of dexamethasone and hydrocortisone in HTC cells has been analyzed by Baxter and Tomkins (1970). The experiments were performed with cells in suspension; minimum sampling times were either 2.5 or 5 minutes. At a concentration of glucocorticoids in the medium of 5×10^{-9} M, both association and dissociation of the steroids are very fast, reaching a maximum within 5 minutes. Both processes are temperature dependent. In further studies, it was shown that specific binding of dexamethasone to the receptors is not inhibited, or only slightly inhibited, when cells are pretreated with mercurials that do not penetrate the membrane, such as PCMB and PCMPS (Levinson et al., 1972). The rate of association of dexamethasone with cell-free extracts was also determined in parallel experiments. Saturation of binding occurs within 5 minutes, so that binding of dexamethasone to cell-free extracts might be faster than in intact cells. However, it has already been mentioned that a slower rate of association of steroids in intact cells does not necessarily mean that the plasma membrane constitutes a barrier to the diffusion of steroids into cells.

These results are suggestive, but not conclusive, of transport of glucocorticoids into HTC cells by simple diffusion. Since association of dexamethasone with proteins in cell-free extracts appears to be very rapid (less than 5 minutes) it is possible that some intracellular binding of the glucocorticoids was taking place during the interval of 2.5 minutes in which the cells were first sampled in the influx experiments. In this case, obviously, no clear conclusion could be drawn on the actual process of transport across the plasma membrane.

In recent studies in the laboratory of the author, the transport of glucocorticoids into HTC cells has been reexamined by means of the zero trans procedure, but with sampling times of 8 seconds at 28°C, well within the linear portion of the uptake curve. The rate of transport of both hydrocortisone and dexamethasone was found to be linear over a range of concentrations from 10 to 400×10^{-9} M; corticosterone, testosterone, progesterone, or estradiol do not prevent transport of the two glucocorticoids into HTC cells (E. P. Giorgi and W. P. Stein, unpublished, 1978). These results confirm the hypothesis of Tomkins and associates (Baxter and Tomkins, 1970; Levinson et al., 1972) that glucocorticoids enter HTC cells by a process of simple diffusion.

Thus, while normal liver cells seem to possess a system of facilitated diffusion for glucocorticoids (Rao et al., 1976b, 1977b), this system is lacking in HTC and other hepatoma cell lines. It could be postulated that the loss of a specialized mode of hormone transport in hepatoma cells is a consequence of cellular transformation. As we shall see later, another example of specific membrane modifications during malignant transformation of hepatic cells is known (Section VI,B). Alternatively, the lack of facilitated diffusion for glucocorticoids in hepatoma cells may be attributed to suppression of function, since these cells are grown in medium containing 5% serum, but with no added hormone, and therefore are exposed to very low concentrations of glucocorticoids during growth. This second hypothesis, however, seems rather unlikely, since in these cells there is no suppression of function of the specific glucocorticoid receptors.

2. Mouse Pituitary Adenoma Cells

The line At-T-20/D1 from mouse pituitary adenoma possesses specific glucocorticoid receptors, and production of ACTH (adrenocorticotropic hormone) by the cells is suppressed by the corticoids (Watanabe et al., 1973). Harrison and co-workers have studied the transport of a synthetic glucocorticoid, triamcinolone acetonide, into these cells (Harrison et al., 1974, 1975, 1976, 1977). For the influx experiments, a suspension of cells was exposed to tritium-labeled steroid (1.15×10^{-9} M), with or without addition of a 500-fold concentration of dexamethasone, in order to measure specific binding. The total uptake (i.e., specific and nonspecific) of triamcinolone acetonide was observed to be near maximal after 2.5 minutes, the shortest time of sampling. In efflux experiments, the cells were preincubated with radioactive triamcinolone acetonide and then excess dexamethasone was added. Under these conditions, the efflux curve of triamcinolone acetonide from the cells showed two components: a rapid and a much slower efflux. The fast efflux is saturable ($K_A = 0.6 \times 10^{11}$ M^{-1}) and is displaced by dexamethasone and natural glucocorticoids, but does not vary with temperature. The slow process of efflux is also saturable, is very sensitive to temperature variations, and is inhibited by pretreatment of the cells with phospholipase A_2 and neuraminidase, or by addition of ethanol or dimethylsulfoxide to the medium (Harrison et al., 1974, 1976, 1977). Whereas the slow efflux of triamcinolone acetonide is probably a process of dissociation of the steroid from the intracellular receptors, the nature of the fast efflux is not clear. Although this process seems to exhibit some of the characteristics of transport by facilitated diffusion, such as saturability and specificity, it seems to be temperature independent, and this is not usually thought to be compatible with the hypothesis of facilitated diffusion. It is also improbable that initial rates were measured in these experiments, since sampling time was 2.5 minutes, whereas, as mentioned above, transport of glucocorticoids across the cell membrane is very fast, and linear for approximately 15–50 seconds. Further studies on mouse pituitary adenoma cells are reviewed in Section VI,B.

VI. Binding of Steroids to Membrane Proteins

In the previous section, we presented the evidence for interaction of steroid hormones with carrier systems. However, this is not the only interaction taking place between steroids and membrane proteins. Both nonspecific and highly specific steroid binding have been observed to proteins extracted from cell membranes or to membrane fractions. Although, as discussed at the end of this section, these steroid binding proteins cannot be identified with the carrier systems, because of their situation at the membrane surface, they might be implicated in steroid transport. In order to understand the possible role of these proteins, their mode of interaction with steroid hormones first must be examined briefly.

A. Low Affinity Binding

Steroids are transferred across the membrane of erythrocytes (RBC) very rapidly, and the pattern of permeability suggests that the process is one of simple diffusion (Brinkmann et al., 1972). However, substantial binding of steroids can be obtained upon incubation with proteins from erythrocyte membranes (De-Venuto et al., 1969; Brinkmann and van der Molen, 1972). In the studies of DeVenuto and co-workers (1969), the proteins were extracted from erythrocyte ghosts with butanol and purified by gel electrophoresis. The purified proteins were dialyzed against a solution containing progesterone; the radioactivity associated with the proteins was determined at the end of dialysis. It appears that 49% of the steroid initially in the medium becomes associated with the proteins, and this percentage of bound hormone remains constant over a wide range of concentrations, up to $0.24 \times 10^{-9} M$ per mg protein. Less steroid binds at higher temperature, probably because of a faster dissociation of the steroid—protein complex. The extent of binding of progesterone is similar in isolated proteins and in intact erythrocyte membranes. Brinkmann and van der Molen (1972) found as well that proteins isolated from erythrocytes ghosts have a great combining affinity for steroids as compared with intact membranes, but suggested that this finding is due to binding of the hormones to sites of the proteins that would not be exposed to the steroids in the intact membrane. This interpretation is in agreement with the results of investigations with spin labeled steroids, which demonstrate that the steroid molecules interact with proteins in the membrane only after lipids are removed by lytic agents (Hubbell *et al.*, 1970). Binding of steroids to RBC membrane proteins might therefore have little physiological significance. Brinkmann and van der Molen extended their investigations to other steroids, and reported that binding of pregnenolone, 20α-OH-progesterone, progesterone, estradiol, androstenediol, and testosterone is unsaturable and that the extent of binding decreases with the polarity of the compounds (Fig. 15). These two findings together lead to the conclusion that any interaction that might take place between the steroids and the membrane proteins is nonspecific in character.

The binding of progesterone, hydrocortisone, and aldosterone to crude membrane preparations from rat liver and kidney also appears to be unsaturable and to follow the polarity rule (Davidson *et al.*, 1963). Aldosterone is not displaced by spirolactone, a steroid that suppresses the biological action of aldosterone. These observations are at variance with those of other workers, who, as described later, have found specific binding of glucocorticoids in membrane fractions of rat liver cells (Suyemitzu and Terayama, 1975). But, since in the studies of Davidson and

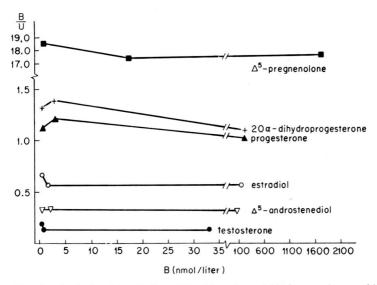

FIG. 15. Scatchard plot of the binding of steroids to hemoglobin-free membranes of human erythrocytes. The membranes were dialyzed for 24 hours at 4°C in phosphate buffer against concentrations of steroids varying from 0.063 to 240 × 10^{-9} M. Protein concentration, 2 mg/ml. On the ordinate, ratio bound to unbound steroid; on the abscissa, bound steroid. (Reproduced from Brinkmann and van der Molen, 1972, with kind permission of North Medical, Amsterdam.)

co-workers, steroids were added to the liver membranes in micromolar concentrations, the presence of specific, high affinity binding of steroids would have been overlooked.

B. High Affinity Binding

1. Binding of Glucocorticoids

a. *Membranes from Rat Liver Cells.* Binding sites with high affinity and specificity for glucocorticoids have recently been identified in membranes of rat liver and hepatoma cells (Suyemitzu and Terayama, 1975; Terayama *et al.,* 1976). In these experiments, the liver was perfused *in vivo* with saline, in order to remove plasma contaminations and, particularly the steroid carrier globulin CBG. Separation of the tissue homogenate by centrifugation in sucrose yielded two membrane fractions; the lighter fraction ($d = 1.13$ to 1.16), upon examination by the electron microscope, appeared to contain mostly fragments of areas of the cell surface directed toward the sinusoid walls and to present many vesicles derived from disrupted microvilli. The heavier fraction ($d = 1.16$ to 1.18) was instead formed from areas in the membrane of cell-to-cell contact and contained fragmented desmosomes and tight junctions. The two membrane fractions were equilibrated with labeled hydrocortisone either by dialysis for 48 hours in the cold, or by incubation for 2 hours at 4°C, followed by centrifugation. Analysis of the data showed that, in the lighter fraction, there is association of hydrocortisone with one set of binding sites; the association constant of these sites is 0.7×10^9 M^{-1}, and the concentration in normal liver cells is 2.5×10^{-9} M. In the heavier fraction, three sets of sites binding hydrocortisone are found: one set has high affinity, $K_A = 0.6 \times 10^9$ M^{-1}; a second set of sites has intermediate affinity, $K_A = 0.3$–0.2×10^9 M^{-1}; and a third set has low affinity, $K_A = 8.0 \times 10^7$ M^{-1}. The concentrations of the high affinity component in the heavier fraction was determined as 3.2×10^{-9} M in normal cells, and 0.23×10^{-9} M in cells of the two malignant hepatoma lines, AH-774 and AH-130; there is a substantial difference also in the concentration of the sites of intermediate affinity: 4.1 and 1.1 $\times 10^{-9}$ M in normal and malignant cells, respectively. The binding of hydrocortisone to both the light and the heavy membrane fraction is prevented by natural corticoids added in excess, but not by the synthetic glucocorticoid dexamethasone, nor by estradiol. Dexamethasone binding by the two membrane fractions is constant over a range of concentrations from 10^{-9} to 10^{-6} M. The binding sites on the membrane therefore appear to be specific for natural glucocorticoids. Thus, the steroid specificity of the membrane binding sites is different from that of the receptors, which have a greater affinity for synthetic glucocorticoids (Litwack *et al.,* 1973). It is also important to note that the concentration of high affinity binding sites in the membrane fractions is 5-fold greater than the concentration of receptors in the cytosol: thus, the possibility that the binding of steroids is due to receptors attached to the membrane fractions can be discarded. The only evidence suggesting the protein nature of the membrane sites is the fact that they are extremely labile and tend to decrease in concentration after prolonged dialysis (Suyemitzu and Terayama, 1975). This point has not been investigated any further.

   b. *Rat Hypophysis.* The presence of membrane proteins with high affinity for glucocorticoids has been demonstrated in another target tissue, the hypophysis. Although these proteins have physicochemical characteristics and steroid specificity closely resembling those of the corticoid carrier globulin CBG, most probably they are situated intracellularly, since they are still present in cells obtained by trypsinization of pituitary gland perfused free of blood (Koch *et al.,* 1976). The kinetics of association and dissociation of corticosterone to these proteins has recently been investigated (Koch *et al.,* 1977, 1978). A crude membrane preparation from pituitary cells sedimenting at 10,000 *g* was incubated with labeled corticosterone (0.75 or 1.5 pmole) in the presence or in the absence of excess unlabeled steroid. These experiments showed that uptake of corticosterone by the membrane preparation reaches a maximum within 2 minutes even at low temperatures. Dissociation of the protein–steroid complex is an extremely rapid process: on addition of excess of unlabeled corticosterone approximately 42% of the hormone is released within 1–2 minutes. Both association and dissociation of the steroid to the membranes seem to be temperature independent. Dissociation from solubilized membrane proteins, on the other hand, is much slower and is not complete after 40 minutes from addition of excess corticosterone. The reason for this difference is not readily apparent. The association constant of the CBG-like proteins at 0°C is $3 \times 10^{10}\ M^{-1}$ and their concentration in the solubilized protein is 29 fmole/mg protein. Binding of tritiated corticosterone to the membrane fraction is inhibited by 50% by excess corticosterone, deoxycorticosterone, or progesterone and by only 10% by dexamethasone or estradiol. Binding of tritiated dexamethasone to the membrane fractions is unsaturable and is not displaced by glucocorticoids, progesterone, and estradiol. Thus, the membrane proteins bind natural glucocorticoids with high affinity, while the synthetic glucocorticoid dexamethasone is bound with low affinity.

   The possibility that proteins similar to those described in normal rat hypophysis are present also in transformed cells from mouse pituitary adenoma has been raised by the most recent work of Harrison and co-workers (1978). For these studies, membrane vesicles were prepared in a 0.84–1.45 sucrose gradient in a rotor centrifuge and recentrifuged at 150,000 *g* for 30 minutes. The integrity and orientation of the vesicles were tested by their ability to transport α-aminoisobutyric acid. Binding of tritiated corticosterone was studied by incubating the vesicles with the steroid ($15 \times 10^{-9}\ M$) for various lengths of time at 22°C. Under these conditions, over 80% of maximum binding is reached within 5 minutes. Corticosterone is displaced almost completely from the membranes by unlabeled hormone at 200-fold concentration and displaced 49 and 32% by cortisone and hydrocortisone, respectively. The synthetic glucocorticoids, dexamethasone and triamcinolone acetonide, in contrast, do not compete for binding. Pretreatment of the vesicles with Pronase or PCMB reduces binding of corticosterone by over 60%, whereas preincubation with phospholipase A_2 completely abolishes the binding (Harrison *et al.*, 1978).

An interesting observation obtained in this study is that incubation of the membrane vesicles in media of different osmolarity has no influence on the uptake at equilibrium of corticosterone. In media of high osmolarity, there is some reduction of the volume of the vesicles, due to the higher osmotic pressure in the external medium and flow of water from the vesicles. In the case of facilitated diffusion, if the change in volume occurs once equilibrium between free steroid in the medium and in the vesicles is reached, the amount of steroid in the vesicles should decrease. The absence of this phenomenon has been interpreted by the authors as suggesting that the binding sites for glucocorticoids are fixed to the membrane and therefore are not involved in transport of the steroid (Harrison *et al.*, 1978). This hypothesis, however, should be confirmed by a demonstration that there is no trapping inside the vesicles of soluble proteins or of other material capable of binding the hormone. If such proteins were present, changes in volume of the vesicles might affect steroid uptake very little.

The relationship of the binding sites for glucocorticoids demonstrated in the membrane fractions of mouse pituitary adenoma cells and the system of facilitated transport of glucocorticoids in intact cells, described by the same authors (see Section V,F), remains to be clarified. The specificity of the proposed transport system appears to be slightly different from that of the membrane binding sites, since corticosterone does not abolish completely transport of triamcinolone acetonide.

2. Binding of Estrogens

Estrogens are bound strongly by some proteins present on the surface of endometrial and liver cells, as demonstrated by means of the very elegant technique of Pietras and Szego (1977). The epithelial cells, isolated from the tissues by digestion with collagenase in Ca^{2+},Mg^{2+}-free medium, were incubated in vessels containing nylon fibers to which an albumin–estradiol hemisuccinate derivative had been covalently bound. After 30 minutes at 22°C, the number of cells attached to the estradiol-coated fibers was counted at the light microscope. As a further check, DNA content was measured in the cells after detachment from the fibers by short incubation with estradiol (100×10^{-9} M). As can be seen from Fig. 16, endometrial cells become adsorbed to the fibers in great numbers; the number of attached liver cells is lower; while attachment of intestinal cells is negligible. Association of the cells with the fibers is reduced by

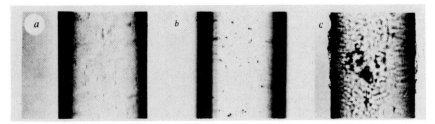

FIG. 16.   Binding of isolated cells to estrogen immobilized by covalent linkage to albumin-derivatized nylon fibers. Cells derived from rat intestinal mucosa (a), liver (b), and endometrium (c) were incubated with the nylon fibers in Petri dishes at a density of 1×10^7 cells/ml for 8 mg fibers. After 30 minutes at 22°C, the washed fibers with the adsorbed cells were transferred to 4 ml fresh medium for microphotography. No binding of any cells was observed when underivatized fibers or fibers coupled only to albumin were used. ×250. (Reproduced from Pietras and Szego, 1977, with kind permission of Macmillan Journals.)

pretreatment with excess estradiol or by lowering the temperature to 4°C, although these treatments have little influence on attachment of intestinal cells. Therefore, the membranes of target tissue cells, endometrium and liver, possess some saturable sites that bind estradiol immobilized on the nylon fibers with high affinity, whereas these sites are absent in the intestinal epithelium, a nontarget tissue.

The presence of proteins that bind estrogens has been confirmed by further experiments on membrane preparations from rat liver cells (Pietras and Szego, 1978). A lighter and a heavier membrane fractions were prepared by sucrose centrifugation and purified by further centrifugation. After incubation of the membrane fractions with tritiated estradiol, with or without addition of excess unlabeled compound, the binding of the hormones in the lighter and heavier fractions of membranes was found to be 40 and 8 times greater than in the tissue homogenate, respectively. The binding is saturable in both fractions, but it has a higher affinity in the lighter membrane fraction, with an association constant of 0.3×10^{10} M^{-1} and a concentration of approximately 500 fmole/mg protein. Estradiol and diethylstilbestrol at a 200-fold excess completely suppress the binding of the hormone, while estradiol-17α (a biologically inactive compound), hydrocortisone, progesterone, and testosterone are ineffective. The lack of competition between testosterone and estradiol indicates that the binding sites on the membrane cannot be identified with SHBG, the carrier plasma globulin, which has a similar affinity for the estrogen and the androgen (Westphal, 1971). The protein nature of the membrane binding sites is suggested by the evidence that pretreatment of the cells with trypsin or heating at 60°C for 1 hour prevent binding of estradiol (Pietras and Szego, 1978).

It is very interesting to note that the binding sites for glucocorticoids and for estrogens are found in the lighter fraction of liver cell membranes, which, as already mentioned, is formed from the surface of the membrane in contact with the sinusoid vessels of the portal system.

C. POSSIBLE ROLE OF MEMBRANE PROTEINS THAT BIND STEROIDS

It might be opportune at this point to stress that binding of steroids to proteins does not necessarily imply that some biological action, such as changes in the function of either ligand or the binding protein, must ensue. Thus, the binding of steroids to proteins isolated from erythrocyte membranes clearly falls into the category of nonspecific adsorption (Brinkmann *et al.*, 1972). Whether the high specific binding of hormones to membrane proteins in endometrial, hepatic, and pituitary cells leads to any change in the conformation of the proteins or in the membrane structure is not known, and one can only speculate on their role at present. It must be stressed that the evidence that these membrane binding sites are entities different from the cytoplasmic receptors is overwhelming. Thus, they might fulfil some role in the exchange of hormones between the cell and the external environment. However, a very high affinity for the hormones has been reported for these membrane proteins, with association constants in the region of 10^9–10^{10} M^{-1}. Since the rate constants of association for steroid–protein interaction are of the order of 10^6–10^8 M^{-1} sec^{-1} (Munck and Brinck-Johnsen, 1968; Heyns and de Moor, 1971; Milgrom *et al.*, 1973), the rate constant of dissociation of the membrane proteins would be expected to be very low, as low as 10^{-1}–10^{-4} sec^{-1}. This would seem incompatible with the fast rates of transport of steroids across the cell membrane (see Tables IV and V in Section IV). Carrier proteins have, in fact, much lower association constants, e.g., $K_A = 0.9 \times 10^6$ M^{-1} for the high affinity system for corticoids in rat liver cells (Rao *et al.*, 1976b, 1977b).

On the other hand, the affinity for the steroids of the membrane proteins is similar to, or slightly higher than, that of the plasma carrier proteins, SHBG and CBG, and it is possible that there is some exchange of steroid between the plasma and the membrane proteins. From the steroid–protein complexes thus formed in the membrane, steroid could dissociate, and enter the cells, if there is a decrease of free hormone inside the cells. In this respect it might be significant that in some cells the sites with the highest affinity for steroids are situated on the surface areas facing the blood vessels, where their role of trapping bound circulating hormone could be best fulfilled. The finding that CBG-bound corticosterone is active in liver (Keller *et al.*, 1969) might well be explained by such a mechanism. It has indeed been suggested that the binding sites on the membrane of pituitary cells are molecules of CBG complexed in the plasma membrane (Koch *et al.*, 1978).

Roles other than transport can also be postulated for these membrane proteins. The possibility that they represent membrane enzymes metabolizing steroids should be considered, especially in regard to the binding sites on liver cell membranes. Alternatively, the proteins might be involved in cell-to-cell communications, whereby the hormone would act as the signal by which the cells recognize each other. This might contribute to the maintenance of the cytoarchitecture of the target tissues. Thus, it would be very important to confirm the finding of Pietras and Szego (1977) that the specific sites are absent in nontarget tissues, like the intestine, or the finding of Terayama and co-workers (1976), that they are deleted in malignant cells. It is well known that loss of normal cell-to-cell apposition is one of the first features of malignant transformation in epithelial tissues.

Finally, it is conceivable that the membrane components with high affinity for steroids might be important in the transfer of the steroids from the membrane to the nucleus. According to the hypothesis of Jackson and Chalkley (1974), the estradiol receptor in uterine cells is originally bound to the plasma membrane; after binding the steroid, the receptor would undergo a conformational change in order to be transferred into the interior of the cell and ultimately to the nucleus. The protein binding steroids on the membrane surface might then be precursors of the receptors. This hypothesis would be compatible with the difference noted in steroid specificity between the two types of proteins. Another hypothesis, proposed by Szego (1978), is that transfer of the steroids from the specific sites on the membrane to the nucleus might be a function of the lysosomes, which might move centripetally from the plasma membrane to the nucleus once they have adsorbed the receptor–steroid complex.

VII. Conclusions

From a consideration of the cellular membrane composition and of the physicochemical characteristics of steroid molecules (Section III), it can be predicted that simple diffusion will be the main mode of transport of steroid hormones into cells. This prediction is borne out by the results of studies of steroid transport into cells and tissues. In fact, simple diffusion is the *only* mode of steroid transport in at least one typical target tissue (endometrium) and in some hormone responsive tissues (thymocytes, skin, hepatoma cells) (Sections IV and V). However, the possibility of great variations in membrane lipid composition can also be expected to cause differences in hormone permeability between cells from various tissues. Selective permeability for individual steroids may be recognized from a lack of correspondence between the rate of transport of the compounds into cells and their solubility in organic solvents or artificial lipid membranes. Many examples of this type of selectivity have been described in the previous sections. The permeability of human endometrium to estradiol, of thymocytes to corticosterone, or of prostate to testosterone, is unusually high for such polar compounds. These changes in permeability might represent an adaptation of the plasma membrane aimed to insure an adequate supply of the active hormone to the cells; or, in the case of the prostate, an adequate supply of the precursor of the active hormone to the cells. The increased permeability of term placenta to estrogens, and particularly to estrogen sulfates, might have a similar significance. On the other hand, in cultured cells, permeability to glucocorticoids seems to be inversely related to cell sensitivity to hormonal action (Section IV).

Important differences between artificial lipid membranes and cell membranes are also apparent. The restricted movements of spin labeled steroids in the membranes of erythrocytes or nerve fibers compared with the absolute freedom of movement in lecithin membranes is a case in point (Section III). The influence of the membrane intrinsic proteins on the ordering of the lipids in the biological membrane might explain this difference. Moreover, the modification by various steroids of androgen entry into the prostate is not compatible with interactions of the steroids with the lipid matrix, which makes it necessary to postulate interactions with some membrane proteins (Section V). Thus, it might be an oversimplification to view the transport of steroid hormones into cells as a process of diffusion through the lipid bilayer regulated exclusively by the solubility of the steroids in the membrane lipids, since this is to ignore the whole structure of the membrane.

Mediated diffusion of steroids has been claimed in some target tissues, but a critical evaluation of the evidence does not support such claims, with the possible exception of mediated diffusion of glucocorticoids in rat liver. It is probable that mediated diffusion of estrogens supplements simple diffusion also in placenta (Section V). Therefore, specialized modes of steroid transport may have developed in organs that process large amounts of steroids, as an adaptation to function. It is interesting to note that, in hepatoma cells grown in the absence of hormones, mediated diffusion of glucocorticoids is lost.

The role of membrane proteins binding steroids with high affinity is still a matter of speculation. These proteins seem to be present exclusively in the membrane of target cells and are characterized by preferential binding of natural over synthetic steroids (Section VI). The contribution of these proteins to steroid transport into the cells could be the trapping of bound hormones from the circulation and their slow delivery across the membrane.

It has been clearly demonstrated that passage of hormones across the plasma membrane is not a limiting step to binding of the hormones to the intracellular receptors in endometrium, thymocytes, and hepatoma cells (Bell and Munck, 1973; Levinson et al., 1972; Peck et al., 1973). On the other hand, it is not yet possible to assess whether saturation of specific membrane proteins in endometrium or hypophysis—or of the still hypothetical carriers in the prostate—in the presence of increased concentrations of free hormones in blood, might substantially reduce the rate of transport of hormones into the tissues. Thus, although the evidence to date does not warrant any firm conclusion, it seems unlikely that transport of steroid across the plasma membrane affords a mechanism of regulation of hormone supply to target tissues. Extracellular binding to plasma carrier proteins and intracellular binding plus metabolism would appear to be much more efficient mechanisms for the regulation of cellular hormone levels (Section II).

Information on the mode of transport of steroid hormones into cells is still very limited, and the effects of physiological and pathological variations on cell permeability to steroids remain largely unknown. Malignant transformation might delete specific membrane proteins (Terayama *et al.*, 1976), and at least one genetically determined change in membrane composition affecting permeability to steroids has been described (Sogbesan *et al.*, 1974). This field of investigation is completely open and is worth pursuing, because a better knowledge of the changes in cell permeability to steroids in physiological and pathological states might indeed open a new therapeutic approach for the treatment of diseased hormone-dependent tissues. Specific modification *in vivo* of cell permeability to steroids might indeed not be beyond the realm of possibilities, as shown by the recently introduced method of treatment of lipodistrophic diseases by incorporation of liposomes into the cell membrane (for review, see Gregoriadis, 1978).

ACKNOWLEDGMENTS

I am grateful to Prof. W. D. Stein for his constructive criticism of the manuscript and to Mrs. C. Stein for stylistic help.

REFERENCES

Baird, D. T., Horton, R., Longcope, C., and Tait, J. F. (1969). *J. Clin. Endocrinol. Metab.* **29**, 293–301.

Ballard, P. L., and Tomkins, G. M. (1970). *J. Cell Biol.* **47**, 222–234.

Bangham, A. D., Standish, M. M., and Weissmann, G. (1965). *J. Mol. Biol.* **13**, 253–259.

Barry, B. W., and El Eini, D. I. D. (1976). *J. Pharm. Pharmacol.* **28**, 219–227.

Baulieu, E. E., and Raynaud, J. P., and Milgrom, E. (1970). *Eur. J. Biochem.* **13**, 293–304.

Baxter, J. D., and Tomkins, G. M. (1970). *Proc. Natl. Acad. Sci. U.S.A.* **65**, 709–715.

Baxter, J. D., and Tomkins, G. M. (1971). *Proc. Natl. Acad. Sci. U.S.A.* **68**, 932–937.

Becker, H., Kaufmann, J., Klosterhalfen, H., and Voigt, K. D. (1972). *Acta Endocrinol. (Copenhagen)* **71**, 589–599.

Beckett, A. H., and Pickup, M. E. (1975). *J. Pharm. Pharmacol.* **27**, 226–234.

Bell, P. A., and Munck, A. (1973). *Biochem. J.* **136**, 97–107.

Bidlingmaier, F., Knorr, D., and Neumann, F. (1977). *Nature (London)* **266**, 647–648.

Bretscher, M. S., and Raff, M. C. (1975). *Nature (London)* **258**, 43–49.

Brinkmann, A. O., and van der Molen, H. J. (1972). *Biochim. Biophys. Acta* **274**, 370–381.

Brinkmann, A. O., Mulder, E., and van der Molen, H. J. (1972). *J. Steroid Biochem.* **3**, 610–615.

Brockerhoff, H. (1974). *Lipids* **9**, 645-650.

Butler, K. W., Smith, I. C. P., and Schneider, H. (1970). *Biochim. Biophys. Acta* **219**, 514-517.

Cowan, R. A., Cowan, S. K., and Grant, J. K. (1975). *J. Endocrinol.* **61**, 121-131.

Davidson, E. T., DeVenuto, F., and Westphal, U. (1963). *Proc. Soc. Exp. Biol. Med.* **113**, 387-391.

Davson, F., and Danielli, J. F. (1970). "The Permeability of Natural Membranes." Hafner, Darien, Connecticut.

Demel, R. A., Bruckdorfer, K. R., and van Deneen, L. L. M. (1972). *Biochim. Biophys. Acta* **255**, 321-330.

DeVenuto, F., Ligon, D. L., Friedrichsen, D. H., and Wilson, H. L. (1969). *Biochim. Biophys. Acta* **193**, 36-47.

Dewey, C. W. (1959). *Am. J. Physiol.* **197**, 423-438.

Diamond, J. M., and Wright, E. M. (1969). *Annu. Rev. Physiol.* **31**, 581-646.

Eik-Nes, K., Schellman, J. A., Lumry, R., and Samuels, L. T. (1954). *J. Biol. Chem.* **206**, 411-419.

Erickson, B. N., Williams, N. N., Hummel, F. C., and Marcey, I. G. (1937). *J. Biol. Chem.* **118**, 15-30.

Franks, N. P., and Lieb, W. R. (1978). *Nature (London)* **274**, 339-342.

Funder, J. W., Feldman, D., and Edelman, I. S. (1973). *Endocrinology* **92**, 954-973.

Giorgi, E. P. (1976). *J. Endocrinol.* **68**, 109-119.

Giorgi, E. P., and Crosignani, P. G. (1969). *J. Endocrinol.* **44**, 219-230.

Giorgi, E. P., and Moses, T. F. (1976). *J. Endocrinol.* **65**, 279-280.

Giorgi, E. P., Stewart, J. C., Grant, J. K., and Scott, R. (1971). *Biochem. J.* **123**, 41-55.

Giorgi, E. P., Stewart, J. C., Grant, J. K., and Reid, J. (1972a). *J. Endocrinol.* **55**, 421-439.

Giorgi, E. P., Stewart, J. C., Grant, J. K., and Shirley, I. M. (1972b). *Biochem. J.* **126**, 107-121.

Giorgi, E. P., Shirley, I. M., Grant, J. K., and Stewart, J. C. (1973). *Biochem. J.* **132**, 465-474.

Giorgi, E. P., Moses, T. F., Grant, J. K., Scott, R., and Sinclair, J. (1974). *Mol. Cell. Endocrinol.* **1**, 271-284.

Godeau, J. F., Schorderet-Slatkine, S., Hubert, P., and Baulieu, E. E. (1978). *Proc. Natl. Acad. Sci. U.S.A.* **75**, 2353-2357.

Goldstein, J. L., and Brown, S. M. (1977). *Annu. Rev. Biochem.* **46**, 897-930.

Gorski, J., and Gannon, F. (1976). *Annu. Rev. Physiol.* **38**, 425-450.

Gower, D. B. (1975). *In* "Biochemistry of Steroid Hormones" (H. L. Makin, ed.), pp. 149-184. Blackwell, Oxford.

Graff, J. C., Wohlhueter, R. M., and Plagemann, P. G. W. (1977). *J. Biol. Chem.* **252**, 4185-4190.

Graham, J. M., and Green, C. (1969). *Biochem. Pharmacol.* **18**, 493-502.

Gregoriadis, G. (1978). *Nature (London)* **275**, 695-699.

Gurpide, E., and Welch, M. (1969). *J. Biol. Chem.* **244**, 5159-5171.

Gurpide, E., MacDonald, P. G., vandeWiele, R., and Lieberman, S. (1963). *J. Clin. Endocrinol. Metab.* **23**, 346-357.

Härkönen, P., Isotalo, A., and Santii, R. (1975). *J. Steroid Biochem.* **6**, 1405-1413.

Harper, M. E., Pike, A., Peeling, W. B., and Griffiths, K. (1973). *J. Endocrinol.* **60**, 117-125.

Harrison, R. W., Fairfield, S., and Orth, D. N. (1974). *Biochem. Biophys. Res. Commun.* **61**, 1262-1267.

Harrison, R. W., Fairfield, S., and Orth, D. N. (1975). *Biochemistry* **14**, 1304-1307.

Harrison, R. W., Fairfield, S., and Orth, D. N. (1976). *Biochim. Biophys. Acta* **444**, 487-496.

Harrison, R. W., Fairfield, S., and Orth, D. N. (1977). *Biochim. Biophys. Acta* **466**, 357-365.

Harrison, R. W., Balasubramanian, K., Yeakley, J., Fant, M., Svec, F., and Fairfield, S. (1978). *In* "Steroid Receptor Systems" (W. W. Leavitt, ed.). (in press).

Hayes, S. D., and Saunders, L. (1966). *Biochim. Biophys. Acta* **116**, 184-185.

Heap, R. B., Symons, A. M., and Watkins, J. C. (1970). *Biochim. Biophys. Acta* **218**, 482–495.

Heyns, W., and de Moor, P. (1971). *J. Clin. Endocrinol. Metab.* **32**, 147–154.

Hillier, S. G., Cole, E. N., Groom, G. V., Boyns, A. R., and Cameron, E. H. D. (1973). *Steroids* **22**, 227–238.

Hoffmann, W., Forbes, T. R., and Westphal, U. (1969). *Endocrinology* **85**, 778–781.

Huang, C.-H. (1976). *Nature (London)* **259**, 242–244.

Hubbell, W. L., and McConnell, H. M. (1969). *Proc. Natl. Acad. Sci. U.S.A.* **63**, 16–22.

Hubbell, W. L., and McConnell, H. M. (1971). *J. Am. Chem. Soc.* **93**, 314–326.

Hubbell, W. L., Metcalfe, J. C., Metcalfe, S. M., and McConnell, H. M. (1970). *Biochim. Biophys. Acta* **219**, 415–427.

Isaacs, W. A., and Hayhoe, F. G. J. (1967). *Nature (London)* **215**, 1139–1142.

Jackson, V., and Chalkley, R. (1974). *J. Biol. Chem.* **249**, 1627–1636.

Jensen, E. V., Suzuki, T., Kawashima, T., Stumpf, W. E., Jungblut, P. W., and De Sombre, E. R. (1968). *Proc. Natl. Acad. Sci. U.S.A.* **59**, 632–638.

Jimenez de Asua, L., O'Farrell, M. K., Clingan, D., and Rudland, P. S. (1977). *Proc. Natl. Acad. Sci. U.S.A.* **74**, 3845–3849.

Kapstein, P., Treiber, L., Wenderger, F., and Oertel, G. W. (1967). *Hoppe-Seyler's Z. Physiol. Chem.* **348**, 401–404.

Keller, N., Richardson, U. I., and Yates, F. E. (1969). *Endocrinology* **84**, 49–62.

King, R. J. B. (1976). *Essays Biochem.* **12**, 41–76.

King, R. J. B., and Mainwaring, W. I. P. (1974). "Steroid-Cells Interactions." Butterworth, London.

Koch, B., Lutz, B., Briaud, B., and Mialhe, C. (1976). *Biochim. Biophys. Acta* **444**, 497–507.

Koch, B., Lutz-Bucher, B., Briaud, B., and Mialhe, C. (1977). *J. Endocrinol.* **73**, 399–400.

Koch, B., Lutz-Bucher, B., Briaud, B., and Mialhe, C. (1978). *J. Endocrinol.* **79**, 215–222.

Kowarski, A., Shalf, J., and Migeon, C. J. (1969). *J. Biol. Chem.* **244**, 5269–5272.

Lacko, L., Wittke, B., and Geck, P. (1975). *J. Cell. Physiol.* **86**, 673–680.

Lasnitzki, I., and Franklin, H. C. (1972). *J. Endocrinol.* **54**, 333–342.

Lasnitzki, I., and Franklin, H. R. (1975). *J. Endocrinol.* **64**, 289–297.

Lasnitzki, I., Franklin, H. R., and Wilson, J. D. (1974). *J. Endocrinol.* **60**, 81–90.

Lawrence, D. K., and Gill, E. W. (1975). *Mol. Pharmacol.* **11**, 280–286.

Levinson, B. B., Baxter, J. D., Rousseau, G. G., and Tomkins, G. M. (1972). *Science* **175**, 189–190.

Liao, S. (1975). *Int. Rev. Cytol.* **41**, 87–172.

Litwack, G., Filler, R., Rosenfield, S. A., Lichtash, N., Wishman, C. A., and Singer, S. (1973). *J. Biol. Chem.* **248**, 7481–7486.

Mainwaring, W. I. P. (1977). "The Mechanism of Action of Androgens." Springer-Verlag, Berlin and New York.

Mainwaring, W. I. P., and Peterken, B. M. (1971). *Biochem. J.* **125**, 285–295.

Melathi, K., and Gurpide, E. (1977). *J. Steroid Biochem.* **8**, 141–145.

Masiak, S. J., and LeFevre, P. G. (1974). *Arch. Biochem. Biophys.* **162**, 442–447.

Mercier-Bodard, C., Marcheret, M., Perat, M., Picard, M.-T., and Baulieu, E. E. (1976). *J. Clin. Endocrinol. Metab.* **43**, 374–386.

Milgrom, E., and Baulieu, E. E. (1970). *Endocrinology* **87**, 276–287.

Milgrom, E., Atger, M., and Baulieu, E. E. (1973). *Biochim. Biophys. Acta* **320**, 267–283.

Mills, T. M., and Spaziani, E. (1968). *Biochim. Biophys. Acta* **150**, 435–445.

Munck, A., and Brinck-Johnsen, T. (1968). *J. Biol. Chem.* **243**, 5556–5565.

Munck, A., and Leung, K. (1977). *In* "Receptors and Mechanism of Action of Steroid Hormones" (J. R. Pasqualini, ed.), Part II, pp. 312–397. Dekker, New York.

Osterhout, W. J. V. (1935). *Proc. Natl. Acad. Sci. U.S.A.* **21**, 125–132.

Overell, B. G., Condon, S. E., and Petrov, V. (1960). *J. Pharm. Pharmacol.* **12,** 150-155.

Pasqualini, J. R., ed. (1977). "Receptors and Mechanism of Action of Steroid Hormones," 2 vols. Dekker, New York.

Peck, E. J., Jr., Burgner, J., and Clark, J. H. (1973). *Biochemistry* **12,** 4596-4603.

Peterson, R. P., and Spaziani, E. (1969). *Endocrinology* **85,** 932-940.

Pietras, R. J., and Szego, C. M. (1977). *Nature (London)* **265,** 69-72.

Pietras, R. J., and Szego, C. M. (1978). *Endocrinology* **102,** Suppl., 76.

Plagemann, P. G. W., and Erbe, W. (1976). *Biochem. Pharmacol.* **25,** 1489-1494.

Rao, G. S., Schulz-Hagen, K., Rao, M. L., and Breuer, H. (1976a). *J. Steroid Biochem.* **7,** 1123-1129.

Rao, M. L., Rao, G. S., Höller, M., Breuer, H., Schattenberg, P. J., and Stein, W. D. (1976b). *Hoppe-Seyler's Z. Physiol. Chem.* **357,** 573-584.

Rao, M. L., Rao, G. S., and Breuer, H. (1977a). *Biochem. Biophys. Res. Commun.* **77,** 566-573.

Rao, M. L., Rao, G. S., Eckel, J., and Breuer, H. (1977b). *Biochim. Biophys. Acta* **500,** 322-332.

Rask, R., and Peterson, P. A. (1976). *J. Biol. Chem.* **251,** 6360-6366.

Richards, C. D., and Hesketh, T. R. (1975). *Nature (London)* **256,** 176-182.

Riggs, T. R., Pan, M. W., and Fang, H. W. (1968). *Biochim. Biophys. Acta* **150,** 92-103.

Rosen, V., Jung, I., Baulieu, E. E., and Robel, P. (1975). *J. Clin. Endocrinol. Metab.* **41,** 761-770.

Sachs, L., Inbar, M., and Shinitzky, M. (1974). *In* "Control of Proliferation in Animal Cells" (B. Clarkson and R. Baserga, eds.), pp. 283-296. Cold Spring Harbor Lab., Cold Spring Harbor, New York.

Sackmann, E., and Träuble, H. (1972). *J. Am. Chem. Soc.* **94,** 4482-4491.

Sandberg, A. A., Slaunwhite, W. R., Jr., and Carter, A. C. (1960). *J. Clin. Invest.* **39,** 1914-1928.

Scheuplein, R. J., Blank, I. H., Branner, G. J., and MacFarlane, D. (1969). *J. Invest. Dermatol.* **52,** 63-70.

Scholtan, W., Schlossmann, K., and Rosenkrantz, H. (1968). *Arzneim.-Forsch.* **18,** 767-775.

Seeman, P. (1972). *Pharmacol. Rev.* **24,** 583-655.

Sessa, G., and Weissmann, G. (1968). *Biochim. Biophys. Acta* **150,** 173-180.

Sha'afi, R. I., and Gary-Bobo, C. M. (1973). *Prog. Biophys. Mol. Biol.* **26,** 103-146.

Simpson-Morgan, M. W., and Sutherland, R. L. (1976). *J. Physiol. (London)* **257,** 123-126.

Singer, S. J., and Nicolson, G. L. (1972). *Science* **175,** 720-731.

Sips, R. (1948). *J. Chem. Phys.* **16,** 490-498.

Slaunwhite, W. R., Jr., Lockie, G. N., Back, N., and Sandberg, A. A. (1962). *Science* **135,** 1062-1072.

Snart, R. S., and Wilson, M. J. (1967). *Nature (London)* **215,** 964.

Sogbesan, A. O., Adadevoh, B. K., and Dada, O. A. (1974). *Acta Endocrinol. (Copenhagen)* **77,** 794-800.

Solomon, A. K. (1960). *Sci. Am.* **203,** 146-156.

Stein, W. D. (1967). "The Movement of Molecules across the Cell Membranes." Academic Press, New York.

Stevens, R. W., and Green, C. (1972). *FEBS Lett.* **27,** 145-148.

Suyemitzu, T., and Terayama, H. (1975). *Endocrinology* **96,** 1499-1508.

Szego, C. M. (1978). *In* "Structure and Function of the Gonadotrophins" (K. W. McKerns, ed.), pp. 431-472. Plenum, New York.

Szego, C. M., Seeler, B. J., Steadman, R. A., Hill, D. F., Kimura, A. K., and Roberts, J. A. (1971). *Biochem. J.* **123,** 523-538.

Tait, J. F., and Burstein, S. (1964). *In* "The Hormones" (G. Pincus, K. V. Thinman, and E. B. Astwood, eds.), pp. 57-130. Academic Press, New York.

Terayama, H., Okamura, W., and Suyemitzu, T. (1976). *In* "Control Mechanisms in Cancer" (W. E. Criss, T. Ono, and J. R. Sabine, eds.), pp. 83-97. Raven Press, New York.

Tremblay, R. R., Forest, M. G., Shalf, J., Martel, J. G., Kowarski, A., and Migeon, C. J. (1972). *Endocrinology* **91**, 556–561.

Tseng, L., Stolee, A., and Gurpide, E. (1972a). *Endocrinology* **90**, 390–404.

Tseng, L., Stolee, A., and Gurpide, E. (1972b). *Endocrinology* **90**, 405–414.

Urry, D. W. (1972). *Ann. N.Y. Acad. Sci.* **195**, 108–125.

Vandenheuvel, F. A. (1963). *J. Am. Oil Chem. Soc.* **40**, 455–471.

Wassermann, W. J., and Masui, Y. (1975). *Exp. Cell Res.* **91**, 381–388.

Watanabe, H., Orth, D. N., and Toft, D. O. (1973). *J. Biol. Chem.* **248**, 7625–7630.

Weissmann, G., and Keiser, H. (1965). *Biochem. Pharmacol.* **14**, 537–546.

Weissmann, G., Sessa, G., and Weissmann, S. (1965). *Biochem. Pharmacol.* **15**, 1537–1551.

Weissmann, G., Collins, T., Evers, A., and Denham, P. (1976). *Proc. Natl. Acad. Sci. U.S.A.* **73**, 510–514.

Westphal, U. (1971). "Steroid-Protein Interactions." Springer-Verlag, Berlin and New York.

Williams, D., and Gorski, J. (1971). *Biochem. Biophys. Res. Commun.* **45**, 258–264.

Williams, D., and Gorski, J. (1973). *Biochemistry* **12**, 297–306.

Williams, G. H., and Underwood, R. H. (1974). *In* "Methods of Hormone Radioimmunoassay" (B. M. Jaffe and H. R. Behrman, eds.), pp. 371–392. Academic Press, New York.

Wolosin, J. M., Ginsburg, H., Lieb, W. R., and Stein, W. D. (1978). *J. Gen. Physiol.* **71**, 93–100.

INTERNATIONAL REVIEW OF CYTOLOGY, VOL. 65

Structural Aspects of Brain Barriers, with Special Reference to the Permeability of the Cerebral Endothelium and Choroidal Epithelium

B. VAN DEURS

Institute of Anatomy, The Panum Institute, University of Copenhagen, Copenhagen, Denmark

I. Introduction

The brain represents a highly specialized part of the vertebrate body with respect to its extracellular fluid environment. The composition of this fluid is different from that of the blood and of noncerebral tissue fluids. Consequently, a barrier between blood and brain exists. This concept of a "hematoencephalic" barrier was established in experiments with intravenous injection of (acid) dyes, which stained most organs but not the brain (Ehrlich, 1887; Lewandowsky, 1900; Goldmann, 1909, 1913; Stern and Gautier, 1921; Wislocki and Leduc, 1952). Later it was shown that this barrier in principle was a barrier to protein, or dye-protein complexes (Rawson, 1943; Tschirgi, 1950). This was verified by light microscopy in experiments with injection of, for instance, fluorescent conjugates (Klatzo et al., 1962) and Evans blue (Clasen et al., 1970).

In these studies of brain barriers, it was established that while the brain tissue was not stained by the dyes, the circumventricular organs and the choroid plexus accumulated intravenously administered material (see, e.g., Wislocki and Leduc, 1952). It could thus be concluded that certain regions of the central nervous system do not possess a barrier between blood and tissue. Lack of a barrier was also demonstrated in early electron microscopic studies applying the now classic "tracer" technique with silver nitrate in the drinking water of the experimental animals, which showed that in, e.g., the choroid plexus, in contrast to the brain parenchyma, the capillaries were permeable to silver, and accordingly "where connective tissue surrounds the capillary endothelium, silver storage occurs, but where the endothelium rests on an abutment of glia, silver is barred" (Dempsey and Wislocki, 1955; see also, for example, Hashimoto and Hama, 1968; Dretzki, 1971). However, the special composition of the cerebrospinal fluid (CSF) compared to serum shows that the choroid plexus does represent a barrier between blood and CSF.

The production and regulation of the CSF, as well as the barrier mechanisms between the blood and the various compartments of the brain, have been the subject of numerous physiological studies (for reviews, see Davson, 1967, 1972; Cserr, 1971; Rapoport, 1976b). Also, from a structural point of view, the interest in brain barriers has been considerable, for more than two decades mainly developed by electron microscopy. The cerebral endothelium and the choroidal epithelium, which represent barriers between blood and extracellular fluid in the neuropil (the blood-brain barrier) and between blood and ventricular CSF (the blood-CSF barrier), respectively, have, in particular, been of interest to electron microscopists using various tracers (mainly enzymatic, lipid-insoluble macromolecules, see Table I) and, more recently, also freeze-fracture replicas. From a discussion of the occurrence of wide perivascular spaces and connective tissue elements, or glial sheaths, as criteria for brain barriers (see Bodenheimer and Brightman, 1968), the presence of tight junctions between the endothelial and epithelial cells came into focus (e.g., Brightman, 1965a,b; Brightman and Reese, 1969; Brightman et al., 1975a,b; Reese and Karnovsky, 1967; Crone and Thompson, 1970; Connell and Mercer, 1974; Dermietzel, 1975a,b; Møllgård and Saunders, 1975; Møllgård et al., 1976; Dermietzel et al., 1977; Bouchaud and Bouvier, 1978; Bouvier and Bouchaud, 1978; Tani et al., 1977; van Deurs and Koehler, 1978, 1979). Also, vesicular uptake and transcellular transport of

TABLE I
ENZYMATIC TRACERS USED IN BRAIN BARRIER STUDIES

Enzyme	MW[a]	Diameter[b]	pI[a]
Horseradish peroxidase (HRP)	39,800	50–60 Å	9.8
Cytochrome c	12,800	$30 \times 34 \times 34$ Å	10.0
Microperoxidase[c] (MP)	1,880	20 Å	4.85

[a] The values for molecular weight (MW) and isoelectric point (pI) are from Plattner et al. (1977).

[b] The values for molecular diameter(s) are from Simionescu et al. (1975).

[c] This microperoxidase is a hemeundecapeptide. A hemenonapeptide (MW 1630) and a hemeoctapeptide (MW 1550) have also been produced (Plattner et al., 1977; Simionescu et al., 1975), but not used in studies of brain barriers.

tracers in the normal cerebral endothelium (Westergaard and Brightman, 1973; van Deurs and Amtorp, 1978), in cerebral endothelium under various experimental conditions (Westergaard, 1975a; Beggs and Waggener, 1976; van Deurs, 1976b; van Deurs and Amtorp, 1978; Westergaard et al., 1976, 1977; Nag et al., 1977; Petito et al., 1977; Povlishock et al., 1978), and in the choroidal epithelium (Becker et al., 1967; Brightman, 1967, 1975; Brightman et al., 1970b; Davis and Milhorat, 1975; van Deurs, 1978a,b; van Deurs et al., 1978) became of special interest. During the same years, studies on permeability properties in noncerebral endo- and epithelia also developed progressively. The scope of this work has, therefore, been to review (ultra-) structural studies of these aspects of brain barriers, particularly in mammals, in relation to similar research on noncerebral tissue. The reader interested in brain barriers of lower vertebrates and invertebrates should consult, for instance, Brightman et al. (1970b), Hashimoto (1972), Bundgaard et al. (1979), and Lane (1978).

II. Cerebral Blood Vessels

A. LOCALIZATION OF THE BLOOD–BRAIN BARRIER

The cerebral capillary endothelium is of the continuous type and can, in principle, be classified together with capillaries of the skin, connective tissue, skeletal, cardiac, and smooth muscles, and lung alveolar capillaries (Bennett et al., 1959; Majno, 1965). In early electron microscopic studies, problems attended the attempt to locate the barrier in the brain to blood-borne macromolecules partly because tracer techniques were inadequate (Dempsey and Wislocki, 1955; van Breemen and Clemente, 1955; Wislocki and Ladman, 1958;

Hager, 1961; Tani and Evans, 1965; Clawson *et al.*, 1966; Dretzki, 1971). It was, for instance, uncertain whether the barrier was located in endothelial or periendothelial structures (Dempsey and Wislocki, 1955; Bodenheimer and Brightman, 1968; Dretzki, 1971).

Shortly after the introduction of horseradish peroxidase (HRP, see Table I) as a tracer for electron microscopy (Graham and Karnovsky, 1966), the blood–brain barrier was located at the ultrastructural level in the cerebral endothelium (Reese and Karnovsky, 1967). After intravenous injection of HRP into mice, the tracer was found in the lumen of the cerebral vessels, and in a few vesicles in the endothelium. No HRP was seen outside the endothelium. It was concluded that the barrier to this protein was due to two characteristic properties of the cerebral endothelium: (1) the endothelial tight junctions (Figs. 1 and 2), which prevented any extravasation of tracer via the intercellular spaces (Fig. 4), and (2) the few endothelial vesicles (measuring about 700 Å in diameter; initially described by Palade, 1953) as compared with noncerebral endothelia (see Section II,B), which did not appear to be involved in transendothelial transport from blood to brain (Figs. 4 and 5) (Reese and Karnovsky, 1967). A hint that the cerebral endothelium differs from that of noncerebral capillaries (Section II,B) also comes from ultrastructural cytochemical studies by Torack and Barrnett (1964), Torack *et al.* (1961), and Marchesi and Barrnett (1963), who located nucleoside phosphatase mainly in the basement membrane of the cerebral capillaries and not within the endothelial cells (in association with vesicles). This is in contrast to what they found in capillaries of the choroid plexus, for instance (vessels that probably exhibit some vesicular transport activity; Section IV,A), and noncerebral capillaries. Torack and Barrnett (1964) correlated this lack of enzymatic activity in the cerebral capillary endothelium with the lack of vesicular transport activity characteristic of the blood–brain barrier, and suggested that the nucleoside phosphatases were involved in control of transport mechanisms across endothelia. Várkonyi and Joó (1968) and Joó (1968) reported that administration of nickel chloride increased the permeability of the cerebral endothelium, and that this enhanced permeability was due to inhibition of ATPase activity. Butyrylcholinesterase activity has been located in the cerebral endothelium (in endothelial vesicles), whereas the enzyme was not seen in capillaries outside the brain (Joó and Csillik, 1966; Joó *et al.*, 1967).

B. PERMEABILITY OF NONCEREBRAL ENDOTHELIUM

Before going further in the description of the cerebral endothelium, it is necessary to look at the permeability properties of continuous, noncerebral capillary endothelia. It is well established from physiological work that the permeability of these endothelia to lipid-soluble molecules (including O_2 and CO_2), which permeate unrestricted the cell membrane and thus use the entire endothelial sur-

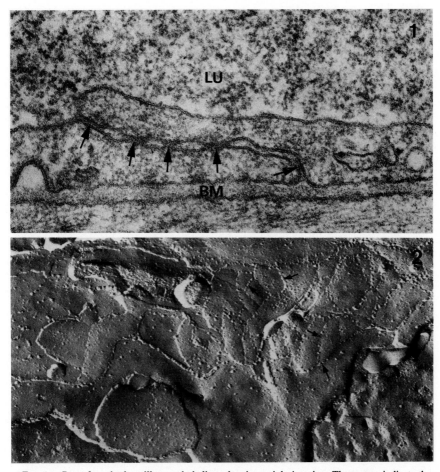

FIG. 1. Part of cerebral capillary endothelium showing a tight junction. The arrows indicate the "fusion points" of adjacent membranes, presumably corresponding to the junctional strands seen in freeze-fracture preparations. LU, Vascular lumen; BM, basement membrane. ×94,000.

FIG. 2. Freeze-fracture micrograph of cerebral endothelium (E-face) showing a tight junction represented by a network of strands (arrows). The strands are built of particles. The vessel type (arteriole, capillary, or venule) could not be determined. ×58,000.

face for diffusion, is very high, and that the permeability to lipid-insoluble (hydrophilic) molecules is much lower, decreasing considerably with increasing molecular weight (in the range of 10,000–100,000 MW) as evaluated by measuring the lymph–plasma ratio of graded dextrans (Garlick and Renkin, 1970). The permeability to large molecules (> 100,000 MW) such as plasma protein and dextran is very restricted (see, for example, Garlick and Renkin, 1970; Renkin,

1978). This gave rise to the "pore theory." According to this, a small pore exists for rapid exchange of water and small lipid-insoluble molecules, occupying a minor fraction of the total capillary surface (less than 0.1%) represented by water-filled channels 70-90 Å in diameter, or slits 40-50 Å in width. Also, a large pore about 250-700 Å in diameter was postulated, being much more infrequent than the small one and allowing passage of large lipid-insoluble molecules (larger than 80-90 Å) (Pappenheimer, 1953; Pappenheimer *et al.*, 1951; Landis and Pappenheimer, 1963; Grotte, 1956; Mayerson *et al.*, 1960). These pores should, because of their dimensions, in principle, be detectable in the electron microscope. However, there were and still are difficulties in identifying these pores at the ultrastructural level (see, for example, Luft, 1973).

Studying cardiac and muscular capillaries of mice, Karnovsky (1967) found that intravenously injected HRP penetrated the endothelial junctions (Fig. 3). The adjacent endothelial membranes at the junctions were often separated by a gap, approximately 40 Å in width, and Karnovsky (1967) suggested that the endothelial "tight" junction was a macula occludens (a discontinuous tight junction) rather than a zonula occludens. The nature of the junctions was also verified by use of lanthanum (Karnovsky, 1967). Apparently, vesicles (Fig. 3) participated in the extravasation of HRP, but the junctions were considered as the morphological equivalent of the small pore system (Karnovsky, 1967; see also Karnovsky, 1970).

Bruns and Palade (1968b) used the considerably larger ferritin molecule (110 Å diameter, MW about 500,000) as a probe for the large pore system. Endothelial vesicles most likely transported ferritin across the endothelium, whereas the junctions were not penetrated. Accordingly, vesicles apparently represented the structural basis for the large pore. The vesicles had the right size for the large pore, but appeared to be much more frequent than expected from physiological data. However, the time a vesicle needs to cross the endothelium (Bruns and Palade, 1968a,b; see also Section III,E) and/or the circumstance that not all vesicles may be involved in transport (at the same time), may account for this discrepancy.

Indeed, it was well warranted to use a smaller tracer than HRP in order to separate the small and large pore. Exogenous myoglobin (MW 17,800, $25 \times 34 \times 42$ Å diameter) was used as probe molecule for the pore systems by N. Simionescu *et al.* (1973). Vesicles seemed to carry myoglobin injected intravenously into rats across the endothelium of muscle capillaries. It was concluded that the vesicles were the structural equivalent of the large pore system, and at least partly, of the small pore system, too. These observations were followed up in studies with the smaller hemepeptide tracer MP (Table I) (N. Simionescu *et al.*, 1975). Microperoxidase crossed the endothelium preferentially or exclusively by means of vesicles and (transient) transendothelial channels formed by vesicles or vesicle chains opening simultaneously at the luminal and abluminal endothelial

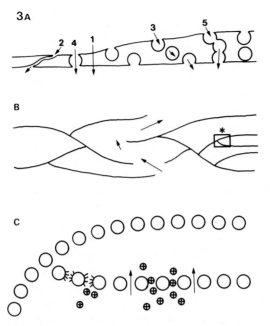

FIG. 3. (A)   Possible pathways across a continuous capillary endothelium (e.g., in muscle and brain): (1) Directly through the membranes of the endothelial cells (lipid-soluble molecules, and perhaps small nonpolar solutes); (2) Through the junctions (the paracellular route) between adjacent endothelial cells (water and smaller lipid-insoluble molecules). (3) Vesicular transport and (4) and (5) through channels formed by one or more vesicles (lipid-insoluble molecules, probably of all sizes). In the normal cerebral endothelium, probably only pathways (1) and (2) operate to any degree of significance, and the junctions are most likely very tight as compared to muscle capillaries. However, in situations with experimental opening of the blood–brain barrier, pathway (2) may be "leaky" and (3)–(5) may be "formed." Under extreme experimental conditions, the junctions (2) may be widened to become freely permeable even to large plasma proteins. Fenestrations, which may occur in certain brain tumors, have not been included. (B) Discontinuities in junctional strands representing relatively large "channels" for permeation of larger hydrophilic molecules. Such channels may be more frequent, for example, in muscle capillaries than in the brain. It should be noted that such channels would not be visible in thin sections. (C) A part of the junctional strands in Fig. 3B (asterisk) at a higher magnification. Here narrow "pores" between the individual tight junctional proteins of the strands are seen. They may represent the permeation routes for ions, and the permeability properties of the strands are thus determined by electrical charges of the tight junctional proteins. Whereas the pores to the left are "closed," those at the right are "open" to the indicated cations.

surface (Fig. 3; see also Fig. 23). These channels were considered to be the equivalent of the small pore since they were continuous, water-filled structures with strictures of ~ 100 Å (N. Simionescu et al., 1975).

In studies of muscle capillaries in mice, Williams and Wissig (1975) found that the junctions constituted a barrier for the transendothelial movement of HRP, whereas vesicular transport occurred. The results were not considered to provide clear-cut evidence that endothelial junctions were the site of the small pore (Williams and Wissig, 1975, compare with the results of Karnovsky, 1967, described earlier), but partly supported the findings of N. Simionescu *et al.*, (1973, 1975). However, Wissig and Williams (1978) have recently published results with MP (obtained in experiments similar to those of N. Simionescu *et al.*, 1975) indicating that some endothelial junctions were permeable to MP and probably represented the small pore. According to Wissig and Williams, vesicles may participate, but they do not transfer substantial amounts of MP. Furthermore, these investigators found no evidence for transendothelial channels. These discrepancies between the studies of N. Simionescu *et al.* (1975) and Wissig and Williams (1978) emphasize the problems attending attempts to obtain "clearcut" functional evidence (in this case: for transendothelial pathways) at the electron microscopic level. An explanation of the discrepancies was given by N. Simionescu *et al.* (1978b), who, by using well-defined microvascular segments (bipolar microvascular fields, N. Simionescu *et al.*, 1978a), observed that the fastest extravasation of MP occurs via junctions in pericytic venules (see also N. Simionescu *et al.*, 1978c), and that vesicles and channels were responsible for the extravasation in capillaries. It was, therefore, suggested (N. Simionescu *et al.*, 1978c) that Wissig and Williams (1978) might have included venules in their "capillary" material. Increased permeability of noncerebral microvasculature is known to occur, especially in the venular end of the microvascular bed, in relation to release of vasoactive amines and opening of cell junctions (see Majno *et al.*, 1961; N. Simionescu *et al.*, 1978c).

Hopefully, more reliable and convincing ultrastructural approaches will be available in the future. This is important also in the study of the possible pathways across the cerebral endothelium (see Sections II,C and D, and III), because this field, in the present author's opinion, is intimately related to the knowledge of capillary permeability in noncerebral tissue.

C. Permeability of Cerebral Endothelium to Hydrophilic Molecules

In electron microscopic micrographs of thin sections, endothelial tight junctions appear as a series of "membrane fusions," sometimes with distinct intervening spaces (Fig. 1). The number of fusion points varies and is often difficult to establish because the intercellular spaces are somewhat tortuous, but, in the brain, this number seems to be between two and six (cf. Tani *et al.*, 1977). Reese and Karnovsky (1967) calculated the ratio of the total width of tight junctions (at the points of "fusion") to the average total width of single cell membranes in cerebral capillaries. They found a value of 1.6–1.8, irrespective of the type of fixative used in the tissue preparation. In comparison, the value for heart capillary junctions was 2.4. Thus, Reese and Karnovsky (1967) characterized the junction in heart capillaries as "close" (see also Karnovsky, 1967, 1970), but that of the brain as "closed," meaning that the cerebral capillary junctions may be tighter than those in noncerebral tissue.

The general features of the cerebral tight junction have been established in many studies with HRP (Brightman, 1967; Brightman and Reese, 1969; Brightman et al., 1970a,b), cytochrome c (see Table I) (Milhorat et al., 1973), and microperoxidase (MP; see Table I) (Feder et al., 1969; Brightman et al., 1970b; Feder, 1970, 1971; Reese et al., 1971; van Deurs and Amtorp, 1978), which are all stopped by the junctions (Figs. 4 and 5). Capillaries "tight" (no extravasation) to HRP have also been reported in the endoneurium of peripheral nerves (Olsson and Reese, 1969), whereas vessels in spinal ganglia are leaky to HRP (Jacobs et al., 1976; Olsson, 1968).

Freeze-fracture studies on noncerebral blood vessels have provided important information about the nature of the endothelial junctions. The tight junction of endothelia is a more or less continuous belt around the cells, consisting of aligned intramembranous particles or simply of shallow ridges and grooves (junctional "strands") varying in number and complexity from segment to segment of the microvascular tree (M. Simionescu et al., 1975, 1976; N. Simionescu et al., 1978c; Staehelin, 1975; Yee and Revel, 1975; van Deurs, 1979).

Freeze-fracture information on the cerebral endothelial junctions is relatively limited (Connell and Mercer, 1974; Tani et al., 1974, 1977; Dermietzel, 1975a; Yamamoto et al., 1976). The reason for this may be that the membranes of the cerebral capillaries in junctional regions only infrequently become fractured (Brightman et al., 1975a; Brightman, 1977; Tani et al., 1977; B. van Deurs, unpublished observations). Another problem with respect to freeze-fracture of cerebral endothelial tight junctions is the precise identification of the endothelium as belonging to arterioles, capillaries, or venules (ACV-units). A precise and reliable identification of these vessel types is necessary and probably requires freeze-fracture of isolated ACV-units (M. Simionescu et al., 1975a; N. Simionescu et al., 1978a,b,c). These microvascular segments are known to exhibit different junctional architecture (M. Simionescu et al., 1975; N. Simionescu et al., 1978c) and different permeability properties (N. Simionescu et al., 1975, 1978a,b,c), at least in muscle tissue. In freeze-fracture replicas, the cerebral endothelial tight junctions (Fig. 2) typically resemble those of noncerebral vessels. Thus, Connell and Mercer (1974) and Yamomoto et al. (1976) described the cerebral "capillary" tight junctions as a network of "ridges" composed of approximately 100 Å particles. Dermietzel (1975a) also reported continuous ridges (fibrils, which are not seen in noncerebral capillaries) in the tight junctions of the cerebral capillaries. Similarly, Tani et al. (1977) observed junctional strands (2-6) formed by ridges, as well as by particles, and

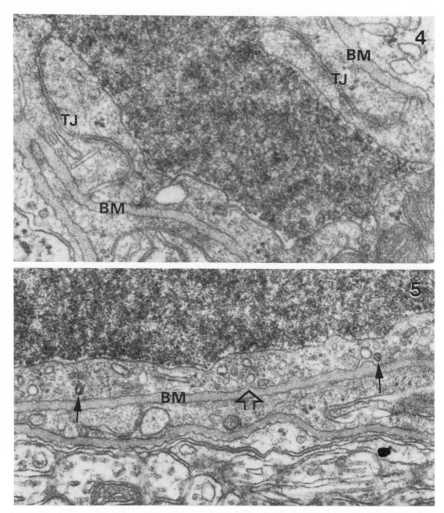

Figs. 4 AND 5.   Parts of cerebral capillaries fixed by ventriculocisternal perfusion of aldehydes 15 minutes after intravenous injection of MP. The tracer (reaction product) is distinct in the vascular lumen, but has not reached the endothelial basement membrane (BM). In Fig. 4, two tight junctions (TJ) are seen, which are not penetrated by the tracer. In Fig. 5, some vesicles are present in the endothelial cell, two of which are tracer-labeled (small arrows), but no vesicles are discharging the tracer into the basement membrane. The large arrow indicates a vesicle without tracer opening onto the abluminal endothelial surface. ×60,000 and ×48,000, respectively.

additionally, a high degree of variability in the architecture of the cerebral endothelial tight junction.

At present, it is not possible to give any reasonable structural explanation of why the cerebral endothelial tight junctions might be "tighter" than those of many other noncerebral blood vessels (e.g., muscle capillaries) (cf. Reese and Karnovsky, 1967). It may be speculated, however, that the system of junctional strands in, for instance, the muscular capillary endothelium, is less continuous than in the cerebral endothelium, or provided with wider "channels," which can be permeated by molecules up to a certain size determined by the width of the discontinuity in the strand (Fig. 3B) (see, for example, Wissig and Williams, 1978). This is obviously difficult to establish when freeze–fracture replicas of cerebral capillary endothelium are rarely obtained. Another aspect is that, for instance, electrical charges of the individual protein molecules of the tight junctional "strands" (which are interposed between the adjacent cells, see also van Deurs and Luft, 1979) probably determine the permeability of the strands to ions (Fig. 3C). Such differences in electrical properties from tissue to tissue would not be possible to visualize directly in the electron microscope, although they may be very important for the tightness and selectivity of the junction. Other aspects of the interpretation of freeze–fracture replicas of tight junctions are mentioned in Section IV,C,2.

When using the term "endothelial vesicle" or micropinocytic vesicle, the reference usually means the typical, uncoated version. However, coated vesicles also occur in the cerebral endothelium, but little information about these organelles is available. According to Joó (1971, 1972), they are frequent in the cerebral endothelium after cAMP administration, or after nickel chloride or mercuric chloride intoxification. They may be involved in selective protein uptake (Joó, 1972). Outside the brain, coated vesicles are similarly present as a minor population of endothelial vesicles (Stehbens, 1965; Bruns and Palade, 1968a), whereas they are predominant in sinus-lining endothelia (Wisse, 1972; Bankston and DeBruyn, 1974; DeBruyn et al., 1975).

As reported by Reese and Karnovsky (1967), the number of vesicles in cerebral capillaries is low. In freeze–fracture preparations, Connell and Mercer (1974) found that the number of endothelial vesicles was 5 per μm^2 in cerebral capillaries. This value was confirmed in a freeze–fracture study by Tani et al. (1977) who, however, emphasized that it applied to the luminal endothelial front. On the abluminal front, occasionally high numbers of vesicles (30–40 per μm^2) were reported. In contrast, the number of vesicles as evaluated in freeze–fracture preparations (mean number of vesicular openings at blood front and tissue front) is 78 per μm^2 in capillaries of the diaghragm, and 89 per μm^2 in myocardial capillaries (M. Simionescu et al., 1974) (see also Fig. 22). There seems to be a very uneven distribution of vesicles in the cerebral endothelium (Tani et al., 1977; van Deurs and Amtorp, 1978). In experiments with intravenously injected

MP (van Deurs and Amtorp, 1978), it was found that 58% of the cross-sectioned vessels contained 5 or fewer MP-labeled vesicles, whereas 12% of the vessels contained 20 or more labeled vesicles; those with 30 or more vesicles always were arterioles (or larger venules). No transport activity, however, was seen in normo- and hypotensive rats (that is, no MP was seen in the endothelial basement membrane) (Fig. 6): an observation that indeed cannot be established in freeze–fracture preparations. Most of these labeled vesicles were connected with the luminal surface. When MP was perfused ventriculocisternally, the tracer diffused readily between the ependymal cells into the extracellular space of the neuropil (Figs. 34 and 35), and was seen in endothelial basement membranes and in some endothelial vesicles. In 52% of the examined cross-sectioned vessel profiles, 5 or fewer vesicles were seen, whereas 14% exhibited 20 or more. In this case, most labeled vesicles were connected with the abluminal surface (van Deurs and Amtorp, 1978). No transport of the tracer from brain to blood (so-called retro-grade transport) could be established in these experiments nor in similar (B. van Deurs, unpublished) HRP experiments (Fig. 7). This is in agreement with other reports (Brightman, 1967; Brightman *et al.*, 1970a; Delorme *et al.*, 1975; Weindl and Joynt, 1973). However, some evidence in favor of an occasional minor vesicular brain-to-blood transport of macromolecules in the cerebral en-dothelium has been presented (Becker *et al.*, 1968; del Cerro, 1974; Wagner *et al.*, 1974; van Deurs, 1977).

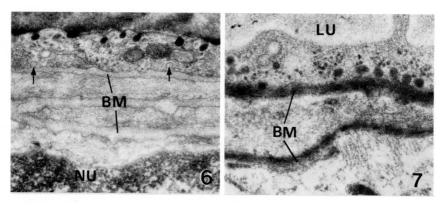

Fig. 6.   Cerebral arteriole 30 minutes after intravenous injection of horseradish peroxidase. The tracer labels a number of vesicles connected with the luminal endothelial surface, but is not seen in the basement membrane (BM) outside the endothelium. Two unlabeled vesicles at the abluminal surface are indicated (arrows). NU, Nucleus of a smooth muscle cell. ×48,000.

Fig. 7.   Cerebral capillary (or venule) 30 minutes after ventriculocisternal perfusion of horse-radish peroxidase. The tracer is present in the endothelial basement membrane and in a number of endothelial vesicles mostly connected with the abluminal endothelial surface. BM, Basement mem-brane; Lu, lumen. ×48,000.

Apparently, there exists a population of "vesicles" (most of which are surface-connected) in segments of the cerebral microvasculature, which, under normal conditions, is not involved in transport or conceivably involved in a very minimal transfer (see below) of macromolecules (van Deurs and Amtorp, 1978), in contrast to what is generally assumed to occur in normal noncerebral vessels.

Physiologically, the cerebral endothelium under normal conditions is very "tight" to protein compared with noncerebral endothelia (see, e.g., Renkin, 1978, Table 3). Straus (1958) made direct quantitative measurements of the HRP content in brain and other tissue after i.v. administration of the tracer molecule and found no HRP in the brain. Similarly, Milhorat et al. (1973) were unable to detect intravenously administered cytochrome c (Table I) in the CSF up to 9 hours after the injection (evaluated spectrophotometrically). However, if longer periods (24–48 hours) are used in order to establish an equilibrium between blood and brain extracellular fluid, intravenously administered protein does penetrate the blood–brain barrier (Cutler et al., 1967). It should be noticed that not only the cerebral endothelium, but also, for instance, the choroidal epithelium, could be involved in this phenomenon (see Section IV,C). Regarding the cerebral endothelial junctions, the presence of a very few "pores" (see preceding, and Fig. 3B) in these could explain a very limited extravasation (molecular sieving) of protein, the rate presumably depending on the molecular size. Such pores have not been shown electron microscopically in the brain, but are discussed in relation to muscle capillaries (e.g., Wissig and Williams, 1978; for a similar discussion related to epithelia, see Section IV,C).

Another possible structural explanation of this obviously very limited penetration of protein from blood into the brain was presented by Westergaard and Brightman (1973), who showed by light and electron microscopy that some extravasation of HRP occurs in segments of arterioles both on the brain surface and within the brain. It was found that the endothelial junctions were never penetrated by HRP. However, HRP-labeled endothelial vesicles were present in segments of arterioles, and in these segments, the basement membrane was also HRP-labeled. It was concluded that a vesicular blood-to-brain transport was responsible for the extravasation. Also, the large ferritin molecule (MW 500,000) was apparently transported across the endothelium by vesicles (Westergaard and Brightman, 1973). Later it was suggested by Nørtved and Westergaard (1974) that a similar transport takes place in segments of venules of the normal brain. This occasional vesicular transport in the normal brain, which must be minimal, has been confirmed [e.g., by van Deurs (1976b), Westergaard et al. (1977), and Povlishock et al. (1978)], whereas it could not be established in other studies [van Deurs and Amtorp (1978), Beggs and Waggener (1976), Petito et al. (1977), and Nag et al. (1977)]. The reason for this discrepancy remains obscure, although minor variations in the physiological state of the experimental animal and in the experimental procedure might be considered. Assuming that vesicular transendothelial transport, when it does occur, is bidirectional (i.e., vesicles are carrying material simultaneously in the luminal and abluminal directions: a reasonable assumption with respect to membrane turnover) (Shea and Bossert, 1973; N. Simionescu *et al.*, 1973; Williams and Wissig, 1975), this limited vesicular blood-to-brain transport in segments of the normal cerebral microvasculature may correspond to the possible retrograde vesicular transport mentioned earlier.

D. Permeability of Cerebral Endothelium to Lipids, Small Lipid-Insoluble Molecules, and Ions

Information on the structural basis for transport and diffusion of lipid-soluble molecules, small lipid-insoluble molecules, and ions across the cerebral endothelium is relatively sparse. Unlike most proteins and other hydrophilic molecules, lipids probably have ready access to the cerebral parenchyma. Ultrastructural experiments have been reported with injection into the blood of polyunsaturated linolenic acid, which itself is strongly osmophilic and thus visible in the electron microscope after a routine preparation procedure (Brightman, 1977). Linolenic acid apparently moved directly across the endothelial cell membrane and was found as droplets of various sizes throughout the endothelial cells and in adjacent smooth muscle cells in the case of arterioles (Brightman, 1977).

In a study of the cerebral endothelium, Richards (1978) concluded, based on a cytochemical visualization of 5-hydroxydopamine, that the endothelial junctions prevent small molecules such as transmitters from passing from blood to brain extracellular fluid.

Milhorat *et al.* (1975b) found that the endothelial tight junctions of cerebral capillaries were permeable to calcium ions visualized electron microscopically with a calcium phosphate precipitate after *in vivo* administration of calcium into the blood stream. "Colloidal" lanthanum, injected intravenously together with a fixative, could also be located by electron microscopy in the intercellular spaces of the cerebral capillary endothelium (Milhorat *et al.*, 1975b), and Casley-Smith (1969) reported permeability of the cerebral endothelial junctions to ions, based on a method with Prussian blue reaction precipitates. In contrast, Brightman and Reese (1969) found that the cerebral endothelial junctions were impermeable to colloidal lanthanum. Also, Bouldin and Krigman (1975) reported that the tight junctions of the cerebral endothelium were not penetrated by "ionic" lanthanum (administered *in vivo* before fixation). "Colloidal" and "ionic" lanthanum seem to give the same results with respect to the permeability of a cell junction (Martinez-Palomo and Erlij, 1973), but the behavior of these probes is unclear. These observations point to the problem of establishing ultrastructurally a paracellular route for ions across the cerebral (as well as noncerebral) endothelium.

E. The Role of Periendothelial Structures in the Blood–Brain Barrier

To what extent periendothelial structures such as the sheath of astrocytic endfeet, phagocytic pericytes, and the endothelial basement membrane surrounding the blood vessels, and the general lack of a perivascular connective tissue space are of significance in the blood–brain barrier is still uncertain (Maynard *et al.*, 1957; Hager, 1961; Wolff, 1963; Bodenheimer and Brightman, 1968; Brightman *et al.*, 1970a; Pappas, 1970; Ford, 1976; van Deurs, 1976b).

Although the astrocytic endfeet form a rather continuous sheath around the cerebral capillaries and once were believed to represent the structural basis of the blood–brain barrier (see review in Davson, 1972, p. 369), these do not halt diffusing macromolecules, since they are connected with gap junctions rather than (continuous) tight junctions (Brightman and Reese, 1969). This is also established in numerous tracer studies (see Section III).

Phagocytic pericytes, described in detail by van Deurs (1976b), are present in the endothelial basement membrane of arterioles in particular (Fig. 8), but also of capillaries and venules. These cells are able to endocytose ferritin, HRP, and MP and incorporate the tracer into their lysosomal apparatus (Brightman, 1965b; Olsson and Hossmann, 1970; Cancilla *et al.*, 1972; Wagner *et al.*, 1974; van Deurs, 1976b; van Deurs and Amtorp, 1978). Similar phagocytic pericytes are well established in histamine or serotonin-treated noncerebral tissue (Majno, 1965; Majno and Palade, 1961), whereas the uptake of extravasated ferritin in ''normal'' animals ''was limited and did not lead to any extensive accumulation of tracer molecules within lytic vacuoles'' (Bruns and Palade, 1968b).

With prolonged time after tracer injection and extravasation in experiments with opening of the endothelial blood–brain barrier (see Section III,B), and when the tracer is no longer present in the extracellular space, the pericytes can be heavily loaded (Cancilla *et al.*, 1972), indicating that the pericytes may be of some significance for the homeostasis (Fig. 8). Furthermore, the endothelial part of the blood–brain barrier seems less efficient in immature animals (see, e.g., Saunders, 1977), and in these cases the phagocytic pericytes may be an important part of a periendothelial barrier (Kristensson and Olsson, 1973).

Brightman (1965b) found, in studies with ferritin injected into the CSF compartments of the brain, that the endothelial basement membrane functioned as a filter to this large tracer. He suggested, therefore, that the basement membrane also might be the primary periendothelial barrier for movement of colloidal particles from blood to brain. While Bruns and Palade (1968b) and Clementi and Palade (1969) could not establish any ''filter'' function of the basement membrane to ferritin in (continuous) muscle capillaries and (fenestrated) intestinal capillaries, respectively, Johansson (1978) concluded in a recent study with interstitially microinjected ferritin that the muscle capillary basement membrane represented a substantial barrier to macromolecules of the size of ferritin.

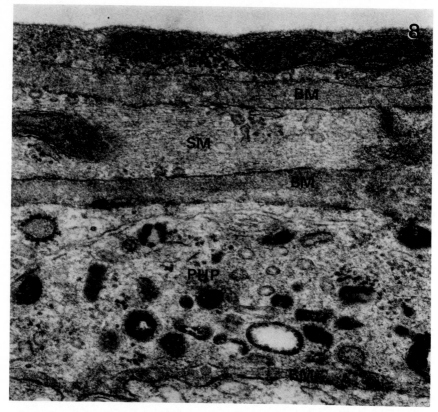

Fig. 8. A phagocytic pericyte (PHP) is seen containing numerous horseradish peroxidase-labeled vesicles. The vessel (an arteriole) is from an experiment with Aramine-induced acute hypertension where intravenously injected horseradish peroxidase extravasates. However, the basement membranes (BM) below the endothelium (EN) and surrounding the smooth muscle cell (SM), as well as the one below the phagocytic pericyte are not tracer-labeled, or only very slightly so, suggesting that the phagocyte has "cleansed" its environment of extravasated tracer. ×64,000.

III. Opening of the Blood–Brain Barrier

A great deal of our knowledge concerning the structural aspects of the permeability properties of the cerebral endothelium, and consequently of the blood–brain barrier, comes from numerous studies in which the barrier in some experimental or pathological situation is broken down. While there is no doubt that many different experimental and pathological situations cause extravasation of tracer molecules into the brain (Figs. 9A and B), there are distinct discrepancies in the interpretation of the results with respect to the mechanism or mechanisms underlying the extravasation (e.g., Brightman, 1977; Westergaard, 1977).

In some cases, it has been reported that the endothelial cells of the cerebral vessels are completely filled with protein tracer after intravenous administration. Even though it has been suggested that this might reflect a protein route directly across the endothelial cell membranes, it is generally considered as an artifact, i.e., passive tracer filling of the endothelial cells after rupture of the membrane caused by the fixative (Hirano *et al.*, 1969, 1970a; Brightman *et al.*, 1970a; Hansson *et al.*, 1975; Beggs and Waggener, 1976; Persson *et al.*, 1978). It is therefore of greater interest to discover which of the characteristic features of the cerebral endothelium in the normal blood–brain barrier (tight junctions and a few, apparently nontransporting vesicles, see Section II,A) can be reversibly altered or modified in some way as to be responsible for experimentally induced breakdown of the barrier. Thus, diffusion through open junctions (Section III,A), transport by means of vesicles (Section III,B), and diffusion through vesicle "chains" (channels) connected with the luminal and abluminal cell membrane (Section III,C) are likely explanations for transient transendothelial "pores" in the cerebral endothelium. In addition, occasional formation of fenestrae (Section III,D) in the endothelium may be considered. The following paragraphs deal with these possible transendothelial "pores" in the cerebral endothelium in relation to experimental opening of the blood–brain barrier to lipid-insoluble molecules (proteins and peptides).

A. OPENING OF TIGHT JUNCTIONS

Among the many "methods" existing to increase the permeability of the blood–brain barrier, the application of hypertonic solutions in particular is associated with the opening of junctions in the brain. Also opening of, for example, hepatocyte junctions with hypertonic disaccharides (Goodenough and Gilula, 1974) and of the choroidal epithelial junctions with hypertonic urea and sucrose (Bouchaud and Bouvier, 1978; see Section IV,C) have been described.

Increased permeability of cerebral vessels to fluorescein-labeled albumin was demonstrated after injection of hypertonic glucose in the carotid artery (Klatzo *et al.*, 1965). Experimental opening of the tight junctions to HRP in the cerebral endothelium was later reported in an ultrastructural study by Brightman *et al.* (1973). These authors infused hyperosmotic urea solutions into one internal carotid artery of rabbits (Rapoport, 1970; Rapoport *et al.*, 1971, 1972) and observed a pronounced focal ("spotted") extravasation of Evans blue–albumin, as well as of HRP. By electron microscopy, the latter tracer was localized in the intercellular spaces between the endothelial cells, that is, in the "pools" between the "fusion points" of the junctions. This was observed mostly in capillaries, but also in some arterioles. However, at the fusion points, the outer leaflets of the

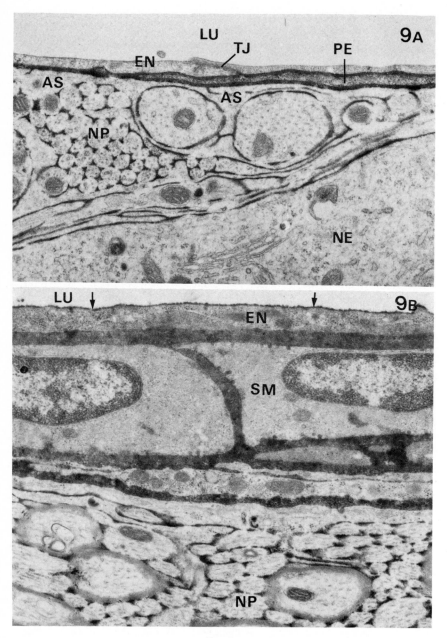

FIG. 9.

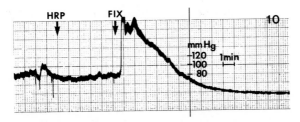

Fɪɢ. 10. Typical arterial blood pressure curve for a rat being fixed by ventriculocisternal perfusion of aldehydes. Approximately 5 minutes after the start of the HRP perfusion, the perfusion of fixative began, and almost immediately the blood pressure increased to about 200 mm Hg, whereafter it decreased to about 40 mm Hg.

adjacent endothelial membranes still seemed to be in very close contact ("fused"). The results of Brightman *et al.* (1973) appear to provide rather convincing evidence in favor of osmotic opening of the junctions and of a paracellular penetration of protein from blood to the vascular basement membrane. The opening of the junctions was considered reversible, since the blood-brain barrier was reestablished after some hours (Rapoport *et al.*, 1971). An explanation was given (Rapoport, 1970; Rapoport *et al.*, 1972; Brightman *et al.*, 1973) that hyperosmolar concentrations of, for example, urea withdraw water from the endothelial cells, resulting in shrinkage of the cells and consequently of opening of the junctions. The authors stated, however, that the endothelium did not appear shrunken, but that this might be due to rehydration of the cells during fixation. Some support for hyperosmotic cell shrinkage and consequently, for opening of the junctions, comes from physiological work in which the enhanced permeability of tracers (urea, sucrose, inulin, labeled albumin) decreased with increasing molecular size (Thompson, 1970). Another explanation of the apparent opening of the junctions might be histamine release, which can cause contraction of endothelial cells. Brightman *et al.* (1973) argued that this was an unlikely explanation, since large doses of histamine, serotonin, or norepinephrine perfused ventriculocisternally in mice did not cause HRP labeling of the "pools"

Fɪɢ. 9. (A) Part of a cerebral capillary or venule and surrounding neuropil (NP) after extravasation of HRP induced by acute hypertension. The lumen (LU) of the blood vessel is devoid of HRP because of the vascular perfusion fixation. The endothelium (EN) does not exhibit tracer-labeled vesicles, and the endothelial tight junction (TJ) appears free of tracer. However, the pericyte (PE) is surrounded by HRP, and the astrocytic layer (AS) has been penetrated by the tracer, which clearly labels the extracellular space of the neuropil. NE, Part of a neuron. ×20,000. (B) This micrograph shows a situation comparable to that in (A). However, in (B), an arteriole is shown, and the smooth muscle cells (SM) are clearly outlined by HRP, which is also present in the neuropil (NP). Furthermore, although the lumen (LU) of the arteriole is devoid of tracer, some HRP adheres to the luminal surface (arrows) of the endothelial cell (EN). ×20,000.

between the junctional fusion points (see also Westergaard, 1975a,b). At present, no evidence exists that HRP itself induces opening of endothelial junctions in the brain of mammals, although this tracer probably does so (by release of vasoactive amines) in noncerebral tissue of some species (see Majno *et al.*, 1961; Cotran and Karnovsky, 1967; Clementi, 1970; Deimann *et al.*, 1976; Straus, 1977; N. Simionescu *et al.*, 1978a). Westergaard (1977) has argued that the HRP in the ''pools'' between the fusion points could have reached this position by transport in micropinocytic vesicles (Westergaard and Brightman, 1973).

The blood–brain barrier has been osmotically opened by hypertonic solutions of lactamide and arabinose to circulating antibody to measles (Hicks *et al.*, 1976). In this study it was assumed, without the use of electron microscopy, that the extravasation of antibody (MW approximately 180,000) was due to widened tight junctions.

Olsson *et al.* (1975) reported that extravasation of HRP in X-ray damaged cerebral capillaries was due to open junctions. In experiments with allergic encephalomyelitis, Lampert and Carpenter (1965) and Hirano *et al.* (1970b) found that Thorotrast and HRP penetrated the cerebral endothelium via open junctions, although vesicles probably participated in the extravasation (Hirano *et al.*, 1970b). In cerebral edema induced by freeze-injury, opening of junctions to HRP was observed (Baker *et al.*, 1971), but the investigators also suggested that vesicles and direct passage through the endothelial cell membranes were involved in the observed extravasation of HRP.

Other situations such as hypertension have also been reported to open tight junctions between cerebral endothelial cells. Rapoport (1976a,b) induced acute hypertension in rats by perfusion of isotonic saline into the brain via the common carotid artery, or by intravenous perfusion of Aramine (metaraminol bitartrate). He observed an opening of the blood–brain barrier to Evans blue–albumin and suggested, based on macroscopic evaluation of the tracer leakage into the brain, that this was due to widening of the endothelial tight junctions (Brightman *et al.*, 1973). assuming that the increased intraluminal pressure in the blood vessels stretches the endothelial cells. In the acute hypertension experiments with perfusion of isotonic saline into carotid artery, Rapoport (1976a) obtained arterial blood pressures as high as 490 mm Hg, and it does not seem unlikely that such a severe rise in blood pressure actually causes opening of the tight junctions. The Aramine-induced acute hypertension caused less extensive opening of the blood–brain barrier than the isotonic saline hypertension, and Rapoport (1976a) found that the threshold for barrier opening by Aramine was about 160 mm Hg in rats, a value close to those obtained in some electron microscopic, Aramine-induced, hypertension experiments (Westergaard *et al.*, 1977; van Deurs and Amtorp, 1978). At the light microscopic level, extravasation of protein was shown in Aramine-induced hypertension by Johansson *et al.* (1970). In an electron microscopic study of acute hypertension, Hansson *et al.* (1975) observed a pronounced extravasation of HRP, and numerous HRP-labeled vesicles, but were uncertain whether, in addition to vesicles, also transendothelial channels (see later) and open junctions might be responsible for the extravasation observed in arterioles, capillaries, and venules.

In other ultrastructural experiments (Giacomelli *et al.*, 1970), rats were made hypertensive by partially constricting the left renal artery until blood pressure exceeded 200 mm Hg (long-term hypertension). Horseradish peroxidase or colloidal carbon was injected intravenously and allowed to circulate for varying times before killing the rats. Giacomelli *et al.* (1970) observed that small arteries and arterioles were highly permeable to HRP, but not to colloidal carbon. Horseradish peroxidase apparently penetrated the endothelial junctions; it was seen throughout the entire length of the junctions, from the lumen to the basement membrane. The penetration of HRP, but not of carbon, suggests that the (opened) junction acts as a "selective" pore ("small pore"). Although numerous endothelial vesicles were also HRP-labeled, they were not considered significantly involved in the transendothelial movement of the tracer or ". . . play at best a secondary role in the formation of edema fluid." Contraction of the endothelial cells (and thus opening of the junctions) might be responsible for the high permeability in these experiments (Giacomelli *et al.*, 1970; see also Rapoport, 1976a,b). In a recent study, Giacomelli *et al.* (1978) studied the effects of carotid artery constriction in severe two-kidney hypertension. The period of hypertension lasted for 1–3 weeks, and for HRP experiments, rats with a blood pressure exceeding 200 mm Hg were used. Also here some evidence for junctional opening was obtained.

B. VESICULAR TRANSPORT

A rather large number of ultrastructural studies indicate that transendothelial vesicular transport is another, or the only, way of extravasation, see Table II. The concept of vesicles playing an important role in transport across the cerebral endothelium is not new. Raimondi *et al.* (1962) and Tani and Evans (1965) reported that pinocytic vesicles transported fluid and ferritin molecules from blood to brain in experiments with brain edema.

Westergaard (1975a, 1978) perfused serotonin through the ventricular system of mice after intravenous administration of HRP. Brains from these mice were compared with a control material, and it was found that, although a little extravasation of HRP could be seen in arteriolar segments of the normal control mice (Westergaard and Brightman, 1973), the extravasation of the tracer was more pronounced in the experimental material. Tracer-labeled endothelial vesicles were found both in venules and capillaries, and especially in arterioles. The endothelial tight junctions were never seen to contain tracer, and no damaged (tracer-filled) endothelial cells were observed. The effect of serotonin on the

TABLE II

ULTRASTRUCTURAL TRACER STUDIES SUPPORTING THE VESICULAR TRANSPORT MECHANISM FOR
EXPERIMENTALLY INDUCED EXTRAVASATION IN THE BRAIN[a]

Method	Tracer injected i.v.[b]	Species	Blood pressure monitored	Reference
Brainstem lesion	HRP[b]	Rat (Wistar–Furth)	−	Manz and Robertson (1972)
Hypertension	Ferritin	Rat (Wistar–King)	−	Eto et al. (1975)
Serotonin	HRP	Mouse (albino)	−	Westergaard (1975a)
Ischemia	HRP	Mongolian gerbil	−	Westergaard et al. (1976)
Compression injury	HRP	Cat	−	Beggs and Waggener (1976)
Acute hypertension	HRP	Rat (Wistar)	+	van Deurs (1976b)
Acute hypertension	HRP	Rat (Wistar)	+	Westergaard et al. (1977)
Acute hypertension	HRP	Rat (Wistar–Furth)	+	Nag et al. (1977, 1979)
Seizures	HRP	Rat (Wistar)	+	Petito et al. (1977)
Seizures	HRP	Rat (Wistar)	+	Bolwig et al. (1977)
Portocaval anastomosis	HRP	Rat (Wistar)	−	Laursen and Westergaard (1977)
Acute hypertension	Microperoxidase	Rat (Sprague–Dawley)	+	van Deurs and Amtorp (1978)
Brain injury	HRP	Cat	+	Povlishock et al. (1978)

[a] In all studies, the endothelial junctions were found to be tight to the tracer. Further details are given in the text.
[b] i.v., Intravenous; HRP, horseradish peroxidase.

permeability varied considerably, and no dose dependency was established. Westergaard (1975a) concluded that serotonin enhanced the vesicular transport of HRP seen to a much lesser extent in the "normal" brain (Westergaard and Brightman, 1973), and suggested that serotonin might affect the plasma membrane of the endothelial cells by elevation of cAMP concentrations. It was also reported that norepinephrine and cAMP perfused through the cerebral ventricles induce vesicular transport (Westergaard, 1975b). Later, Westergaard (1977) suggested that the most likely mechanism behind the serotonin-induced extravasation is local hypertension (see following) caused by serotonin-induced vasoconstriction of arterioles.

That cAMP might be involved in transendothelial vesicular transport in the brain has been suggested in other permeability studies (Beggs and Waggener, 1976; Petito et al., 1977). Support for cAMP playing a role in formation of cerebral endothelial vesicles also comes from works of Joó (1972) and Joó et al.

(1975). Joó (1972) showed that a significant cAMP-induced increase in the number of endothelial vesicles (uncoated and coated) can be produced. He compared the number of endothelial vesicles in the cerebral endothelium after cAMP administration with data from noncerebral endothelia with pronounced vesicular activity, and found that the cAMP-induced vesicular activity in the brain operates with approximately one-quarter of the efficiency of normal noncerebral capillaries.

Westergaard *et al.* (1976) studied extravasation of HRP in the brain by inducing ischemia in Mongolian gerbils. This species lacks arterial communication between the cerebral and the vertebral system in about 50% of cases. This makes it especially well suited for ischemia studies by occlusion of the common carotid artery in only one side. Horseradish peroxidase was injected intravenously into animals with neurological signs after artery occlusion for various periods, and allowed to circulate for 5 minutes before vascular perfusion. Although severe postischemic damage of cells of the neuropil was often observed, the endothelial cells appeared undamaged. The intercellular junctions were apparently not penetrated by HRP, and the endothelial cells contained HRP-labeled vesicles of varying size and shape. It was concluded that transendothelial vesicular transport was responsible for the observed postischemic extravasation. Westergaard (1977) later suggested that a mechanism responsible for the permeability of the blood–brain barrier in these experiments might be the release of serotonin from blood platelets in the occluded half of the brain. The serotonin may cause hypertension (by vasoconstriction of arterioles) resulting in increased vesicular transfer.

Similarly, protocaval anastomosis was found to induce vesicular transport of HRP across the cerebral endothelium (mainly arterioles) of rats (Laursen and Westergaard, 1977).

Manz and Robertson (1972) reported impermeable tight junctions and vesicular blood-to-brain transport of HRP in experiments with brainstem lesions of thiamine-deficient rats.

Compression injury of the spinal cord of cats was also found to induce vesicular transport in capillaries (Beggs and Waggener, 1976). These authors injected HRP immediately before cord injury and allowed the tracer to circulate from 90 seconds to 4 hours. They examined spinal cord segments up to 9 cm away from the injury and found, using a quantitative approach, a correlation between the intensity of vesicular transport activity and the degree of vascular leakage, which was less pronounced with increasing distance from the injury. The junctions between the endothelial cells were never penetrated by the tracer. Beggs and Waggener (1976) concluded that vesicular transport may play an important role in posttraumatic edema. Damaged capillary endothelial cells with HRP free in the cytoplasm, situated adjacent to intact cells, were observed in zones of maximal cord injury, but apparently the HRP remained within the damaged cell and did not leak directly into the basement membrane. At 9 cm from the zone of injury, the number of HRP-labeled vesicles was still increased (up to 3-fold) relative to controls, but here almost no leakage of HRP was evident. Beggs and Waggener (1976) suggested, therefore, that the transport also depends on factors other than the initiation of vesicle formation. That vesicles may be present in the cerebral endothelium without any appreciable transport activity is supported by observations of van Deurs and Amtorp (1978). It was suggested by Beggs and Waggener (1976) that cAMP might be of importance for vesicular transport activity. Furthermore, Beggs and Waggener (1976) and Beggs *et al.* (1975) found that, in addition to cAMP, microtubules may be involved, since the number of microtubules appeared to be increased in regions with high vesicular activity. Also vesicle-to-microtubule attachments were reported (Beggs *et al.*, 1975).

Further support for transendothelial vesicular transport of HRP in cerebral arterioles, capillaries, and venules comes from a study of mechanical brain stem injury in cats (Povlishock *et al.*, 1978). As early as 3 minutes after injury, HRP-labeled abluminal vesicles were seen, apparently discharging their contents into the endothelial basement membrane. The junctions appeared completely tight. Interestingly, Povlishock *et al.* (1978) observed a rise of 75 mm Hg in arterial blood pressure for 60 seconds in response to the brain injury.

Hypertension is of special interest in the study of vesicular transport across the cerebral endothelium, and rather numerous reports exist in this topic. Although, for example, long-term hypertension may cause extravasation by means of opened junctions (Giacomelli *et al.*, 1970, 1978; see Section III,A) several investigators have found that vesicular blood-to-brain transport is the only structural explanation for extravasation of protein tracer in acute hypertension experiments (Eto *et al.*, 1975; van Deurs, 1976b; Nag *et al.*, 1977; Westergaard *et al.*, 1977; van Deurs and Amtorp, 1978). This may indicate that different mechanisms of extravasation are involved in acute hypertension and long-term hypertension. Furthermore, the results of Giacomelli *et al.* (1970, 1978) do not preclude that vesicular transport precedes, or occurs simultaneously, with the opening of junctions.

In an electron microscopic study of hypertensive encephalopathy in rats, Eto *et al.* (1971) observed an increased number of vesicles in the endothelium of capillaries and venules. More recently, Eto *et al.* (1975) obtained evidence that vesicles transported ferritin molecules across the endothelium in hypertensive rats, whereas the junctions were intact.

Acute hypertension was induced in rats by intravenous infusion of Aramine (van Deurs, 1976b; Westergaard *et al.*, 1977). This caused a rapid rise in arterial blood pressure, from about 100 mm Hg to about 170–200 mm Hg, lasting for a few minutes (see also van Deurs and Amtorp, 1978). All the experimental animals were artificially respired, their body temperature maintained at a constant level, and the arterial blood pressure continuously monitored from the femoral artery. After a resting period of 20–45 minutes with stable blood pressure, HRP was infused slowly during 4–12 minutes. After the start of the HRP-infusion, Aramine was infused intravenously during 1.5–8 minutes. The blood pressure increased almost immediately. In a second series of experiments, the rats were infused with Aramine for 8–14 minutes, and HRP then infused for 6–11 minutes. In both experimental groups, a pronounced extravasation of HRP was observed. Since relatively many HRP-labeled vesicles occurred in segments of the microvasculature, the junctions never appeared penetrated, and damaged endothelial cells did not occur, it was concluded that vesicular transport was responsible for the extravasation. In a third series of experiments (Westergaard *et al.*, 1977), the alpha-blocker Regitin (phentolamin) was injected simultaneously with Aramine. HRP was then infused as described above. This procedure did not cause any increase in the blood pressure, and no extravasation of HRP occurred. Based on these observations, it was concluded that the extravasation caused by Aramine-induced hypertension was not a drug-effect of Aramine, but due to the elevated hydrostatic pressure. Finally, it was also investigated whether Aramine, Regitin, and HRP had any influence on blood gasses. However, no significant changes in arterial pH, p_{CO_2}, or p_{O_2} were observed; most important, the p_{O_2} was, although rather low (due to experimental intervention), always within acceptable limits (Westergaard *et al.*, 1977).

Evidence in favor of a vesicular transport in the Aramine-induced acute hypertensive rat brain has also been obtained from experiments with MP (van Deurs and Amtorp, 1978). The results in this study were essentially the same as those of previous studies described earlier (van Deurs, 1976b; Westergaard *et al.*, 1977).

Another kind of experimental acute hypertension is that induced by angiotensin. Also this method seems to cause vesicular transport in the cerebral endothelium (Nag *et al.*, 1977). Focal cerebral edema and regions with pronounced extravasation of HRP were seen in the rats. Many HRP-labeled endothelial vesicles were present already after 90 seconds in segments of arterioles and in the basement membrane of these vessels. After 8 minutes, HRP extravasation was also seen in capillaries and venules, but was still most pronounced in arterioles. No damage of endothelial cells or disruption of endothelial junctions was seen.

Hirano *et al.* (1970a) have reported that electroshock may cause an opening of the blood–brain barrier, but they were uncertain whether vesicular transfer or leaky junctions were responsible for the observed extravasation of HRP. Experimental seizures have more recently been brought into focus as a way of introducing arterial hypertension and vesicular transport activity in the cerebral endothelium. Lorenzo *et al.* (1972) investigated the permeability of cerebral endothelium to protein during drug-(pentylenetetrazol-) induced seizures. They found regional increase in the extravasation of ^{125}I-labeled albumin. Lorenzo *et al.* (1972) considered the alteration in permeability reversible and obtained evidence that a significant amount of ^{125}I-labeled albumin returned to the bloodstream. They supported the view that the extravasation was due to a vesicular transport rather than to opening of the endothelial junctions, but emphasized that this problem could be resolved only in a fine structure study. An interesting observation in the studies of Lorenzo et al. (1972, 1975) was an increase in arterial blood pressure (as well as in venous and CSF pressure) from approximately 130 mm Hg to 230 mm Hg within 5 minutes after the onset of seizures. Later, Lorenzo et al. (1975) reinvestigated the Metrazol-(pentylenetetrazol-) induced seizure effect on the blood–brain barrier by use of HRP. At the light microscopic level, their observations confirmed their previous work (Lorenzo et al., 1972). However, at the ultrastructural level, they reported that in general there were only a few HRP-labeled vesicles in capillary endothelial cells, but that some HRP could be seen in junctions. In larger vessels, they found more labeled endothelial vesicles, but also here the junctions appeared labeled. They concluded, therefore, that vesicles (at least in capillaries) may not be the principal route for transendothelial passage of protein during seizures.

At the ultrastructural level, especially the work of Petito et al. (1977) is of interest with respect to seizures, hypertension, and HRP extravasation. These investigators provided evidence that a number (20–30) of electroconvulsant shocks caused a vesicular transport of HRP from blood to brain, especially in arterioles and capillaries, and to a lesser extent in venules. Most interestingly, Petito et al. (1977) also found that the electroshocks and the extravasation of HRP was related to an abrupt rise in systemic blood pressure. In experiments where hypertension was prevented by cervical cordotomy, only very little or no extravasation occurred. In the experiments of Petito et al. (1977), no evidence in favor of penetrated junctions (Lorenzo et al., 1975) was obtained. Bolwig et al. (1977) reported similar results from experiments with epileptic seizures. They found only little opening of the blood–brain barrier after a single electroshock. After 10 electroshocks, they found a pronounced extravasation apparently due to vesicular transport, accompanied by hypertension. Transsection of the spinal cord prevented both hypertension and the extravasation. Bolwig et al. (1977) concluded that the acute arterial hypertension was the mechanism behind the breakdown of the blood—brain barrier (Westergaard et al., 1977), and that vesicular transport was the structural basis for this breakdown. Exactly how increased arterial blood pressure causes vesicular transport activity is still uncertain. Recently, Simionescu and Simionescu (1978) have reported that micropinocytosis in noncerebral endothelial cells is independent of blood flow and hydrostatic pressure.

Brain edema has been induced by ultraviolet irradiation, and a focal extravasation of intravenously injected HRP was observed, which could be due to vesicular transport by means of both micropinocytic and larger (>1400 Å) vesicles (Sasaki et al., 1977). The authors found the highest pinocytic activity in venules. Only intact junctions were reported to be present. In experiments with mercury intoxication, Ware *et al.* (1974) observed increased vesicular transport of HRP in cerebral vessels without disruption of the tight junctions, and in experiments with air embolism, Persson *et al.* (1978) also obtained some evidence in favor of vesicular transport in cerebral endothelia.

C. Transendothelial Channels

N. Simionescu *et al.* (1975) suggested the existence of patent transendothelial channels across the endothelium of muscle capillaries formed by vesicles or vesicle chains opening simultaneously at the luminal and abluminal endothelial surfaces (Fig. 3A). This concept was supported by very convincing illustrations, including analysis of tilted specimens. The channels were considered of importance in transendothelial passage of MP, representing the structural equivalents of the physiologist's small pores (N. Simionescu *et al.*, 1975; see also Section II,B). The existence of such channels in muscle capillaries could not be established in earlier work using single sections, as well as tridimensional reconstruction (Palade and Bruns, 1968), and has recently been questioned by Wissig and Williams (1978).

Regarding the brain, transendothelial channels were initially demonstrated in the cerebral endothelium of sharks, where these structures were penetrated by HRP, whereas junctions were not (Hashimoto, 1972). Westergaard (1975a) described HRP-labeled endothelial tubules spanning from the luminal to the abluminal membrane of the cerebral endothelium. However, according to Westergaard (1975a), serial sections did not support their identification as transendothelial channels. Nor were Laursen and Westergaard (1977) or Westergaard *et al.* (1977) able to demonstrate transendothelial channels as evaluated on single sections. Channels or tubules in the endothelial cells of the brain have also been reported by Hansson *et al.* (1975) in experiments with acute hypertension, and by Povlishock *et al.* (1978) in experiments with mechanical brain injury. Povlishock *et al.* (1978) believed that tubules represented fused vesicles, but these investigators were not able to demonstrate continuity of the tubules with the luminal or abluminal endothelial membrane. Nag *et al.* (1977) have reported HRP-labeled channels and poorly defined rows of vesicles extending through the width of the endothelial cells, but they did not deal with the possibility of transendothelial pores. Persson *et al.* (1978) reported the presence of a few transendothelial channels in experiments with air embolism.

However, Beggs and Waggener (1976) showed by means of serial sections that transendothelial channels or vesicle chains connecting the luminal and abluminal cell membranes can occur in the cerebral endothelium under certain circumstances. It is important to notice that these channels (which Beggs and Waggener referred to as "facilitated pathways") often appeared very tangled, and therefore in many cases, can be established only by superimposition of several micrographs (Beggs and Waggener, 1976), and not by evaluation of single sections. Recently, Brightman (1977) described experiments with infusion of hyperosmotic solutions. After infusion of arabinose, the tissue was fixed by vascular perfusion of aldehydes, and 1 hour *after* fixation, HRP was infused into the aorta. A distinct leakage of HRP into the brain was observed, which could not be due to vesicular transfer of tracer molecules across the endothelium (assuming that this is an active, energy-requiring process, which must be stopped by fixation). Consequently, the leakage could be due to (1) open junctions, (2) transendothelial channels, or (3) damaged endothelial cells. Only a few open junctions were observed, and Brightman (1977) considered the transendothelial channels, which might be formed (before fixation) by confluent vesicles, as the most likely medium of transfer.

In general, these channels seem to be relatively infrequent in the brain under various experimental conditions, and their role in transendothelial movement of fluid and solids may be minor compared to vesicles, as evaluated from the relative frequency of these endothelial structures.

D. Formation of Fenestrae

Characteristic for a number of brain tumors is the fenestration of the cerebral capillary endothelium. Examples are endothelia of intracerebral fibroma (Hirano *et al.*, 1975), metastatic renal carcinoma in the brain (Hirano and Zimmerman, 1972), human hemangioblastoma (Tani *et al.*, 1974), and other tumors (Brightman and Robinson, 1976; Vick and Bigner, 1972; Hirano and Matsui, 1975). In addition to fenestrae, open (leaky) junctions, endothelial discontinuities, and increased number of endothelial vesicles have been reported in neoplastic cerebral capillaries (Long, 1970; Vick and Bigner, 1972; Hirano and Matsui, 1975), features which all may be responsible for a leaky blood–brain barrier. Such neoplastic capillaries have been shown to be permeable to bloodborne HRP (Brightman *et al.*, 1971; Brightman and Robinson, 1976; Brightman, 1977). These observations represent an interesting "plasticity" of blood vessels (transformation from nonfenestrated to fenestrated endothelium), perhaps in response to certain external factors and disease. That vesicles (in continuous endothelia) and fenestrae may represent related structures were indicated in the study of Palade and Bruns (1968); see also Fig. 23, in which both fenestrae and the transendothelial channel formed by fused vesicles are "closed" by apparently identical diaphragms.

E. Technical Considerations

The main matter of controversy in the discussion of which structural features of the cerebral endothelium are responsible for the extravasation occurring in many experimental and pathological situations seems to be whether transport is vesicular (Section III,B) (as a transient phenomenon), as indicated in a number of studies (see Table II), or whether the endothelial junctions can be (reversibly) opened (Section III,A), or indeed, whether the two pathways can exist simultaneously. Thus, the discussion of the structural equivalents to the "small" and "large" pores of noncerebral capillaries (see Section II,B) is apparently not directly applicable to cerebral vessels, since junctions and vesicles seem to be "alternative pathways" in many studies. As stated by Karnovsky (1970): "the concept of a 'uniform pore' is useful in physiological thinking but requires correlation with possibly quite different structures, depending on the capillary type." This may be true, in particular, for the cerebral microvasculature.

In the numerous studies of experimental opening of the blood–brain barrier, arterioles, capillaries, and venules appear to be involved in the increased permeability, although to a varying degree, but no generalizations can be made with respect to which microvascular segment is most susceptible to experimental intervention.

1. *Fixation Problems*

It is a common feature of all the studies (Table II) supporting a vesicular transport across the cerebral endothelium that (1) the endothelial junctions did not appear penetrated by tracer (Fig. 11); (2) tracer-labeled vesicles occurred [in varying positions (Section III,E,3) and numbers (Section III,E,4)] in the endothelial cells (Figs. 12–17); and (3) the tracer was typically seen in the endothelial basement membrane and often also in the extracellular space of the adjacent neuropil (Figs. 9A,B, 11, 13–17). In the evaluation of these aspects, an important factor to be considered is the fixation technique. By far the greatest part of the brain tissue used in the ultrastructural studies of blood–brain barrier permeability is fixed by vascular perfusion of aldehydes. This application of fixative washes out blood and tracers in the vascular lumen, and typically no reaction product is seen at the blood side of the endothelium in, for example, HRP studies (Fig. 9A). This would make statements like "the junctions were not penetrated by tracer" somewhat questionable (Fig. 9A). However, the vascular perfusion fixation of the brain is always incomplete (even with preperfusion of physiological saline), leaving more or less numerous vascular segments with some blood cells and free or endothelial surface-bound tracer (Figs. 9B, 11–14) (van Deurs, 1976b; Reese and Karnovsky, 1967; Westergaard *et al.*, 1977; Hashimoto and Hama, 1968). This situation may be considered as good as immersion fixation, since it creates images where junctions exhibit tracer on both the luminal and the abluminal surface, and therefore may justify the conclusion that the junction was not penetrated by the tracer (Fig. 11). In the evaluation of the tightness of a junction, it should also be considered whether the section is perpendicular to the adjacent membranes all the way down from the luminal to the abluminal surface; the individual fusion points should be visible (Figs. 1 and 11).

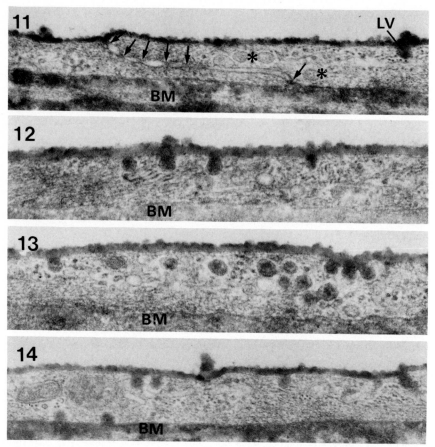

FIGS. 11-14. These are all micrographs from one particular arteriole in an experiment with acute hypertension. Extravasation of intravenously injected HRP in the vessel is obvious. Figure 11 shows a tight junction: the "fusion points" (arrows) are not penetrated by the tracer, although this is present on the luminal surface, as well as in the basement membrane (BM). LV indicates a single luminal vesicle with HRP, and the asterisks show some unlabeled vesicles. In Fig. 12, only luminal vesicles with HRP are seen; the basement membrane (BM) appears devoid of tracer in this segment of the vessel. In Fig. 13, both luminal and apparently "free" vesicles with HRP are seen, and in Fig. 14, some basal vesicles with HRP are present and the basement membrane (BM) is distinctly tracer-labeled. ×67,000. (From van Deurs, 1976b.)

In some studies, ventriculocisternal perfusion fixation with aldehydes has been used (van Deurs and Amtorp, 1978; Feder, 1971). This technique, in which the fixative is perfused at a constant rate from cannulae in the lateral ventricles to an outflow cannula inserted into the cisterna magna (see van Deurs and Amtorp, 1978), is in fact an *in situ* immersion fixation. It has advantages over vascular perfusion fixation: (1) the blood and tracer are left in the vascular lumen (Figs. 4

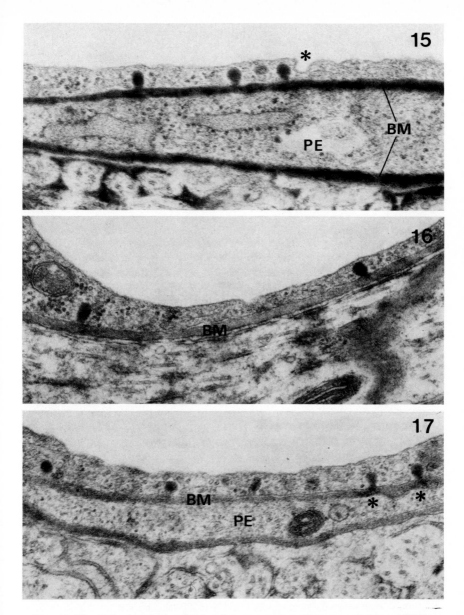

Figs. 15–17. Parts of cerebral capillaries after extravasation of intravenously injected HRP has been induced by acute hypertension. The tissue is fixed by vascular perfusion of aldehydes, which explains why the lumens are devoid of HRP. In Fig. 15, a typical situation is seen: some vesicles connected with the basal endothelial surface are heavily tracer-labeled, and so is the basement membrane (BM) surrounding the pericyte (PE). The asterisk indicates an unlabeled luminal vesicle. In contrast, in Fig. 16, two heavily labeled vesicles are connected with the abluminal endothelial cell membrane, although the underlying basement membrane (BM) appears almost unlabeled. In Fig. 17, some HRP-labeled basal vesicles are also seen, and in two cases (asterisks), they seem to discharge HRP into the basement membrane (BM) between the endothelial cell and the pericyte (PE). ×60,000.

and 5), and (2) the vessel wall is not stretched by exposure to a high hydrostatic pressure. The fixation quality is good (in general better than after immersion fixation), and the fixation conditions may be more physiological than those of immersion and vascular perfusion fixation. On the other hand, the ventriculocisternal perfusion fixation preserves satisfactorily only parts of the cerebral parenchyma close to the ventricles, limiting the use of this technique considerably (van Deurs and Amtorp, 1978). As the experimental rat is living (artificially respirated, etc.) at the beginning of fixation, the fixation technique appears ideal, and it might be expected that the brain would be gradually fixed at increasing distances from the ventricles with increasing time, the blood still flowing in vessels of the "unfixed" parts of the brain. However, this is not the case. It is characteristic of the ventriculocisternal perfusion fixation that the arterial blood pressure of the experimental animal rises considerably immediately after the start of the fixative perfusion, and then decreases within some minutes to a level between 0 and 40 mm Hg (Fig. 10) (van Deurs and Amtorp, 1978). Sooner or later the blood pressure will be zero in all cases (mostly within 10–20 minutes). This is, at least partly, the reason why only the brain parenchyma close to the cerebral ventricles is "well-fixed." The fixation technique is particularly suited to the choroid plexus (Section IV, van Deurs et al., 1978; van Deurs, 1978a,b). For more information about fixation of brain tissue, see, for example, Gonzalez-Aguilar (1969), Karlsson and Schultz (1965), Palay et al. (1962), Schultz and Karlsson (1965), van Harreveld and Khattab (1968), and Williams and Jew (1975).

A main source of evidence in favor of vesicular transendothelial transport is the rapid (30–90 second) sequential tracer labeling of luminal vesicles,[1] free vesicles and abluminal vesicles (and basement membrane) with increasing circulation time of the tracer in the blood stream, as described for muscle capillaries by N. Simionescu et al. (1973, 1975) who used in situ immersion fixation. However, the fixative in vascularly perfused animals mostly washes away any tracer present in luminal vesicles (Fig. 15). Therefore, only basal and free tracer-labeled vesicles are seen in the cerebral endothelium in most blood–brain barrier studies, and quantitative estimates of the number of labeled vesicles will be inaccurate (Beggs and Waggener, 1976). Also, it can be impossible to establish with perfusion fixation the characteristic pattern of sequential tracer labeling with increasing time of tracer circulation (N. Simionescu et al., 1975) seen with immersion fixation.

---

[1] In order to describe the distribution of labeled endothelial vesicles, the following terminology will be used: luminal vesicles are vesicles opening at the luminal surface of the endothelium; free vesicles are not opening onto any surface in the particular (400–700 Å thick) section (which does not exclude that they can open onto a surface at an other plane of sectioning); and basal vesicles are vesicles opening at the abluminal (basal) endothelial surface.

2. Choice of Tracer

Another technical aspect to be considered, especially in relation to endothelial junctions, is the choice of tracer. As shown in Table II, HRP is by far the most commonly used tracer in permeability studies of the brain. The reason for this is probably that HRP is a widely used tracer in permeability studies of noncerebral tissue, and is a protein of relatively low molecular weight (Table I), allowing the judgment that if HRP in a particular situation cannot penetrate a junction, then larger proteins (e.g., albumin) cannot do so either. Finally, HRP is very easy to visualize enzymatically (Graham and Karnovsky, 1966) without obvious damage to fine structure. However, it has some disadvantages. It has been shown that HRP administered intravenously can result in arterial hypotension and extravasation of blood proteins, probably because of release of histamine (or other vasoactive amines) and opening of the endothelial tight junctions in noncerebral tissue (Majno and Palade, 1961; Majno et al., 1961; Cotran and Karnovsky, 1967; Clementi, 1970; Deimann et al., 1976; Straus, 1977; N. Simionescu et al., 1978c). Treatment with histamine-antagonists prevents the hypotension and extravasation. This toxic effect of HRP has been reported to be especially pronounced in certain rat strains (Wistar; Sprague–Dawley), whereas Wistar-Furth rats do not react significantly to HRP (Deimann et al., 1976). However, the possible toxicity of HRP does not seem to have influenced the results of permeability studies in the brain. Although various (sometimes very high) doses of HRP have been administered to, for instance, Wistar and Sprague-Dawley rats (Table II), controls clearly show that HRP does not extravasate in the normal brain. Thus, it is unlikely that the extravasation of HRP in several experimental situations should be due to a toxic effect of HRP. Additionally, in several studies on brain vascular permeability, the arterial blood pressure has been monitored during the experiments (Table II), and no hypotension caused by HRP infusion has been reported (see, for example, Nag et al., 1977; Petito et al., 1977; Westergaard et al., 1977). In experiments by the author with intravenous injection of HRP in Sprague-Dawley rats, it was found that "in three of the four rats this caused a significant decrease in blood pressure" (van Deurs, 1978a), whereas preliminary experiments in the author's laboratory (M. Møller and B. van Deurs, unpublished observations) have shown that some Wistar rats may not be sensitive (i.e., do not exhibit any hypotension) to HRP. In a study of the cochlear capillaries in the guinea pig, no significant HRP-induced histamine release was found, and pretreatment with antihistamine did not alter the results (Winther, 1971). However, Vegge and Haye (1977) found a dose-dependent extravasation of HRP in guinea pigs. This was not affected by treatment with antihistamine, whereas acetylsalicylic acid prevented extravasation. In general it must, therefore, be concluded that the toxic effect of HRP in different species, strains, and individuals is far from elucidated satisfactorily and, as stated by Deimann et al. (1976), monitoring of blood pressure is an indispensable requirement in most studies with HRP (and perhaps with other hemeprotein and peptide tracers, too; see van Deurs and Amtorp, 1978; van Deurs, 1978b).

3. Localization of Tracer

Also, the localization of HRP (and other tracers) under various experimental conditions needs some comment. In many cases, in experiments with intravenously administered HRP, it is concluded from micrographs with HRP in abluminal endothelial vesicles, in the basement membrane, and in the extracellular space of the neuropil (Fig. 15), that the tracer arrived at the basement membrane following vesicular transport from the blood across the endothelium and thereafter spread into the neuropil. However, movement of tracer in the opposite direction also has to be considered. For instance, if the junctions in some vessels leak, HRP can, after extravasation in these vessels, diffuse into the extracellular space of the neuropil, and then label the basement membrane and basal vesicles in other vessels in a "retrograde" fashion. The presence of tracer-labeled (basal) vesicles may not necessarily be indicative of vesicular transport (van Deurs and Amtorp, 1978).

Some evidence in favor of movement of tracer from blood via endothelial vesicles toward basement membranes and eventually to the extracellular space of the neuropil, assuming that the junctions are tight, is found in micrographs in which only the endothelial basement is labeled, in addition to endothelial vesicles, whereas extracellular structures at some distance from the endothelium are unlabeled (see, for example, Westergaard et al., 1977, Fig. 8a; van Deurs and Amtorp, 1978, Fig. 14). It would be even more convincing if such observations could be correlated with the time of HRP circulation in the bloodstream, making it possible to visualize a progressing front of extravasated tracer with increasing time. However, such data are difficult to obtain, as will be explained later.

Very convincing evidence in favor of a blood-to-brain transport by means of vesicles is found in micrographs where tracer-labeled vesicles are fused with the abluminal endothelial cell membrane adjacent to an unlabeled basement membrane, or the basement membrane contains tracer only immediately beneath the opening (discharging) vesicle (Figs. 16 and 17) (Karnovsky, 1967; N. Simionescu et al., 1973; Beggs and Waggener, 1976; Povlishock et al., 1978). Also here, a detailed time sequence analysis is, in principle, well warranted, but difficult to perform. It should be noticed, however, that not even the situation with a heavily tracer-labeled vesicle opening onto, for instance, an unlabeled or slightly labeled basement membrane is beyond question, since we know little about the effects of fixation on tracer proteins bound to membranes (i.e., of a vesicle), situated "freely" in extracellular fluid, and located in the basement membrane. Similarly, little information exists about diffusion of tracers and enzymatic reaction products during and after fixation.

It has been calculated that movement of a single endothelial vesicle (a process often assumed to be caused by Brownian motion and presumably bidirectional, although it is difficult to imagine "random diffusion" of a macromolecular structure such as an endothelial vesicle in the cytoplasm containing highly organized systems of filaments, structural proteins, etc.) is a process lasting for a few seconds to minutes (Shea and Karnovsky, 1966; Shea *et al.*, 1969; Tomlin, 1969; Karnovsky and Shea, 1970; Casley-Smith and Chin, 1971; Green and Casley-Smith, 1972; Shea and Bossert, 1973; Rubin, 1977). These theoretical calculations must, however, be evaluated with the utmost care, as they rely on a number of uncertain parameters. In their works on microvascular permeability to myoglobin and MP in muscles, N. Simionescu *et al.* (1973, 1975) found that the time required for the progressive development in endothelial tracer labeling (first luminal vesicles, then free vesicles, and finally also abluminal vesicles and basement membrane) is on the order of 1 minute. A similar period of time (less than 90 seconds) for vesicular transport to be detectable in the brain has also been reported (Beggs and Waggener, 1976; Nag *et al.*, 1977). In either case, the brief period of time involved makes experimentation difficult. For instance, slow and careful tracer infusion followed by initiation of vascular perfusion fixation through the heart, or removal of tissue for immersion fixation, are lengthy processes as compared with the permeability times mentioned above. Furthermore, chemical fixation for electron microscopy must be considered as a very slow process compared to physiological phenomena and probably represents the morphologist's main problem at present (cf. Luft, 1972), although it is often disregarded. In addition, at least in the brain, the vesicle population is very unevenly distributed (van Deurs and Amtorp, 1978), and a great variation in transport activity from segment to segment of the microvasculature apparently exists under experimental conditions. For instance, the entire "sequence" of labeling patterns of the endothelium can sometimes be seen within different parts of the same vessel in the same electron microscopic section (Figs. 12–14), and vessels with completely unlabeled basement membranes occur frequently even after relatively long tracer circulation (van Deurs, 1976b). This phenomenon may be due to incomplete mixing of the tracer in the plasma (N. Simionescu *et al.*, 1973). In the experiments with hypertension described in Section III,B (van Deurs, 1976b; Westergaard *et al.*, 1977), muscle capillaries from the tongue were also examined (B. van Deurs, unpublished observations) and here, too, capillaries with tracer labeling only of luminal and free vesicles could be seen among vessels with pronounced extravasation (Fig. 18). Similar "unleaky" muscle capillaries have been reported elsewhere (Karnovsky, 1967; Williams and Wissig, 1975). Thus, even though detailed short-time sequence studies are well warranted in the analysis of the blood–brain barrier, several factors apparently make such approaches rather complicated.

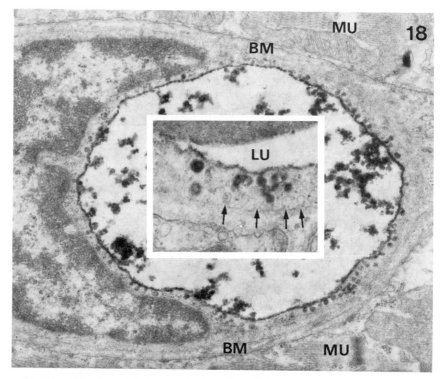

FIG. 18.   This micrograph shows a muscle capillary from the tongue of a hypertensive rat intravenously injected with HRP and fixed by vascular perfusion approximately 10 minutes after the
start of the tracer injection. The fixation has been incomplete, since tracer can be seen in the vascular
lumen and on the luminal endothelial surface. It is obvious that extravasation of HRP has not
occurred in this vascular segment: all tracer-labeled vesicles are luminal. The basement membrane
(BM) between the capillary and muscle cells (MU) is completely devoid of HRP. The inset shows
HRP-labeling of the luminal (LU) endothelial surface and in vesicles or vesicle-chains, but the basal
vesicles (arrows) are completely devoid of tracer. ×25,000; inset, ×48,000.

4. Quantitation

In many experimental situations with a pronounced extravasation of tracer into
the brain, the number of endothelial vesicles in general appears to be relatively
low, as evaluated from published micrographs. Also, most vesicles are basal.
This probably represents the most important single argument against vesicular
transport causing a significant extravasation in the brain. It should be noticed,
however that although the number of cerebral endothelial vesicles (in certain
experimental situations) is lower than in skeletal muscle capillaries, the capillary
surface is larger in the brain (240 cm^2/gm of tissue) than in a skeletal muscle (70
cm^2/gm of tissue) (Crone, 1973). This means that if there are 3.5 times as many

vesicles in muscular endothelial cells as in cerebral endothelia, the extravasated amount of protein per gram of tissue may still be the same.

The enzymatic properties of the tracer (HRP) raise problems with respect to quantitation. Extravasation of smaller or larger amounts of HRP (within certain limits) will most likely produce a reaction product with the same electron density, and accordingly, the same visibility in the electron microscope.

In a number of studies, "increased permeability" by means of vesicular transport has been suggested, although quantification of the vesicles has not been undertaken. Quantitative or semiquantitative data on cerebral endothelial vesicles in normal and experimental situations are found in the studies of Joó (1972), Connell and Mercer (1974), Reyners et al. (1975), Beggs and Waggener (1976), Tani et al. (1977), and van Deurs and Amtorp (1978). Recently, Nag et al. (1979) showed in a quantitative HRP study of experimental acute hypertension that the number of vesicles was significantly increased in permeable arteriolar segments as compared with nonpermeable segments of the same rats and comparable segments of normotensive animals. Furthermore, in normotensive rats HRP did not influence the number of vesicles as compared to rats receiving saline only (Nag et al., 1979). As mentioned earlier (Section III,B), Nag et al. (1977, 1979) did not observe tracer penetration of the endothelial junctions in their hypertension experiments.

5. Vesicular Transport to Lysosomes

Another argument against transendothelial transport by means of vesicles (thus favoring open tight junctions) in the brain has been presented: the process of vesicular transport is an active energy-requiring process susceptible to poisons used in attempts to open the blood–brain barrier. Furthermore, vesicular transport is a relatively slow process and would consequently be an "inefficient" mechanism of extravasation (Brightman et al., 1970a). However, according to Florey (1964), neither cyanide, cooling, dinitrophenol, nor anoxia influence the rate of vesicular uptake of tracers. Recently, Williams and Wagner (1978) reported that micropinocytic activity in isolated (noncerebral) capillaries is insensitive to metabolic inhibitors and independent of the presence of ATP.

It has been suggested by Brightman (1977) that the vesicles in the cerebral endothelium are incorporating substances for metabolic purposes or transport (in the case of foreign material) endocytozed material to lysosomes, rather than performing transendothelial transport. However, lysosomes and lysosome-like bodies are relatively rare in most endothelia, including those of the brain, exceptions being the endothelial cells of postcapillary venules in lymph nodes (van Deurs, 1978c) and the sinus-lining endothelial cells (Bankston and DeBruyn, 1974; DeBruyn et al., 1975; Wisse, 1972). Indeed, a few multivesicular bodies and lysosome-like vacuoles do become tracer-labeled in the cerebral endothelium (see, for example, Clawson et al., 1966; Beggs and Waggener, 1976; van Deurs and Amtorp, 1978), and it is likely that they are formed by fusion of endothelial vesicles (Povlishock *et al.*, 1978), but the relatively sparse number of such structures does not support the view that the cerebral endothelial vesicles are involved mainly in lysosome formation (Brightman, 1977). Furthermore, no information exists to indicate that the number of tracer-labeled lysosomes increases significantly in the cerebral endothelium with increased time of tracer exposure under various experimental conditions, as it does, for instance, in the choroidal epithelium, which exhibits a well developed lysosomal apparatus (see Section IV,D).

F. CONCLUSIONS

In conclusion, several factors have to be evaluated in order to determine the structural basis for the cerebral capillary permeability and the blood–brain barrier under normal, as well as under experimental conditions, and technical problems are obvious.

The concept of a vesicular transendothelial transport (of macromolecules) in the brain is still based mainly on circumstantial evidence, and criticism can be raised. However, during the last years, convincing information has accumulated in favor of such a mechanism, and at present it seems likely that a vesicular transport takes place in the brain under certain circumstances, as it apparently does under normal conditions in most noncerebral tissue. This does not in any way exclude the possibility that opening of cerebral endothelial junctions, for example, in experiments with hypertonic solutions or long-term hypertension, especially where the arterial blood pressure becomes considerably higher than 200 mm Hg (Brightman *et al.*, 1973; Rapoport, 1976b; Giacomelli *et al.*, 1970, 1978), or the formation of transendothelial channels (Beggs and Waggener, 1976; Brightman, 1977) can be the structural basis for extravasation in the brain as well. Moreover, different kinds of "pores" for blood-to-brain movement of material (particularly macromolecules) may be opened simultaneously. Even though electron microscopy has provided valuable insight into the blood–brain barrier and the cerebral endothelium, many factors are still unresolved. As stated by Luft (1973) in a review of the use of the electron microscope in studies of capillary permeability in general "... the past 15 years have shown this instrument to be more a key to Pandora's Box than a magic sword to slay dragons still lurking in dim vascular recesses."

IV. Choroid Plexus

The choroid plexus, described by Tennyson and Pappas (1961) as an enormous ingrowth of highly vascularized pia covered by special ependyma, consists of numerous villi. Each villus is composed of a vascular core, mainly composed of wide-caliber fenestrated capillaries (Maxwell and Pease, 1956), surrounded by connective tissue cells (leptomeningeal cells; Maxwell and Pease, 1956). Exposed to the CSF is an epithelium that includes roughly cuboidal cells with an irregular brush border at the apical surface and some basolateral foldings (Maxwell and Pease, 1956; Millen and Rogers, 1956; Tennyson and Pappas, 1961; Cancilla *et al.*, 1966; see also reviews by Dohrmann, 1970; Milhorat, 1976). Hitherto, clear-cut ultrastructural differences between the choroid plexus of the lateral, third, and fourth ventricles have not been established, but only a few investigators have dealt with this problem (Davis *et al.*, 1973; van Deurs *et al.*, 1978; van Deurs and Koehler, 1978, 1979). In order to localize the blood–CSF barrier in the plexus, it is necessary to consider the endothelium, the connective tissue and the epithelium separately (Figs. 19 and 20).

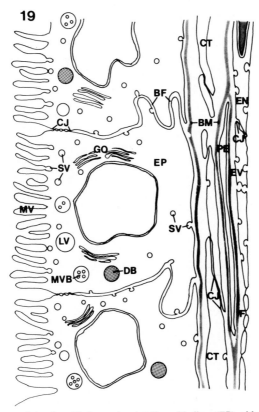

FIG. 19. Diagram of the choroid plexus showing the epithelium (EP) with the apical microvilli (MV) and some basolateral foldings (BF), the layer of connective tissue cells (CT), and the endothelium (EN). These three layers, which are separated by basement membranes (BM), are all provided with cell junctions (CJ) (discussed in the text). Elements of the vacuolar apparatus of the epithelial cells are also shown: DB, dense body; GO, Golgi complex; LV, large vacuole; MVB, multivesicular body; SV, small vesicles, sometimes connected with the epithelial surface (endocytic or exocytic). In the endothelium, fenestrae (F), as well as vesicles (EV), are present. PE, Pericyte.

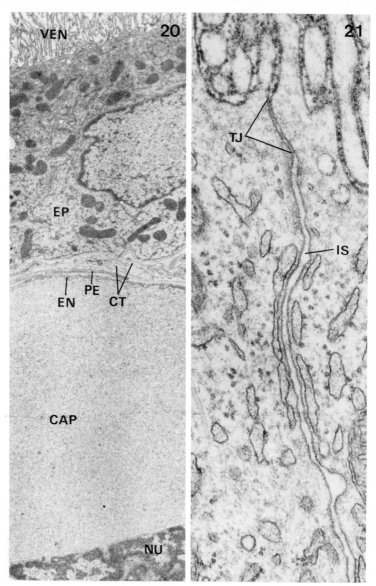

FIG. 20.   Survey electron micrograph of a part of the rat choroid plexus (compare with Fig. 19). At the apical surface of the epithelial cell (EP), microvilli projecting into the cerebral ventricle (VEN) are seen. Below the epithelium are connective tissue (meningeal) cells (CT), a pericyte (PE), and the (fenestrated) endothelium (EN) of a choroidal capillary (CAP). NU, Nucleus of the capillary endothelium. ×8500.

FIG. 21.   Apical part of the rat choroidal epithelium in an experiment with ventriculocisternal perfusion of MP. This tracer is distinct on the microvilli at the ventricular surface, but has not penetrated the tight junction (TJ) and is absent from the intercellular space (IS). Note that no other junctional types than the apical tight junction are present. ×62,900.

A. Permeability of Blood Vessels

The fenestrated capillaries of the choroid plexus (in some lower vetebrates the choroidal capillaries may be nonfenestrated, cf. Carpenter, 1966) are permeable to hydrophilic dyes in the blood stream (Dempsey and Wislocki, 1955). The pathway(s) across the endothelium followed by various ionic and macromolecular tracer molecules is not completely clear and is probably related to the size and nature of the tracer (Pappas and Tennyson, 1962; Dretzki, 1971; Hashimoto and Hama, 1968; Brightman et al., 1970b; Castel et al., 1974; Brightman, 1967, 1975; van Deurs et al., 1978). Several pathways most likely exist: (1) transendothelial transport by means of vesicles, which are present in the thick parts of the endothelium (Figs. 22 and 23), (2) diffusion via transient pores or channels in the endothelium (Fig. 23), (3) diffusion via open or "leaky" cell junctions between adjacent endothelial cells (van Deurs, 1979), and (4) diffusion across fenetrae in the thin parts of the endothelium (Figs. 23 and 24). These four pathways (regarding the first three, see also Section II,B–D) across fenestrated endothelia are also discussed with respect to the pineal gland (Møller et al., 1978). The possibility of vesicular transport in capillaries of circumventricular organs is interesting as it suggests that lack of a barrier to macromolecules is not necessarily correlated to the existence of fenestrae (Møller et al., 1978). The diaphragms of the fenestrae may be proteinaceous (e.g., Palade and Bruns, 1968) and represent a route of extravasation that may be equivalent to the physiologist's "small" pore (Clementi and Palade, 1969).

Freeze–fracture studies of the fenestrated capillaries in the choroid plexus revealed that the "tight" junctions between the endothelial cells did not constitute a continuous network (zonula occludens), but rather a system of discontinuous, blindly ending strands (Dermietzel, 1975a; van Deurs, 1979). On the E-face of the endothelial cell membrane, the junctions appear as shallow grooves devoid of particles, and on the P-faces, as low ridges bearing some particles (van Deurs, 1979). The discontinuous organization of the strands is not supportive of any barrier function, at least for smaller molecules (e.g., MP; see N. Simionescu et al., 1978c; Wissig and Williams, 1978; see Fig. 3A), but may prevent diffusion of larger molecules. Interestingly, these junctions of the choroid plexus capillaries—the vessels identified with certainty in the replicas by the localization of fenestrae (van Deurs, 1979)—resemble those of venules rather than of capillaries as described by M. Simionescu et al. (1975).

The arterioles of the choroid plexus exhibit zonulae occludentes that are probably tighter than those of the capillaries (van Deurs, 1979; Dermietzel, 1975a). The junctional strands (low ridges with some particles on P-faces and grooves with particles on E-faces) form a rather small-meshed, continuous network resembling that described in arterioles of other tissue (M. Simionescu et al., 1975). Gap junctions may be seen in the tight junction meshes (see M.

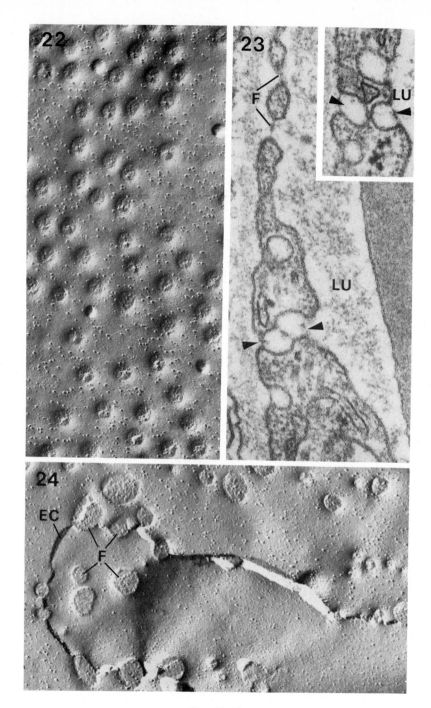

Figs. 22–24.

Simionescu *et al.*, 1975). Vesicles are rather numerous in the choroidal arteriolar endothelium and are probably involved in transendothelial transport of macromolecules. Venules, although not identified with certainty in freeze–fracture replicas (van Deurs, 1979), exhibit a tight junction architecture appearing as a less-developed version of the capillary type, and are presumably highly leaky.

B. PERMEABILITY OF MENINGEAL SHEATHS

Extravasated tracers like HRP, cytochrome c, and MP are localized in the endothelial and epithelial basement membranes, and in the interstitium surrounding the connective tissue cells (Brightman, 1967, 1975; Milhorat *et al.*, 1973; Davis and Milhorat, 1975; van Deurs, 1978a,b; van Deurs *et al.*, 1978). Some micropinocytic vesicles and a few vacuoles with tracer may be seen in the connective tissue cells, but they do not exhibit a well developed lysosomal apparatus. Typical macrophages are also present in the interstitium and can accumulate, for example, intravenously administered Myofer (Dretzki, 1971). Phagocytes in the developing rabbit choroid plexus are common (Tennyson and Pappas, 1964). The investigators suggested that such phagocytes, after migration between the choroidal epithelial cells, could give rise to the epiplexus or Kolmer cells localized on the ventricular surface of the plexus, which are also phagocytic (Fig. 26) (cf. Carpenter *et al.*, 1970; Sturrock, 1978). The function of these macrophages in relation to the blood–CSF barrier seems unclear.

The connective tissue cells or leptomeningeal cells form a rather continuous "sheath" between the endothelium and epithelium (Maxwell and Pease, 1956; Dermietzel, 1975b; van Deurs, 1979). According to Maxwell and Pease (1956), the connective tissue layer is about 85% continuous, but "although this system might serve as a baffle in slowing diffusion, it would not interpose a definitive barrier." In agreement with this observation of Maxwell and Pease (1956), no tracer or freeze–fracture experiments suggest that the connective tissue layer constitutes a barrier between the endothelium and the epithelium (Brightman, 1967, 1975; Davis and Milhorat, 1975; Dermietzel, 1975b; van Deurs, 1978a,b, 1979).

FIG. 22. Freeze–fracture micrograph of an endothelial cell of a rat choroidal capillary. Numerous vesicular openings (or fenestrae) are seen. ×55,000.

FIG. 23. Thin section of a rat choroid capillary. When fenestrae (F) are seen in the thin part of the endothelium, vesicles are present in the thick part. Arrowheads indicate a transendothelial "channel" formed by vesicles (note that the vesicles are still closed by "diaphragms"). The inset shows a luminal and a basal vesicle (arrowheads) that are not fused, although in close contact. LU, Lumen of the vessel. ×93,000.

FIG. 24. Freeze–fracture micrograph of a fenestrated part of the endothelium of the rat choroidal capillary. The fenestrae (F) are positively identified as they connect both membranes of the endothelial cell. EC, Endothelial cell cytoplasm (cross-fractured). ×55,000.

The connective tissue cells often overlap extensively (especially in the basal parts of the plexus), being connected to each other with complex arrangements of cell junctions, which can only be elucidated with difficulty in thin sections. However, freeze–fracture material has revealed the presence of extended tight junction–gap junction complexes between adjacent connective tissue cells of the choroid plexus (Dermietzel, 1975b; van Deurs, 1979). The tight junctional strands are often "isolated" or may form complex geometric patterns, but do not form a continuous network or belt (Fig. 25). The organization of the tight junctional strands appears to be different from that described between the meningeal cells at the brain surface. Here the tight junctional strands form a continuous network and represent a diffusion barrier to HRP and MP (the "arachnoid barrier layer") (Nabeshima and Reese, 1972; Nabeshima et al., 1976). On P-faces of the choroidal connective tissue cells, the junctional strands

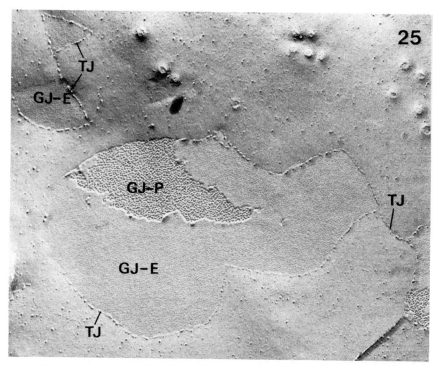

FIG. 25.   Freeze–fracture replica of a meningeal cell membrane from the rat choroid plexus. It is obvious that the tight junction "elements" (TJ) do not form a continuous "barrier" system, but are surrounding and interconnecting gap junctions, which are seen both on E-faces (GJ-E) and on P-faces (GJ-P). ×52,000.

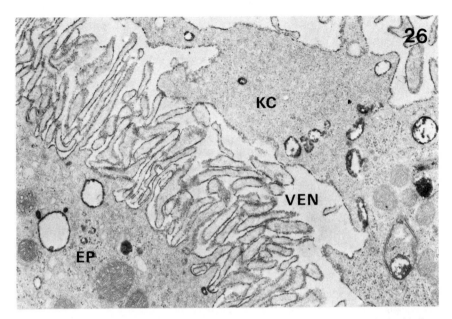

Fig. 26. A Kolmer cell (KC) is seen on the ventricular (VEN) surface of the choroidal epithelium (EP). The tissue has been fixed by ventriculocisternal perfusion of aldehydes 15 minutes after the start of ventriculocisternal perfusion of HRP, which is taken up by the Kolmer cell, as well as by the epithelial cell. ×20,000.

appear as elongated crests with a distinct furrow running along the edge of the crest, occupied by a few particles. On E-faces, elongated impressions are found, bearing some aligned particles-and-bars (van Deurs, 1979). The gap junctions between these cells vary considerably in size and are very often surrounded and interconnected by tight junction elements (Fig. 25) (Dermietzel, 1975b; van Deurs, 1979). A special gap junction type (the "segmented nexus type") was described in the cat choroidal meninges (Dermietzel, 1975b; Dermietzel and Schünke, 1975). The intimate association between tight and gap junctions may be indicative of junctional assembly or dissembly as described in certain developing tissue (e.g., Decker and Friend, 1974; Elias and Friend, 1976), although this aspect is unclear and difficult to evaluate in an apparent "nondeveloping" tissue such as the adult choroid plexus connective tissue. Gap junctions are generally assumed to be involved in electrotonic and metabolic coupling of cells (McNutt and Weinstein, 1973; Staehelin, 1974; Gilula, 1977), but their role between the connective tissue cells of the plexus (and of the "arachnoid barrier layer") is unclear. Synchronized behavior, for example, with respect to electrical properties, has been suggested (Dermietzel, 1975b; Nabeshima et al., 1976).

C. Permeability of the Epithelium: The Blood–CSF Barrier

1. *Tracer Experiments*

From the connective tissue interstitium, extravasated HRP, cytochrome c, and MP are endocytosed by (coated) micropinocytic vesicles (Friend and Farquhar, 1967) at the basolateral surface of the choroidal epithelium (Becker *et al.*, 1967; Brightman, 1967, 1975; Hashimoto and Hama, 1968; Reese *et al.*, 1971; Milhorat *et al.*, 1973; Milhorat, 1976; David and Milhorat, 1975; van Deurs, 1978a,b). Thereafter, the tracers are sequestered in the lysosomal apparatus of the epithelial cells, which will be discussed in detail in Section IV,D.

Eighteen days after intravenous administration of Myofer into rats, this tracer was localized in numerous structures resembling dense bodies in the choroidal epithelium (Dretzki, 1971). Hemosiderin granules were observed in the choroidal epithelium in an early electron microscopic work, which may indicate uptake of blood breakdown products into the epithelial cells (Case, 1959). In a study of vitamin A-deficient rabbits, Witzel and Hunt (1962) observed "weakened" areas between endothelial cells of the choroid plexus (indicative of enhanced permeability) and an apparent increase in number of "inclusion bodies" in the choroidal epithelium. Following intravenous injection of cytochrome c, Davis and Milhorat (1975) reported a significant increase in lysosomal activity of the choroidal epithelial cells as compared to controls. Milhorat (1976) says that "concurrent with the introduction of exogenous material into the cytoplasm, the production of lysosomal enzymes is triggered, as suggested by the apparent increase in the number of small acid phosphatase positive-rich vesicles (primary or protolysosomes) and multivesicular bodies (secondary or heterolysosomes) at the lateral extremes of the Golgi cisterns." It might be argued, however, that this observation is somewhat peculiar (unless very high doses of exogenous protein are introduced into the blood stream), since the choroidal epithelium is also exposed to extravasated protein in normal (nonexperimental) situations. An increase in acid phosphatase activity of the choroidal epithelial cells could not be established with certainty in a study with intravenous administration of HRP, whereas some increase apparently occurred after ventriculocisternal perfusion of a large dose of HRP (van Deurs, 1978a; see also Section IV,D).

In a study of the choroid plexus in *Necturus,* Carpenter (1966) found that the barrier to intravenously injected Thorotrast was the epithelial basement membrane. This may reflect that Thorotrast is a considerably larger molecule than HRP and cytochrome c, and thus easily becomes "trapped" in the epithelial basement membrane. However, HRP (Brightman, 1967, 1975), cytochrome c (Milhorat *et al.*, 1973; Davis and Milhorat, 1975), and MP (Reese *et al.*, 1971; van Deurs, 1978b) penetrate the epithelial basement membrane and the intercellular clefts of the choroidal epithelium from the connective tissue to the apical tight junctions, which effectively prevent further diffusion into the CSF, and thus represent the blood–CSF barrier to lipid-insoluble macromolecules.

The significance of epithelial endocytic uptake of extravasated macromolecules in the blood–CSF barrier seems unclear, although this phenomenon has often been considered of importance for the barrier (Brightman, 1967, 1975; Milhorat et al., 1973; Davis and Milhorat, 1975; Milhorat, 1976). It must be emphasized that vesicles with extravasated material have never been seen to discharge their content at the apical (ventricular) surface of the choroidal epithelium (Brightman, 1967, 1975; Davis and Milhorat, 1975), although tracer-labeled vesicles are occasionally found very close to the apical surface (Brightman, 1975). Thus, vesicular blood-to-CSF transport of macromolecules apparently does not take place.

Using lanthanum, Brightman and Reese (1969) showed that this tracer may penetrate one or two "fusion points" of the choroidal epithelial tight junction, but never the entire junction. Brightman and Reese (1969) suggested that "some tight junctions were not fused but, instead, perforated by a narrow pore." (see Section IV,C,2). However, ionic lanthanum has been reported to pass through the junctions (Castel et al., 1974; Bouldin and Krigman, 1975). The blood–CSF barrier to macromolecules localized at the epithelial tight junctions may therefore be leaky to ions.

The importance of the epithelial tight junctions as a barrier for macromolecules has also been established by ventriculocisternal administration of tracers (Fig. 21; see also Section IV,D). Whereas tracers (ferritin, HRP,MP) readily penetrate the ependymal lining of the cerebral ventricles following a paracellular route and reach the extracellular space of the neuropil (Figs. 34 and 35) (Brightman, 1965a,b; Brightman and Reese, 1969; Brightman et al., 1970b; van Deurs, 1977, 1978b), they never penetrate the tight junctions of the choroid plexus epithelium in "normal" animals (Fig. 21) (Brightman, 1967, 1975; van Deurs 1978a,b; van Deurs et al., 1978). The tight junctions also constitute a barrier to intraventricularly perfused 5-hydroxydopamine visualized at the electron microscopic level by a cytochemical procedure (Richards, 1978). However, in experiments with ventriculocisternal perfusion of hypertonic urea or sucrose, an opening of the choroidal epithelial tight junctions to HRP has been reported (Bouchaud and Bouvier, 1978).

With respect to the barrier properties of the choroid plexus, there may exist a "functional leak" as proposed by Brightman et al. (1970b) and more recently discussed by van Deurs et al. (1978). The barrier between blood and CSF, i.e., the choroidal epithelial tight junctions, may be "circumvented" if, for instance, blood-borne protein extravasates at the base of the plexus, diffuses for a rather short distance into the neuropil, and here readily penetrates the ependyma to reach the CSF in the ventricles. Unfortunately, no clear-cut evidence exists in favor or disfavor of the "functional leak." Obviously, however, this "functional leak" cannot be a leakage in the brain barrier system to any significant degree, but rather a "disturbing" factor in experimentation.

2. Freeze-Fracture

Besides constituting a barrier to macromolecules between blood and brain, the choroid plexus epithelium also functions in the isotonic secretion of the CSF (Wright, 1972, 1975; Quinton et al., 1973; Welch and Araki, 1975; Milhorat, 1976; see also Davson, 1967, 1972; Cserr, 1971; Rapoport, 1976b). Based on light microscopic autoradiographic and physiological experiments, a ouabain-sensitive Na^+,K^+-ATPase is believed to be localized at the apical epithelial surface pumping ions into the CSF (Wright, 1972, 1975; Quinton et al., 1973). In contrast to this, Milhorat et al. (1975a) reported, in an electron microscopic and cytochemical study with the nitrophenylphosphatase technique, that the Na^+,K^+-ATPase was present not at the apical, but at the basolateral choroidal epithelial membrane. Whether this confusing difference in the localization of the ATPase is due to the various techniques remains unclear (DiBona and Mills, 1979). A luminal (apical) localization of the Na-pump is highly remarkable; in principle, all other transporting (absorptive, as well as secretory) epithelia exhibit a basolateral localization (see DiBona and Mills, 1979).

In addition to the Na^+,K^+-pump, a paracellular pathway for passive ion permeation most likely exists between the epithelial cells, i.e., via the tight junctions (Wright, 1972, 1975). Electron microscopic support for this is found in the lanthanum studies of Castel et al. (1974) and Bouldin and Krigman (1975). It has been shown in electrophysiological studies of the choroid plexus in the frog (Wright, 1972, 1975) and cat (Welch and Araki, 1975) that the choroidal epithelium has a relatively high conductance, 5.4 millimho/cm^2 and 6.3 millimho/cm^2, respectively. The choroidal epithelium can, therefore, be classified as "leaky" (cf. Frömter and Diamond, 1972; Diamond, 1974, 1978; Ussing et al., 1974). Claude and Goodenough (1973) have correlated the tightness of various epithelia (based on physiological measurements) with the number of tight junctional strands seen in freeze-fracture replicas. In leaky epithelia, the number of strands (mean values) is about 1–4, and in tight epithelia, about 5–10. Brightman et al. (1975a,b) described the tight junctions of the choroid plexus epithelium of mice in freeze-fracture preparations as a system of interconnected ridges with discontinuities of various widths as seen on the P-face (Figs. 28 and 29). This architecture of the tight junction is different from that seen in most other (tighter) epithelia (Friend and Gilula, 1972; Claude and Goodenough, 1973; Staehelin, 1974), e.g., the small intestine, where the junctional ridges or strands appear as almost continuous and frequently interconnected fibrils on the P-face and where the few discontinuities in the P-face ridges can be accounted for by complementary particles in E-face grooves (van Deurs and Koehler, 1979).

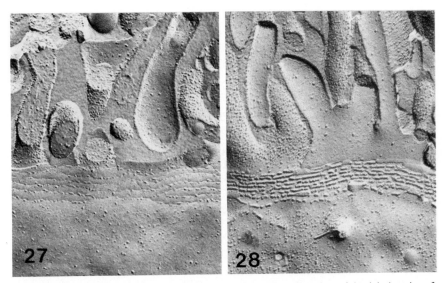

FIGS. 27 AND 28. Freeze-fracture micrographs of an E-face and a P-face of the tight junction of the rat choroidal epithelium. On the E-face (Fig. 27), grooves with numerous particles are seen, and on the P-face (Fig. 28) discontinuous parallel "fibrils" are located. ×45,000.

On the E-face of the choroidal epithelial cells, Brightman *et al.* (1975a) demonstrated (by use of single side replicas) grooves with particles (Figs. 27 and 29). They pointed out that these particles might represent displaced elements from the P-face ridges, whose discontinuities may thus be artifacts produced during tissue preparation (fracturing), or the discontinuities in the ridges could conceivably represent actual "pores" (Brightman and Reese, 1969) in the junction, being a possible structural basis for the paracellular pathway of ions in the CSF secretion.

In recent freeze–fracture studies of the rat choroidal epithelium, a mean number of junctional strands of about 7.5 was found (with statistically nonsignificant differences between two rat strains, and between the choroid plexus obtained from the lateral, third, and fourth ventricles) (van Deurs, 1978d; van Deurs and Koehler, 1978, 1979). This is a surprisingly high number of junctional strands in a leaky epithelium (Claude and Goodenough, 1973), but other examples of an apparent lack of correlation between number of strands and the transepithelial permeability have been reported (Martinez-Palomo and Erlij, 1975; Møllgård *et al.*, 1976; Mazurkiewicz *et al.*, 1977). The discrepancy between number of tight junctional strands and permeability properties of the choroidal epithelium of the rat further stressed the relevance of elucidating the possible presence of "pores" in the strands suggested by Brightman *et al.* (1975a,b). Analysis of complementary (double) replicas revealed that about 43% of the P-face ridges were discontinuous. After correction for "displaced" particles in

FIG. 29.   Freeze–fracture micrograph showing a transition between E- and P-faces of the rat choroid plexus epithelial junction. Note the many particles in the E-face grooves and the numerous discontinuities in the P-face ridges. ×130,000.

the complementary E-face grooves, however, it was found that 24% of the junctional strands were "actual" discontinuities (van Deurs and Koehler, 1979). In contrast, the frequency of discontinuities in the small intestinal tight junction was about 5% (van Deurs and Koehler, 1979). Another possible explanation of the discontinuities in the strands of the choroidal epithelial tight junction was also given: the discontinuities might simply represent material lost during the fracturing (see also Brightman *et al.*, 1975a). In this connection, certain problems in relating the freeze–fracture appearance of tight junctions to the *in vivo* function should be mentioned (see also Section II,C, and Fig. 3B and C). Almost all material used for freeze–fracture has been fixed in glutaraldehyde, and glutaraldehyde seems to produce the characteristic tight junction fibrils by cross-linking "globular" proteins, whereas the strands in unfixed (noncryoprotected) tissue appear as particle chains (van Deurs and Luft, 1979). Moreover, satisfactory matching of complementary replicas only seems possible when using glutaraldehyde-fixed tissue (van Deurs and Koehler, 1979). Another problem in the interpretation of freeze–fracture replicas of tight junctions is, that the fracture reveals structures within the cell membrane, and not between the adjacent cells where the possible paracellular pathway is localized. If the tight junction is represented by a single fibril (or particle chain) between the adjacent cells as suggested by Wade and Karnovsky (1974), a discontinuity in the fibril as seen in a freeze–fracture replica most likely corresponds to a paracellular "pore." However, if the tight junction is composed of two fibrils (or particle chains), one in each membrane (Bullivant, 1978), a discontinuity in a replica does not necessarily influence the permeability of the paracellular pathway. These examples emphasize that replicas (of tight junctions) should be evaluated with care if they are to be correlated with physiological data.

As described for other developing tissue (Schneeberger *et al.*, 1978) and also for human and sheep fetal choroidal epithelium, the number of junctional strands is relatively low (Bohr and Møllgård, 1974; Møllgård *et al.*, 1976). However, a mean of approximately 3.5 tight junctional strands was found in both early (40 days of gestation) and late (125 days of gestation) sheep fetuses, suggesting that a barrier was already developed in the early fetus (Møllgård *et al.*, 1976; Møllgård and Saunders, 1977). Moreover, recent studies of complementary replicas showed that the P-face fibrils are continuous, that is, gaps in the fibrils can be closed by particles in the complementary E-face grooves (Møllgård *et al.*, 1979). In order to explain an apparently unrestricted diffusion of nonelectrolytes, as well as albumin from blood to CSF in the early fetus, and the decrease with fetal development in CSF content of nonelectrolytes and protein (see also Saunders, 1977), Møllgård and Saunders (1975, 1977) and Møllgård *et al.* (1976) suggested the presence of a tubular system of ER-cisterns in the early fetal choroidal epithelium functioning as a transcellular pathway (circumventing the tight junctions). The concept of a tubular system (for sodium transport) has later been extended to include also the adult sheep choroidal epithelium, as well as other transporting epithelia (Møllgård and Rostgård, 1978).

The tight junction of the choroid plexus epithelium of adult rats is of three different geometric types (van Deurs and Koehler, 1978, 1979). Type 1 comprises parallel strands with very few interconnections (55–60% of the total length of the junction) (Figs. 27 and 28), Type 2 comprises relatively parallel strands with frequent interconnections (35–40%) (Fig. 29), and Type 3 comprises complicated, often wavy systems (less than 5%), which may represent a "com-

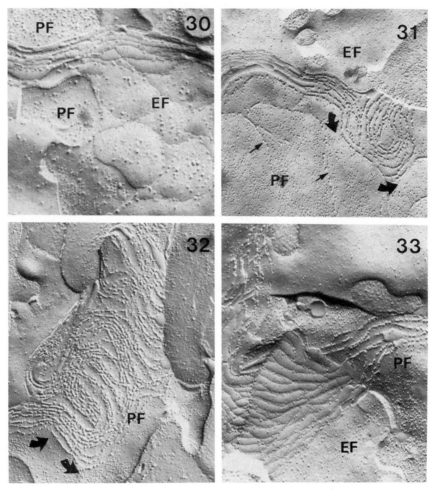

FIGS. 30–33. Variations in the architecture of the tight junctions of the rat choroidal epithelium as revealed by freeze–fracture. EF, E-face; PF, P-face. In Fig. 30, numerous "loops" of junctional strands at the abluminal aspect of the junction are seen. Figures 31 and 32 show a "wavy" organization of the junctional band continuous with bands of usual organization (curved arrows), and Fig. 31 also shows some blindly ending strands (small arrows). Figure 33 shows how the number of junctional strands increases (by bifurcation) where more than two cells are adjacent. Fig. 30, ×76,000; Figs. 31–33, ×43,000.

pressed'' version of Types 1 and 2 (Figs. 31 and 32). Transitions from one type to another occur. Similar ''stretch-unstretch transformations'' of tight junctions have been described in other epithelia (Hull and Staehelin, 1976) and may, in the case of the choroid plexus epithelium, reflect changes in the epithelial surface in relation to fluid transport. Swollen or dilated segments of the intercellular spaces of the choroidal epithelium have been described in relation to various functional stages of the transporting epithelium (Burgess and Segal, 1970; Schultz et al., 1977; Santolaya and Roderiguez Echandia, 1968). Blindly ending strands and loops formed by strands are sometimes seen at the abluminal aspect of the tight junctional band (Figs. 30 and 31). Bouvier and Bouchaud (1978), also studying the rat choroidal epithelium in freeze-fracture, reported that the tight junction, in most cases, consisted of bands of at least five strands. They also found, in fewer regions, a tight junction architecture without any precise orientation of the strands, which formed ''meshworks.'' Therefore, Bouvier and Bouchaud (1978) suggested the presence of two distinctly different forms of tight junctions in the choroid plexus epithelium. However, it is not unlikely that the tight junctional ''meshwork'' in fact corresponds to the Type 3 geometry described earlier, and thus is a structural modulation rather than a different tight-junctional form.

Characteristically, where more than two choroidal epithelial cells are adjacent, the junctional depth and the number of strands increase considerably along the lateral epithelial surface in the apical-basal direction (Brightman, 1977; Bouvier and Bouchaud, 1978). This type of architecture has been shown to be typical of other epithelia (Friend and Gilula, 1972; Staehelin, 1973, 1974) and probably represents additional support at ''weak points'' in the transepithelial barrier. The increase in number of strands at these intersections of more than two cells is sometimes associated with a rather perfect bifurcation of the strands (Fig. 33).

Gap junctions in association with the tight junctional strands are present in many tissues (see, for example, Decker and Friend, 1974; Elias and Friend, 1976; van Deurs, 1979). This has also been reported in the choroidal epithelium of mice (McNutt and Weinstein, 1973; Brightman et al., 1975a), whereas gap junctions could not be established in a study of the rat choroidal epithelium (Bouvier and Bouchaud, 1978). Examining complementary replicas of the choroid plexus epithelium of rats, van Deurs and Koehler (1978) found that, although more or less densely packed particle clusters were sometimes seen in association with the tight junctional strands, the clusters could not be identified as gap junctions, because no groups of pits were present on the complementary fracture face. This does not exclude the presence of gap junctions in the choroidal epithelium of other species. Thus, gap junctions are present in the tight junctional meshes of the cat and also of the fetal chick choroidal epithelium (Dermietzel et al., 1977; Sandri et al., 1977).

Desmosomes have been reported in the choroid plexus epithelium of *Necturus* (Carpenter, 1966).

Since the tight junction in the choroid plexus epithelium, at least in rats, is the only junctional type between adjacent cells (Fig. 21) (van Deurs and Koehler, 1979), the following functions of the tight junction should be considered: (1) leakiness to ions in relation to CSF secretion, (2) tightness to larger molecules for maintenance of the blood–CSF barrier, and (3) adherence of adjacent epithelial cells in all functional states of the epithelium.

Relatively few species have been examined by freeze–fracture to study the nature of the choroidal epithelial tight junctions. While the junctional structure in the cat choroid plexus (Sandri *et al.*, 1977) resembles that of the rat, in chick embryos and sheep fetuses almost continuous tight junctional strands have been demonstrated (Dermietzel *et al.*, 1977; Møllgård *et al.*, 1979), which suggests that generalizations from species to species, or from immature to mature individuals, may be unwarranted.

In experiments with ventriculocisternal perfusion of hypertonic urea and sucrose, Bouchaud and Bouvier (1978) showed a change in the structure of the rat choroidal epithelial tight junctions. An increase in number of discontinuities in the P-face "ridges" was characteristic. Also, Bouchaud and Bouvier (1978) found an increased number of particles in the E-face grooves, but did not use complementary replicas in order to match P-face and E-face material to confirm the increased discontinuity in the strands as compared to normal animals. Swellings of the intercellular spaces corresponding to findings of Wade *et al.* (1973) in experiments with toad urinary bladder exposed to urea were also reported (Bouchaud and Bouvier, 1978). Other experimental studies support the view that discontinuities in the junctional strands may underlie an increased permeability of epithelia (Goodenough and Gilula, 1974; Brightman *et al.*, 1975a; Humbert *et al.*, 1975). An increased hydrostatic pressure within the ventricular system (e.g., in the case of hydrocephalus, cf. Milhorat *et al.*, 1970) may also be considered as a factor causing opening of the epithelial tight junctions in the choroid plexus.

D. Epithelial Endocytosis and Lysosomal Sequestration of Molecules from CSF

While it is well established that ventriculocisternally injected or perfused macromolecules do not penetrate the choroidal epithelial tight junctions (Brightman, 1967, 1975; Becker and Almazon, 1968; van Deurs, 1978a,b; van Deurs *et al.*, 1978), there has been some controversy in the literature as to whether the epithelial cells are able to endocytose the tracers. Small molecules such as iodide (Becker, 1961) and thiocyanate (Welch, 1962) are known to be removed from the CSF by the plexus.

Endocytosis of protein from the apical surface of a polarized epithelium such as the choroidal would be reasonable to expect. Numerous examples have been described, e.g., in the vas deferens (Friend and Farquhar, 1967), trachea

(Richardson *et al.*, 1976), seminal vesicles (Mata and David-Ferreira, 1973; Mata, 1976), yolk sac (Seibel, 1974; Moxon *et al.*, 1976; King, 1977), intestine (Graney, 1968; Cornell *et al.*, 1971; Walker *et al.*, 1972; Orlic and Lev, 1973; Rodewald, 1973), and kidney proximal tubules (Maunsbach, 1966a,b; Graham and Karnovsky, 1966; Graham and Kellermeyer, 1968; Anderson, 1972). However, according to Brightman (1967, 1975) and Bouchaud and Bouvier (1978), only very little endocytosis of tracers takes place at the apical surface of the choroidal epithelium. As stated by Brightman (1975) "when HRP is perfused intraventricularly, only a few apical vesicles pinocytose this protein compared to the many engulfing it from the perivascular space." Ferritin-labeling of apical multivesicular bodies was, however, shown in some micrographs (Brightman, 1971, 1975). Carpenter *et al.* (1967), in experiments with ventricular administration of autologous blood, described hemosiderin granules in the choroidal epithelial cells, but were uncertain whether this represented uptake from the CSF. In contrast, Tennyson and Pappas (1961) clearly demonstrated micropinocytosis of ventriculocisternally perfused Thorotrast in the choroid plexus epithelium. Also, a distinct microcytosis of HRP and MP has been observed at the ventricular surface of the choroidal epithelium (Becker and Almazon, 1968; van Deurs, 1976a, 1978a,b; van Deurs *et al.*, 1978). Ruthenium red labeling of micropinocytic vesicles "budding" from the apical epithelial surface has been reported by Becker and Sutton (1975). This discrepancy with respect to apical endocytosis from the CSF reported in various studies is difficult to explain. It does not appear to be a species-dependent phenomenon.

The most comprehensive microscopic studies of choroidal uptake of macromolecular material from the CSF are those of van Deurs (1978a,b) and van Deurs *et al.* (1978), which will be discussed now. Two principal aspects of endocytosis from the CSF into the rat choroidal epithelium will be dealt with: transport to lysosomes (van Deurs, 1978a,b) (this section) and the possibility of transepithelial vesicular transport (van Deurs *et al.*, 1978) (Section IV,E).

After ventriculocisternal perfusion of HRP and MP, a distinct time-dependent uptake of the tracers was demonstrated in the choroidal epithelium, followed by sequestration within lysosomes. The endocytic uptake of tracer molecules was exclusively by means of coated micropinocytic vesicles, 70–120 nm in diameter (Friend and Farquhar, 1967), formed at the apical epithelial surface membrane between the microvilli (Figs. 36 and 38). The micropinocytic vesicles "discharge" their contents into larger (0.2–0.8 μm) *apical* or *endocytic vacuoles* (Fig. 38). This may happen as fusion of micropinocytic vesicles with each other, and/or by fusion of micropinocytic vesicles with preexisting apical vacuoles. In "control" experiments, with rats not exposed to tracers, micropinocytic vesicles and apical vacuoles (and other components of the vacuolar or lysosomal apparatus) are also present in the choroidal epithelial cytoplasm. The next step in the sequence of events following endocytosis is somewhat unclear, but appar-

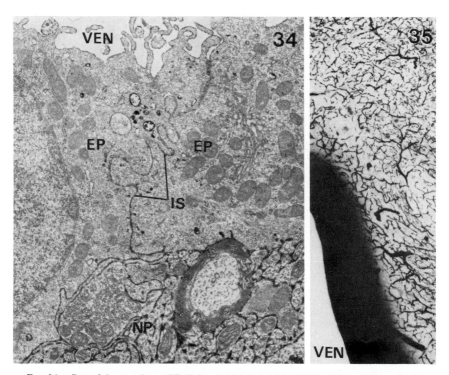

FIG. 34.   Part of the ependyma (EP) lining the third ventricle (VEN) of the rat. The animal has been perfused ventriculocisternally for 15 minutes with MP and thereafter fixed by ventriculocisternal perfusion of aldehydes. The tracer is seen in the intercellular space (IS) between the epithelial cells of the ependyma (EP) and in the extracellular space of the neuropil (NP). ×14,000.

FIG. 35.   Light microscopic picture of a 120 μm-section of a situation comparable with that in Fig. 34, except that HRP instead of MP has been perfused for 15 minutes. Here the penetration of tracer from the third ventricle (VEN) into the neuropil to a depth of approximately 300 μm is obvious. The fact that all blood vessels in this micrograph are "reacted" can be ascribed to the ventriculocisternal perfusion fixation: red blood cells are left in the vessels and react with the diaminobenzidine-H_2O_2 medium during incubation. ×50.

FIGS. 36–38.   Sequence in the choroidal epithelial uptake of macromolecules following ventriculocisternal perfusion of HRP or MP for various times. Fig. 36: After only 5 minutes of tracer perfusion (HRP), labeling of some micropinocytic vesicles (MV) is seen. Fig. 38: After 30 minutes of perfusion (HRP), an additional labeling of apical or endocytic vacuoles (EV) is obvious. Fig. 37: After 60 minutes of perfusion (with MP), multivesicular bodies (MVB) are frequent. GO, Golgi complex. Fig. 36, ×14,000; Fig. 37, ×25,000; Fig. 38, ×30,000.

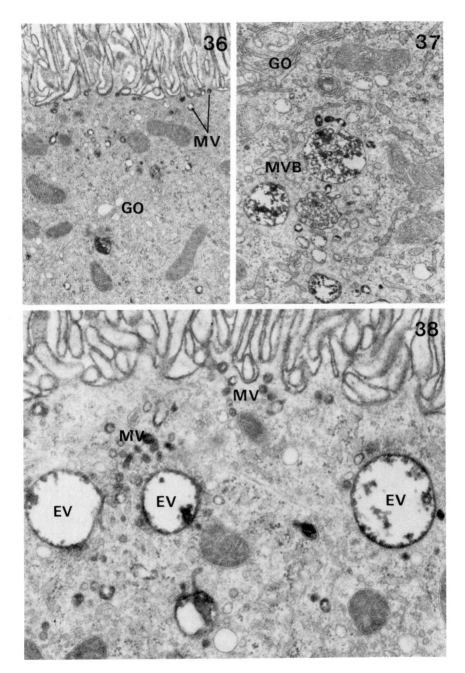

FIGS. 36–38.

ently the apical vacuoles, characterized by tracer labeling mainly or exclusively on the inner aspect of the limiting membrane, develop into *multivesicular bodies* (Fig. 37). It seems likely that these multivesicular bodies are formed by invagination and budding off of parts of the vacuolar membrane, although other explanations may also be given (Hirsch *et al.,* 1968; Friend, 1969). This may be indicative of a kind of membrane regulation to prevent the apical vacoules from growing larger. The size of the apical vacuoles never exceeds a certain limit (approximately 0.8 μm). With increased time, the density of tracer labeling in the matrix of the multivesicular bodies increases, as does the number of small vesicles. Thus, it is possible to distinguish between "light multivesicular bodies," with a few vesicles and relatively little tracer labeling, and "dark multivesicular bodies," with many vesicles and a heavy labeling. Such light and dark multivesicular bodies (0.4-0.7 μm) were also described in material without preceding tracer exposure of the choroidal epithelium, by Cancilla *et al.* (1966).

After 60 minutes of ventriculocisternal perfusion with HRP followed by 2.5 hours of perfusion with artificial CSF without HRP, the reaction product is localized predominantly in vacuoles with a dense, homogeneous content, the so-called *dense bodies* (Fig. 39) (0.2-0.7 μm diameter). Some of these still have the character of (very) dark multivesicular bodies. A few light multivesicular bodies and endocytic vacuoles may also be seen. It is characteristic that these events take place mainly in the apical cytoplasm. Especially endocytic vacuoles and multivesicular bodies are found here, although they may also be seen in other regions of the cell. Dense bodies could be seen throughout the cytoplasm, but mainly in a perinuclear position (see also Becker and Sutton, 1975).

Before going further, it is convenient to look at the electron microscopic localization of acid phosphatase activity. In agreement with the observations of Davis and Milhorat (1975), positive reaction for this hydrolytic enzyme, which is the classic marker enzyme for lysosomes, was demonstrated in Golgi complexes of the choroidal ephithelial cells (Fig. 40). Typically, two central, among 4 to 6, Golgi cisterns were reacted, whereas those on both sides did not exhibit any reaction. Small vesicles close to the Golgi complexes also exhibited some reaction. Dense bodies were mostly very acid phosphatase-positive, although some variation could be seen. Also, at the light microscopic level, acid phosphatase positive bodies, 0.2-0.8 μm in diameter, have been described in the choroid plexus epithelium (Becker and Sutton, 1963, 1975; Cancilla *et al.,* 1966). On the other hand, reaction was demonstrated only in rather few multivesicular bodies, typically in those with many vesicles (Fig. 40) (the dark type). These acid phosphatase-positive multivesicular bodies sometimes contained nonreacted "droplets" in the matrix similar to those seen in a number of dense bodies (Fig. 40). This suggests that the light multivesicular bodies are heterophagosomes, whereas the dark ones are (early) secondary lysosomes. This view is somewhat different from that presented by Milhorat (1976). Based on these results, and on

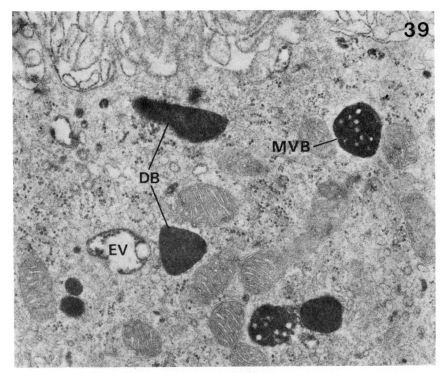

FIG. 39.   After 60 minutes of HRP-perfusion followed by 150 minutes of perfusion with artificial CSF, especially tracer-labeled dense bodies (DB) and heavily labeled multivesicular bodies (MVB) are present in the choroidal epithelium. ×30,000.

those outlined earlier with the labeling sequence of vesicles and vacuoles, the following conclusion is reasonable: the Golgi complexes are the site of synthesis of hydrolytic (lysosomal) enzymes, and small vesicles derived from Golgi cisterns may represent primary lysosomes. These primary lysosomes add hydrolytic enzymes to apical (endocytic) vacuoles, or to the light multivesicular bodies, resulting in the formation of secondary lysosomes (digestive vacuoles) represented by the dense bodies or the dark multivesicular bodies. In this respect, the multivesicular bodies thus represent a heterogeneous group of organelles. This interpretation of the function of the lysosomal apparatus of the choroidal epithelial cells is, in principle, in agreement with that described in several other tissues (de Duve and Wattiaux, 1966; Maunsbach, 1969; Beck *et al.*, 1972). The sequence of events described earlier following ventriculocisternal perfusion of tracers corresponds to that after intravenous injection (van Deurs, 1978a). It will be noticed that after micropinocytosis following intravenous administration of HRP or cytochrome c, many of the vacuoles, multivesicular bodies, and dense

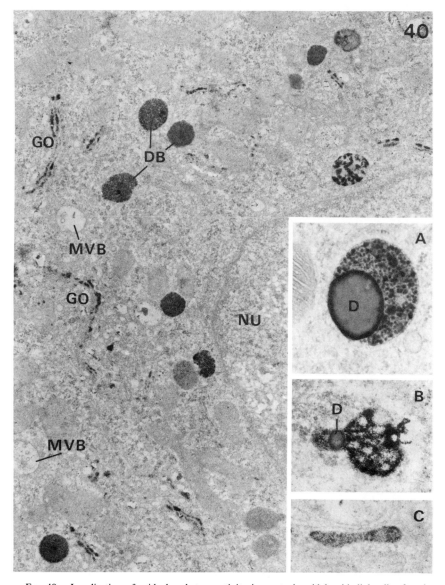

FIG. 40. Localization of acid phosphatase activity in a rat choroidal epithelial cell, after the plexus has been exposed to HRP perfused ventriculocisternally for 60 minutes, followed by 150 minutes of perfusion with artificial CSF before fixation. Reaction product is seen in Golgi complexes (GO) and dense bodies (DB) (compare with Fig. 39). The multivesicular bodies (MVB) are not reacted. NU, Nucleus. ×18,000. Insets A–C show a reacted dense body with a "droplet" (D), a reacted multivesicular body with a similar droplet (D), and a reacted elongated dense body. ×43,000. (From van Deurs, 1978a.)

bodies are also present in the apical half of the cell (see, for example, Brightman, 1967, Figs. 1 and 6; Milhorat, 1976, Fig. 17). However, the uptake of tracers from the apical epithelial surface in general appeared to be at least as pronounced as that occurring at the basolateral membrane (van Deurs, 1978a), in contrast to the findings of Brightman (1967, 1975). In material from long-time experiments with ventriculocisternal perfusion of HRP, where the tracer is distinct at the basolateral surface of the epithelium (see Section IV,E), only relatively little uptake from the basolateral epithelial surface was the rule (see, e.g., van Deurs, 1978a, Figs. 3 and 4). Cancilla *et al.* (1966) reported that more coated vesicles were ''formed'' at the basolateral epithelial membranes than at the apical membranes, being indicative of a higher endocytic activity at the vascular pole than at the ventricular one. However, a number of the vesicles seen in contact with the basolateral membrane may not be ''forming,'' but rather in the process of exocytosis (see Section IV,E).

Structures in the choroidal epithelium resembling autophagic vacuoles are seen only infrequently (Cancilla *et al.,* 1966; van Deurs, 1978a). In an electron microscopic study of the choroid plexus of humans from 6 to 75 years of age, Dohrmann and Bucy (1970) found an apparent increase of cytoplasmic ''inclusions'' with age. Whether these represent ''old'' secondary lysosomes derived from heterophagosomes, autophagic vacuoles, or both, is unclear.

In addition to the characteristic tracer labeling of vesicles and vacuoles belonging to the lysosomal apparatus, some tubular structures or cisterns become labeled both after ventriculocisternal and intravenous administration of tracers (Becker *et al.,* 1967; Milhorat *et al.,* 1973; van Deurs, 1978b). These structures do not represent an alternative medium of endocytosis, since they are never surface-connected, in contrast to what is seen, for example, in the kidney proximal tubules (Graham and Karnovsky, 1966; Graham and Kellermeyer, 1968; Anderson, 1972). Becker *et al.* (1967) suggested that the tracer-labeled cisterns in the choroidal epithelium belonged to the endoplasmic reticulum (ER), but according to van Deurs (1978a), their membranes in general appeared thicker than ER membranes (although it is rather difficult to measure the membrane thickness of peroxidase-labeled organelles). Apparent fusion of micropinocytic vesicles with ER cisterns and Golgi elements was suggested by Cancilla *et al.* (1966), but clear-cut evidence for micropinocytic vesicles discharging their content (HRP or MP) into such cisterns was not obtained by van Deurs (1978a,b). Thus, it is difficult to establish the nature of these cisterns. Tracer-labeled cisterns or tubules have been reported in a variety of epithelia (e.g., Graham and Karnovsky, 1966; Cornell *et al.,* 1971; Orlic and Lev, 1973; Rodewald, 1973; Seibel, 1974), but no general features can be established with respect to their structure, formation, or function. It is, however, not unlikely that at least some of the tracer-labeled cisterns in the choroidal epithelium are formed by lateral fusion of micropinocytic vesicles, since some of the cisterns have a distinct, beaded appearance (Milhorat *et al.* (1973; van Deurs, 1978a). According to this interpretation, they represent a morphological alternative to the predominant vacuolar heterophagosomes (apical or endocytic vacuoles). Also, a few elongated, acid phosphatase-positive "dense bodies" may be seen in the epithelial cells (van Deurs, 1978a).

E. Vesicular Transepithelial Transport of Molecules from CSF

In a light microscopic study of isolated choroid plexuses with administration of fluorescent- and [^{131}I]-labeled protein, Smith *et al.* (1964) demonstrated the tracers in the epithelial cells (presumably in lysosomes), as well as in the connective tissue. The latter observation suggests a transepithelial transport of the proteins, from the apical to the basolateral surface of the epithelium (see also Dodge and Fishman, 1970). Physiological work has also suggested an active removal and transport (by pinocytosis) of small molecules from the CSF in the choroidal epithelium (Pappenheimer *et al.*, 1961; Welch and Sadler, 1966). However, the tracers used in electron microscopic work should not be directly compared with those used in physiological work (van Deurs, 1978b).

At the electron microscopic level, some investigators have reported that after ventriculocisternal administration of HRP, the tracer is not only endocytosed by the choroidal epithelium and sequestered in lysosomes, but also transported across the epithelium from the apical to the basolateral surface by means of vesicles "circumventing" the epithelial tight junctions (Becker and Almazon, 1968; van Deurs, 1976a). In contrast, Brightman (1967, 1975) did not observe any transepithelial transport.

Transepithelial transport of macromolecules in vesicles, which must still be considered controversial, has been reported in several epithelia (Bulger and Trump, 1969; Creemers and Jaques, 1971; Cornell *et al.*, 1971; Orlic and Lev, 1973; Rodewald, 1973; Mata 1976; Moxon *et al.*, 1976), whereas it does not take place, for example, in rat vas dererens (Friend and Farquhar, 1967), nor in the rat kidney proximal tubules (Maunsbach, 1966a,b). Opening of the epithelial tight junctions to allow the transepithelial movement of tracers has to be considered (van Deurs *et al.*, 1978). For instance, in a study of the rat jejunal permeability to HRP administered from the luminal surface, Cooper *et al.* (1978), using intraluminal infusion of hypertonic mannitol, demonstrated an increased vesicular uptake of the tracer into the epithelium, as well as tracer penetration, but were uncertain whether this was exclusively due to opening of the tight junctions, or whether vesicles participated.

The possibility of a transepithelial vesicular CSF-to-blood transport of macromolecules in the choroid plexus epithelium has been reinvestigated (van Deurs *et al.*, 1978) with respect to "sources of error" in earlier experiments (Becker and Almazon, 1968; van Deurs, 1976a). It was characteristic in these experiments (van Deurs *et al.*, 1978) that after 15–30 minutes or more of ventriculocisternal perfusion with HRP, the tracer was distinctly present in some epithelial intercellular spaces and in the connective tissue of the plexus. Horseradish peroxidase was sometimes seen in the choroidal blood vessels, too. Several possible explanations for this location of ventriculocisternally perfused HRP must be considered.

1. It could be speculated that the epithelial tight junctions had been opened during the perfusion experiment, for instance, because of an increased hydrostatic pressure in the ventricular system (see also Bouchaud and Bouvier, 1978). This does not appear likely, however, because the perfusion was performed with a small underpressure (-10 cm of water) in the ventricular system, and reaction product was neither seen in the "pockets" of intercellular spaces nor adsorbed to the membranes between the "fusion points" of the tight junction.

2. It might be argued that the tracer, by some mechanism, had reached the bloodstream and then been carried back to the choroid plexus to extravasate there and thus reach the intercellular spaces. The tracer did actually reach the bloodstream. After 60 minutes of perfusion, HRP was distinct in the lumen of the kidney proximal tubules, and in apical vacuoles of the tubule epithelial cells. Whether the arachnoid villi were collapsed during the ventriculocisternal perfusion (they were supposed to collapse with a negative intraventricular pressure, i.e., a pressure lower than that of the villus veins, cf. Davson, 1967, 1972), or some HRP reached the blood via cisterna magna, the subarachnoid space (where HRP was seen) and the arachnoid villi, is debatable (see also Hochwald and Wallenstein, 1967). It should be mentioned that HRP also reached the pineal gland, probably via the subarachnoid space (M. Møller and B. van Deurs, unpublished). However, both the small intestine and the kidney were examined in the ventriculocisternal perfusion experiments (van Deurs *et al.*, 1978), and only little or no HRP was found outside the fenestrated capillaries of these organs, in contrast to the situation in the choroid plexus. This makes it highly unlikely that the tracer in the connective tissue and the intercellular spaces of the choroid plexus should have been labeled from the blood side.

3. The "functional leak" (see Brightman *et al.*, 1970b; van Deurs *et al.*, 1978, Section IV,C) is of importance in this discussion. Conceivably, some HRP penetrates the ependyma close to the base of the choroid plexus and then diffuses into the plexus connective tissue. It does not seem likely, however, that significant amounts of a large molecule like HRP diffuse out into the villus parts of the plexus within, for example, 30 minutes. If this were true, it also means that large amounts of protein extravasated into the plexus would gain entrance into the CSF by diffusion in the opposite direction. Furthermore, in experiments with the considerably smaller tracer MP perfused ventriculocisternally, MP readily penetrated the ependymal lining and reached the extracellular space of the neuropil,

whereas it was not seen in the choroid plexus except for the most basal parts (van Deurs, 1978b; see also following).

4. Finally, in order to explain the tracer labeling of the intercellular spaces of the choroidal epithelium after ventriculocisternal perfusion of HRP, the possibility of a transepithelial vesicular transport remains to be discussed. After 15–30 minutes or more of perfusion, numerous HRP-labeled micropinocytic vesicles were seen close to the intercellular spaces and sometimes fused with the lateral epithelial membrane. When the "degree" of labeling of the intercellular spaces and the vesicles connected with the lateral membranes is about the same, it is impossible to establish whether the vesicles are involved in the process of exocytosis (discharge of material endotcytosed from the apical surface) or endocytosis (of material in the intercellular spaces). However, in many cases, distinctly labeled vesicles were fused with the lateral epithelial membrane lining intercellular spaces with only little or no HRP. The reaction often appeared only immediately outside the vesicle opening. This situation, illustrated in Figs. 41–46 (see also Section III,E), is difficult to interpret otherwise than that the vesicles are discharging their content by exocytosis. This observation was, therefore, considered to be rather convincing evidence in favor of a transepithelial vesicular transport in the choroidal epithelium, from the CSF to the intercellular spaces, from where the tracer can easily reach the bloodstream (van Deurs et al., 1978). Thus, some vesicles formed at the apical epithelial surface fuse with elements of the lysosomal apparatus, whereas others become involved in a transcellular transport.

In repeating experiments with MP instead of HRP, however, no obvious transepithelial transport from CSF to the intercellular spaces could be established (van Deurs, 1978b). Microperoxidase is only sequestered in lysosomes of the choroidal epithelial cells. This discrepancy between results from HRP and MP experiments may be explained in several ways: (1) MP and HRP may be handled in different ways in the epithelial cells, e.g., because of differences in the isoelectric points (Table I) or other features of the two tracer molecules (Graham et al., 1969; Farquhar, 1978). (2) Transepithelial vesicular transport may require a preceding "overloading" of the lysosomal apparatus, which did not occur in the MP experiments. It is characteristic at least for certain cell types involved in adsorptive pinocytosis (in which the molecules to be endocytosed have to be bound to the cell surface, in contrast to fluid phase or "bulk" endocytosis) that the uptake is easily saturated (Steinman et al., 1974). (3) MP is actually transported across the epithelium, but is impossible to visualize enzymatically, for example, because limited amounts of such a small tracer molecule would diffuse relatively rapidly away. Other explanations may be given, and the above explanations do not exclude one another.

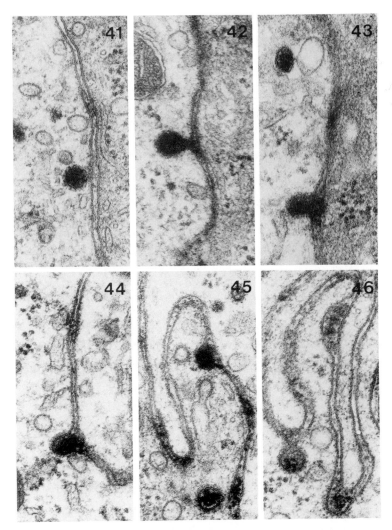

Figs. 41–46.   HRP-labeled, coated micropinocytic vesicles in close contact with or fused with the lateral epithelial membrane of the rat choroid plexus after 60 minutes of ventriculocisternal perfusion of the tracer. The intercellular spaces contain HRP immediately outside the opening vesicles. The micrographs indicate that the vesicles are discharging their contents rather than endocytosing the tracer from the intercellular spaces. ×61,000.(From van Deurs *et al.*, 1978.)

F. Conclusions

The experiments described here with ventriculocisternal perfusion of HRP and MP suggest that the choroid plexus, in addition to its role in CSF secretion and the blood–CSF barrier, is involved in regulation of the composition of the CSF by its capability for active absorption from the ventricular surface. An uptake from the CSF (in case of certain macromolecules followed by lysosomal seques-tration) and a CSF-to-blood transport mechanism might be of special importance for removing metabolites that are synthesized only in the brain and that cannot be removed through the cerebral capillaries. Removal of drugs that have entered the brain should also be considered. It has been shown that penicillin can be actively transported from the CSF to the blood (Walters *et al.*, 1976) and also that several catecholamine transmitter-metabolites are transported through the choroid plexus from CSF to blood by a probenicid-sensitive mechanism (Wolfson *et al.*, 1974). The demonstration of several peptidergic hormones in some of the circumven-tricular organs (Weindl and Sofroniew, 1978) suggests the possibility of secre-tion of hormones to the CSF, and the choroid plexus might be involved in removing such hormones from the fluid. Finally, certain proteins are specifically synthesized and located in the brain, e.g., the S-100 protein, the glial fibrillary acidic protein, and the synaptic-located proteins D1, D2, and D3 (Bock, 1978). During the synaptic turnover, some of the synaptic proteins diffuse to the CSF where they can be demonstrated by immunochemical techniques (Jørgensen and Bock, 1975). The choroid plexus might be of importance in removing such brain-specific proteins from the CSF.

V. Summarizing Comments

This review of ultrastructural studies of the blood–brain and blood–CSF bar-riers related to similar research on noncerebral endothelia and epithelia gives rise to the following summarizing comments.

1. The endothelium of the cerebral microvasculature is, under "normal" conditions, tight to proteins and peptides, or conceivably, a very limited blood-to-brain (and brain-to-blood) transport by means of endothelial vesicles takes place in segments of the microvasculature. This barrier to lipid-insoluble mac-romolecules can be ascribed to the tight junctions between the endothelial cells. Uncertainties exist, however, with respect to the permeability properties of the cerebral endothelial junctions to ions and small lipid-insoluble molecules. So far, freeze–fracture studies of the tight junctions of the cerebral endothelium have provided only little relevant information, and do not explain the apparent dif-ference in tightness between cerebral and noncerebral endothelial junctions.

2. Many experimental or pathological situations, for example, acute hyper-

tension, definitely cause an ''opening'' of the blood–brain barrier. One possible pathway across the cerebral endothelium for larger hydrophilic molecules appears to be vesicular transport comparable with that occurring in normal noncerebral tissue. This may be indicative of the presence of a labile ''pore system'' in the cerebral microvasculature, which can be reversibly opened, but normally is closed. Transendothelial channels may also be responsible for the extravasation. In addition, especially intravenous administration of hypertonic solutions and long-term hypertension (where the arterial blood pressure exceeds 200 mm Hg to a considerable extent) seem to cause an opening of the tight junctions between the endothelial cells. However, it is difficult to draw direct parallels to the ''small pore–large pore'' concept of noncerebral capillary permeability. Among periendothelial structures the basement membrane and phagocytic pericytes may be of some importance in the blood–brain barrier.

3. Extravasation of protein and peptides in the choroidal blood vessels is, in contrast to the situation in cerebral vessels, pronounced under normal conditions, and macromolecules also readily penetrate the connective tissue ''sheaths'' between the choroidal endothelium and epithelium, to reach the epithelial intercellular spaces. Here, the tracers are definitely stopped by the apical tight junctions, which consequently represent the structural basis for the blood–CSF barrier to larger molecules. These junctions may, on the other hand, be leaky to ions, a leakiness that is considered of importance in the CSF secretion of the plexus. Freeze–fracture studies suggest that narrow hydrophilic pores are present in the junctional strands of the choroidal epithelium at least in certain species.

4. Following ventriculocisternal perfusion of protein or peptide tracers, these molecules are endocytosed (by micropinocytosis) at the apical (ventricular) surface of the choroidal epithelium. Thereafter, the tracers are sequestered within a well developed lysosomal apparatus. Whereas the epithelial tight junctions are not penetrated by protein and peptides from the CSF, some transepithelial, vesicular transport from the apical to the basolateral epithelial surface may take place, thus ''circumventing'' the tight junctions. After this transport process, tracers will reach the choroidal blood stream. To what extent this also applies to, for example, homologous proteins and small lipid-insoluble molecules remains uncertain, but the absorptive ability of the choroidal epithelium strongly suggests a role in the regulation of the (macro)molecular composition of the CSF.

These ultrastructural features of brain barriers have been revealed by freeze–fracture (a technique that has its own inherent problems of interpretation) and especially by using thin-sectioned, tracer-injected material. The tracers used (mostly enzymatic macromolecules) are to some extent an ''involuntary'' choice, because of the requirements for electron microscopic visualization, and probably limit the general value of information obtained. It is therefore well warranted to use, to a much greater extent, for example, nonenzymatic tracers of different sizes and electric charges in order to obtain more quantitative and directly comparable results with respect to cellular uptake, membrane turnover, and localization of transendothelial and transepithelial "pores." Also, the use of autoradiography and isotope-labeled homologous proteins, which are well established techniques in other fields of research, would be useful in the study of the choroid plexus in particular. Alternative approaches are also needed to further elucidate the vesicular transport activity in the cerebral endothelium (as well as in noncerebral endothelia) and the possibility of opening of the junctions. Furthermore, additional experimental data are warranted, for instance, concerning in- and outflow amounts of tracers in ventriculocisternal perfusion experiments, and with respect to the importance of hydrostatic and colloidal osmotic pressure in blood and CSF in order to obtain a convergence of ultrastructural and physiological methodology. Finally, more species need to be included in the structural investigations to avoid generalizations. Also, more effort to correlate brain barrier mechanisms in mammals with those of lower vertebrates and invertebrates would be interesting and probably lead to a better understanding of the structural aspects of brain barriers in general.

ACKNOWLEDGMENTS

I wish to thank Dr. J. H. Luft and Dr. M. Møller for their suggestions and criticisms during the preparation of the manuscript. I also thank Miss. K. Pedersen for technical assistance, Mrs. B. Risto for the photographic work, and Mrs. K. Krogh for typing the manuscript.

REFERENCES

Anderson, W. A. (1972). *J. Histochem. Cytochem.* **20,** 672.
Baker, R. N., Cancilla, P. A., Pollock, P. S., and Frommes, S. P. (1971). *J. Neuropathol. Exp. Neurol.* **30,** 668.
Bankston, P. W., and DeBruyn, P. P. H. (1974). *Am. J. Anat.* **141,** 281.
Beck, F., Lloyd, J. B., and Squier, C. A. (1972). *In* "Lysosomes. A Laboratory Handbook" (J. T. Dingle, ed.), p. 200. North-Holland Publ., Amsterdam.
Becker, B. (1961). *Am. J. Physiol.* **201,** 1149.
Becker, N. H., and Almazon, R. (1968). *J. Histochem. Cytochem.* **16,** 278.
Becker, N. H., and Sutton, C. H. (1963). *Am. J. Pathol.* **43,** 1017.
Becker, N. H., and Sutton, C. H. (1975). *In* "The Choroid Plexus in Health and Disease" (M. G. Netsky and S. Shuangshoti, eds.), p. 113. University Press of Virginia, Charlottesville.
Becker, N. H., Novikoff, A. B., and Zimmerman, H. M. (1967). *J. Histochem. Cytochem.* **15,** 160.
Becker, N. H., Hirano, A., and Zimmerman, H. M. (1968). *J. Neuropathol. Exp. Neurol.* **27,** 439.
Beggs, J. L., and Waggener, J. D. (1976). *Lab. Invest.* **34,** 428.
Beggs, J. L., Waggener, J. D., and Miller, W. (1975). *Proc., Electron Microsc. Soc. Am.* **33,** 398.
Bennett, H. S., Luft, J. H., and Hampton, J. C. (1959). *Am. J. Physiol.* **196,** 381.
Bodenheimer, T. S., and Brightman, M. W. (1968). *Am. J. Anat.* **122,** 249.

Bock, E. (1978). *J. Neurochem.* **30**, 7.

Bohr, V., and Møllgård, K. (1974). *Brain Res.* **81**, 314.

Bolwig, T. G., Hertz, M. M., and Westergaard, E. (1977). *Acta Neurol. Scand.* **56**, 335.

Bouchaud, C., and Bouvier, D. (1978). *Tissue & Cell* **10**, 331.

Bouldin, T. W., and Krigman, M. R. (1975). *Brain Res.* **99**, 444.

Bouvier, D., and Bouchaud, C. (1978). *Biol. Cell.* **31**, 109.

Brightman, M. W. (1965a). *J. Cell Biol.* **26**, 99.

Brightman, M. W. (1965b). *Am. J. Anat.* **117**, 193.

Brightman, M. W. (1967). *Prog. Brain Res.* **29**, 19.

Brightman, M. W. (1971). *Tex. Rep. Biol. Med.* **29**, 245.

Brightman, M. W. (1975). *In* "The Choroid Plexus in Health and Disease" (M. G. Netsky and S. Shuangshoti, eds.), p. 86. University Press of Virginia, Charlottesville.

Brightman, M. W. (1977). *Exp. Eye. Res.* **25**, Suppl., 1.

Brightman, M. W., and Reese, T. S. (1969). *J. Cell Biol.* **40**, 648.

Brightman, M. W., and Robinson, J. S. (1976). *In* "Head Injuries. Proceedings of the Second Chicago Symposium on Neural Trauma" (R. L. McLaurin, ed.), p. 107. Grune & Stratton, New York.

Brightman, M. W., Klatzo, I., Olsson, Y., and Reese, T. S. (1970a). *J. Neurol. Sci.* **10**, 215.

Brightman, M. W., Reese, T. S., and Feder, N. (1970b). *In* "Capillary Permeability" (C. Crone and N. A. Lassen, eds.), p. 468. Munksgaard, Copenhagen.

Brightman, M. W., Reese, T. S., Vick, N. A., and Bigner, D. D. (1971). *J. Neuropathol. Exp. Neurol.* **30**, 139 (abstr.).

Brightman, M. W., Hori, M., Rapoport, S. I., Reese, T. S., and Westergaard, E. (1973). *J. Comp. Neurol.* **152**, 317.

Brightman, M. W., Prescott, L., and Reese, T. S. (1975a). *Brain-Endocr. Interact.* **2**, 146.

Brightman, M. W., Shivers, R. R., and Prescott, L. (1975b). *In* "Fluid Environment of the Brain" (H. F. Cserr, J. D. Fenstermacher, and V. Fencl, eds.), p. 3. Academic Press, New York.

Bruns, R. R., and Palade, G. E. (1968a). *J. Cell Biol.* **37**, 244.

Bruns, R. R., and Palade, G. E. (1968b). *J. Cell Biol.* **37**, 277.

Bulger, R. E., and Trump, B. F. (1969). *J. Morphol.* **127**, 205.

Bullivant, S. (1978). *Electron Microsc., Proc. Int. Congr., 9th, 1978* Vol. III, p. 659.

Bundgaard, M., Cserr, H., and Murray, M. (1979). *Cell Tissue Res.* **198**, 65.

Burgess, A., and Segal, N. B. (1970). *J. Physiol. (London)* **208**, 88P.

Cancilla, P. A., Zimmerman, H. M., and Becker, N. H. (1966). *Acta Neuropathol.* **6**, 188.

Cancilla, P. A., Baker, R. N., Pollock, P. S., and Frommes, S. P. (1972). *Lab. Invest.* **26**, 376.

Carpenter, S. J. (1966). *J. Comp. Neurol.* **127**, 413.

Carpenter, S. J., McCarthy, L. E., and Borison, H. L. (1967). *Neurology* **17**, 993.

Carpenter, S. J., McCarthy, L. E., and Borison, H. L. (1970). *Z. Zellforsch. Mikrosk. Anat.* **110**, 471.

Case, N. M. (1959). *J. Biophys. Biochem. Cytol.* **6**, 527.

Casley-Smith, J. R. (1969). *Experientia* **25**, 845.

Casley-Smith, J. R., and Chin, J. C. (1971). *J. Microsc. (Oxford)* **93**, 167.

Castel, M., Sahar, A., and Erlij, D. (1974). *Brain Res.* **67**, 178.

Clasen, R. A., Pandolfi, S., and Hass, G. M. (1970). *J. Neuropathol. Exp. Neurol.* **29**, 266.

Claude, P., and Goodenough, D. A. (1973). *J. Cell Biol.* **58**, 390.

Clawson, C. C., Hartmann, J. F., and Vernier, R. L. (1966). *J. Comp. Neurol.* **127**, 183.

Clementi, F. (1970). *J. Histochem. Cytochem.* **18**, 887.

Clementi, F., and Palade, G. E. (1969). *J. Cell Biol.* **41**, 33.

Connell, C. J., and Mercer, K. L. (1974). *Am. J. Anat.* **140**, 595.

Cooper, M., Teichberg, S., and Lifshitz, F. (1978). *Lab. Invest.* **38**, 447.

Cornell, R., Walker, W. A., and Isselbacher, K. J. (1971). *Lab. Invest.* **25**, 42.

Cotran, R. S., and Karnovsky, M. J. (1967). *Proc. Soc. Exp. Biol. Med.* **126**, 557.

Creemers, J., and Jacques, P. J. (1971). *Exp. Cell Res.* **67**, 188.

Crone, C. (1973). In "The Inflamatory Process" (B. W. Zweifach, L. Grant, and R. T. McCluskey. eds.), Vol. 2, p. 95 Academic Press, New York.

Crone, C., and Thompson, A. M. (1970). In "Capillary Permeability" (C. Crone and N. A. Lassen, eds.), p. 447. Munksgaard, Copenhagen.

Cserr, H. F. (1971). *Physiol. Rev.* **51**, 273.

Cutler, R. W. P., Devel, R. K., and Barlow, C. F. (1967). *Arch. Neurol. (Chicago)* **17**, 261.

Davis, D. A., and Milhorat, T. H. (1975). *Anat. Rec.* **181**, 779.

Davis, D. A., Lloyd, B. J., and Milhorat, T. H. (1973). *Anat. Rec.* **176**, 443.

Davson, H. (1967). "Physiology of the Cerebrospinal Fluid." Churchill, London.

Davson, H. (1972). In "The Structure and Function of Nervous Tissue" (G. H. Bourne, ed.), Vol. 4, p. 321. Academic Press, New York.

DeBruyn, P. P. H., Michelson, S., and Becker, R. P. (1975). *J. Ultrastruct. Res.* **53**, 133.

Decker, R. S., and Friend, D. S. (1974). *J. Cell Biol.* **62**, 32.

de Duve, D., and Wattiaux, R. (1966). *Annu. Rev. Physiol.* **28**, 435.

Deimann, W., Taugner, R., and Fahimi, H. D. (1976). *J. Histochem. Cytochem.* **24**, 1213.

del Cerro, M. (1974). *J. Comp. Neurol.* **157**, 245.

Delorme, P., Gayet, J., and Grignon, G. (1975). *Cell Tissue Res.* **157**, 535.

Dempsey, E. W., and Wislocki, G. B. (1955). *J. Biophys. Biochem. Cytol.* **1**, 245.

Dermietzel, R. (1975a). *Cell Tissue Res.* **164**, 45.

Dermietzel, R. (1975b). *Cell Tissue Res.* **164**, 309.

Dermietzel, R. and Schünke, D. (1975). *Am. J. Anat.* **143**, 131.

Dermietzel, R., Meller, K., Tetzlaff, W., and Waelsch, M. (1977). *Cell Tissue Res.* **181**, 427.

Diamond, J. M. (1974). *Fed. Proc., Fed. Am. Soc. Exp. Biol.* **33**, 2220.

Diamond, J. M. (1978). *Fed. Proc., Fed. Am. Soc. Exp. Biol.* **37**, 2639.

DiBona, D. R., and Mills, J. W. (1979). *Fed. Proc., Fed. Am. Soc. Exp. Biol.* **38**, 134.

Dodge, P. R., and Fishman, M. A. (1970). *N. Engl. J. Med.* **283**, 316.

Dohrmann, G. J. (1970). *Brain Res.* **18**, 197.

Dohrmann, G. J., and Bucy, P. C. (1970). *J. Neurosurg.* **33**, 506.

Dretzki, J. (1971). *Z. Anat. Entwicklungsgesch.* **134**, 278.

Ehrlich, P. (1887). *Ther. Monatsh.* **1**, 88.

Elias, P. M., and Friend, D. S. (1976). *J. Cell Biol.* **68**, 173.

Eto, T., Omae, T., and Yamamoto, T. (1971). *Arch. Histol. Jpn.* **33**, 133.

Eto, T., Yamamoto, T., and Omae, T. (1975). *Arch. Histol. Jpn.* **38**, 299.

Farquhar, M. G. (1978). *J. Cell Biol.* **78**, R 35.

Feder, N. (1970). *J. Histochem. Cytochem.* **18**, 911.

Feder, N. (1971). *J. Cell Biol.* **51**, 339.

Feder, N., Reese, T. S., and Brightman, M. W. (1969). *J. Cell Biol.* **43**, 35a (abstr.).

Florey, H. W. (1964). *Q. J. Exp. Physiol. Cogn. Med. Sci.* **49**, 117.

Ford, D. H. (1976). *Rev. Neurosci.* **2**, 1.

Friend, D. S. (1969). *J. Cell Biol.* **41**, 269.

Friend, D. S., and Farquhar, M. G. (1967). *J. Cell Biol.* **35**, 357.

Friend, D. S., and Gilula, N. B. (1972). *J. Cell Biol.* **53**, 758.

Frömter, E., and Diamond, J. M. (1972). *Nature (London)* **235**, 9.

Garlick, D. G., and Renkin, E. M. (1970). *Am. J. Physiol.* **219**, 1595.

Giacomelli, F., Wiener, J., and Spiro, D. (1970). *Am. J. Pathol.* **59**, 133.

Giacomelli, F., Rooney, J., and Wiener, J. (1978). *Exp. Mol. Pathol.* **28**, 309.

Gilula, N. B. (1977). In "International Cell Biology" (B. R. Brinkley and K. R. Porter, eds.), p. 61. Rockefeller Univ. Press, New York.

Goldmann, E. E. (1909). *Beitr. Klin. Chir.* **64**, 192.
Goldmann, E. E. (1913). *Abh. Preuss. Akad. Wiss., Phys.-Math. Kl.* **1**, 1.
Gonzales-Aguilar, F. (1969). *J. Ultrastruct. Res.* **29**, 76.
Goodenough, D. A., and Gilula, N. B. (1974). *J. Cell Biol.* **61**, 575.
Graham, R. C., and Karnovsky, M. J. (1966). *J. Histochem. Cytochem.* **14**, 291.
Graham, R. C., and Kellermeyer, R. W. (1968). *J. Histochem. Cytochem.* **16**, 275.
Graham, R. C., Limpert, S., and Kellermeyer, R. W. (1969). *Lab. Invest.* **20**, 298.
Graney, D. O. (1968). *Am. J. Anat.* **123**, 227.
Green, H. S., and Casley-Smith, J. R. (1972). *J. Theor. Biol.* **35**, 103.
Grotte, G. (1956). *Acta Chir. Scand., Suppl.* **211**, 1.
Hager, H. (1961). *Acta Neuropathol.* **1**, 9.
Hansson, H.-A., Johansson, B., and Blomstrand, C. (1975). *Acta Neuropathol.* **32**, 187.
Hashimoto, P. H. (1972). *Am. J. Anat.* **134**, 41.
Hashimoto, P. H., and Hama, K. (1968). *Med. J. Osaka Univ.* **18**, 331.
Hicks, J. T., Albrecht, P., and Rapoport, S. I. (1976). *Exp. Neurol.* **53**, 768.
Hirano, A., and Matsui, T. (1975). *Hum. Pathol.* **6**, 611.
Hirano, A., and Zimmerman, H. M. (1972). *Lab. Invest.* **26**, 465.
Hirano, A., Becker, N. H., and Zimmerman, H. M. (1969). *Arch. Neurol. (Chicago)* **20**, 300.
Hirano, A., Becker, N. H., and Zimmerman, H. M. (1970a). *J. Neurol. Sci.* **10**, 205.
Hirano, A., Dembitzer, H. M., Becker, N. H., Levine, S., and Zimmerman, H. M. (1970b). *J. Neuropathol. Exp. Neurol.* **29**, 432.
Hirano, A., Llena, J. F., and Chung, H. D. (1975). *Acta Neuropathol.* **32**, 175.
Hirsch, J. G., Fedorko, M. E., and Cohn, Z. A. (1968). *J. Cell Biol.* **38**, 629.
Hochwald, G. M., and Wallenstein, M. (1967). *Am. J. Physiol.* **212**, 1199.
Hull, B. E., and Staehelin, L. A. (1976). *J. Cell Biol.* **68**, 688.
Humbert, F., Grandchamp, A., Pricam, C., Perrelet, A., and Orci, L. (1975). *J. Cell Biol.* **69**, 90.
Jacobs, J. M., MacFarlane, R. M., and Caranagh, J. B. (1976). *J. Neurol. Sci.* **29**, 95.
Johansson, B., Li, C.-L., Olsson, Y., and Klatzo, I. (1970). *Acta Neuropathol.* **16**, 117.
Johansson, B. R. (1978). *Microvasc. Res.* **16**, 362.
Joó, F. (1968). *Nature (London)* **219**, 1378.
Joó, F. (1971). *Br. J. Exp. Pathol.* **52**, 646.
Joó, F. (1972). *Experientia* **28**, 1470.
Joó, F., and Csillik, B. (1966). *Exp. Brain Res.* **1**, 147.
Joó, F., Várkonyi, T., and Csillik, B. (1967). *Histochemie* **9**, 140.
Joó, F., Rakonczay, Z., and Wollemann, M. (1975). *Experientia* **31**, 584.
Jørgensen, O. S., and Bock, E. (1975). *Scand. J. Immunol.* **4**, Suppl. 2, 25.
Karlsson, U., and Schultz, R. L. (1965). *J. Ultrastruct. Res.* **12**, 160.
Karnovsky, M. J. (1967). *J. Cell Biol.* **35**, 213.
Karnovsky, M. J. (1970). *In* "Capillary Permeability" (C. Crone and N. A. Lassen, eds.), p. 341. Academic Press, New York.
Karnovsky, M. J., and Shea, S. M. (1970). *Microvasc. Res.* **2**, 353.
King, B. F. (1977). *Am. J. Anat.* **148**, 447.
Klatzo, I., Miquel, J., and Otenasek, R. (1962). *Acta Neuropathol.* **2**, 144.
Klatzo, I., Wiśniewski, H., and Smith, D. E. (1965). *Prog. Brain Res.* **15**, 73.
Kristensson, K., and Olsson, Y. (1973). *Acta Neurol. Scand.* **49**, 189.
Lampert, P., and Carpenter, S. (1965). *J. Neuropathol. Exp. Neurol.* **24**, 11.
Landis, E. M., and Pappenheimer, J. R. (1963). *In* "Handbook of Physiology" Sect. 2, Vol. II, p. 961. Washington, D. C.
Lane, N. J. (1978). *Electron Microsc., Proc. Int. Congr., 9th, 1978* Vol. III, p. 673.
Laursen, H., and Westergaard, E. (1977). *Neuropathol. Appl. Neurobiol.* **3**, 29.
Lewandowsky, M. (1900). *Z. Klin. Med.* **40**, 480.

Long, D. M. (1970). *J. Neurosurg.* **32,** 127.

Lorenzo, A. V., Shirahige, I., Liang, M., and Barlow, C. F. (1972). *Am. J. Physiol.* **233,** 268.

Lorenzo, A. V., Hedley-Whyte, E. T., Eisenberg, H. M., and Hsu, D. W. (1975). *Brain Res.* **88,** 136.

Luft, J. H. (1972). *Proc., Electron Microsc. Soc. Am.* **30,** 132.

Luft, J. H. (1973). In "The Inflamatory Process" (B. W. Zweifach, L. Grant, and R. T. McCluskey, eds.), Vol. 2, p. 47. Academic Press, New York.

McNutt, N. S., and Weinstein, R. S. (1973). *Prog. Biophys. Mol. Biol.* **26,** 47.

Majno, G. (1965). In "Handbook of Physiology" (W. F. Hamilton and P. Dow, eds.), Sect. 2, Vol. III, p. 2293. Am. Physiol. Soc., Washington, D.C.

Majno, G., and Palade, G. E. (1961). *J. Biophys. Biochem. Cytol.* **11,** 571.

Majno, G., Palade, G. E., and Schoefl, G. I. (1961). *J. Biophys. Biochem. Cytol.* **11,** 607.

Manz, H. J., and Robertson, D. M. (1972). *Am. J. Pathol.* **66,** 565.

Marchesi, V. T., and Barrnett, R. J. (1963). *J. Cell Biol.* **17,** 547.

Martinez-Palomo, A., and Erlij, D. (1973). *Pfluegers Arch.* **343,** 267.

Martinez-Palomo, A., and Erlij, D. (1975). *Proc. Natl. Acad. Sci. U.S.A.* **72,** 4487.

Mata, L. R. (1976). *J. Microsc. Biol. Cell.* **25,** 127.

Mata, L. R., and David-Ferreira, J. F. (1973). *J. Microsc.* **17,** 103.

Maunsbach, A. B. (1966a). *J. Ultrastruct. Res.* **15,** 197.

Maunsbach, A. B. (1966b). *J. Ultrastruct. Res.* **16,** 1.

Maunsbach, A. B. (1969). In "Lysosomes in Biology and Pathology" (I. Neuberger and E. L. Tatum, eds.), p. 115. North-Holland Publ., Amsterdam.

Maxwell, D. S., and Pease, D. C. (1956). *J. Biophys. Biochem. Cytol.* **2,** 467.

Mayerson, H. S., Wolfram, C. G., Shirley, H. H., and Wasserman, K. (1960). *Am. J. Physiol.* **198,** 155.

Maynard, E. A., Schultz, R. L., and Pease, D. C. (1957). *Am. J. Anat.* **100,** 409.

Mazurkiewicz, J. E., Addis, J. S., and Barrnett, R. J. (1977). *J. Cell Biol.* **75,** 71a (abstr.).

Milhorat, T. H. (1976). *Int. Rev. Cytol.* **47,** 255.

Milhorat, T. H., Mosher, M. B., Hammock, M. K., and Murphy, C. F. (1970). *N. Engl. J. Med.* **283,** 286.

Milhorat, T. H., Davis, D. A., and Lloyd, B. J. (1973). *Science* **180,** 76.

Milhorat, T. H., Davis, D. A., and Hammock, M. K. (1975a). *Brain Res.* **99,** 170.

Milhorat, T. H., Davis, D. A., and Hammock, M. K. (1975b). *J. Neurosurg.* **42,** 315.

Millen, J. W., and Rogers, G. E. (1956). *J. Biophys. Biochem. Cytol.* **2,** 407.

Møller, M., van Deurs, B., and Westergaard, E. (1978). *Cell Tissue Res.* **195,** 1.

Møllgård, K., and Rostgård, J. (1978). *J. Membr. Biol.* **40,** 71.

Møllgård, K., and Saunders, N. R. (1975). *J. Neurocytol.* **4,** 453.

Møllgård, K., and Saunders, N. R. (1977). *Proc. R. Soc. London, Ser. B* **199,** 321.

Møllgård, K., Malinowska, D. H., and Saunders, N. R. (1976). *Nature (London)* **264,** 293.

Møllgård, K., Lauritzen, B., and Saunders, N. R. (1979). *J. Neurocytol.* **8,** 139.

Moxon, L. A., Wild, A. E., and Slade, B. S. (1976). *Cell Tissue Rev.* **171,** 175.

Nabeshima, S., and Reese, T. S. (1972). *J. Neuropathol. Exp. Neurol.* **31,** 176.

Nabeshima, S., Reese, T. S., Landis, D. M. D., and Brightman, M. W. (1976). *J. Comp. Neurol.* **164,** 127.

Nag, S., Robertson, D. M., and Dinsdale, H. B. (1977). *Lab. Invest.* **36,** 150.

Nag, S., Robertson, D. M., and Dinsdale, H. B. (1979). *Acta Neuropathol.* **46,** 107.

Nørtved, J., and Westergaard, E. (1974). *Anat. Rec.* **178,** 428 (abstr.).

Olsson, Y. (1968). *Acta Neuropathol.* **10,** 26.

Olsson, Y., and Hossmann, K.-A. (1970). *Acta Neuropathol.* **16,** 103.

Olsson, Y., and Reese, T. S. (1969). *Anat. Rec.* **163,** 319.

Olsson, Y., Klatzo, I., and Carsten, A. (1975). *Neuropathol. Appl. Neurobiol.* **1**, 59.
Orlic, D., and Lev., R. (1973). *J. Cell Biol.* **56**, 106.
Palade, G. E. (1953). *J. Appl. Physiol.* **24**, 1424.
Palade, G. E., and Bruns, R. R. (1968). *J. Cell Biol.* **37**, 633.
Palay, S. L., McGee-Russell, S. M., Gordon, S., and Grillo, M. (1962). *J. Cell Biol.* **12**, 385.
Pappas, G. D. (1970). *J. Neurol. Sci.* **10**, 241.
Pappas, G. D., and Tennyson, V. M. (1962). *J. Cell Biol.* **15**, 227.
Pappenheimer, J. R. (1953). *Physiol. Rev.* **33**, 387.
Pappenheimer, J. R., Renkin, E. M., and Borrero, L. M. (1951). *Am. J. Physiol.* **167**, 13.
Pappenheimer, J. R., Heisey, S. R., and Jordan, E. F. (1961). *Am. J. Physiol.* **200**, 1.
Persson, L. I., Johansson, B. B., and Hansson, H.-A. (1978). *Acta Neuropathol.* **44**, 53.
Petito, C. K., Schaefer, J. A., and Plum, F. (1977). *Brain Res.* **127**, 251.
Plattner, H., Wachter, E., and Gröbner, P. (1977). *Histochemistry* **53**, 223.
Povlishock, J. T., Becker, D. P., Sullivan, H. G., and Miller, J. D. (1978). *Brain Res.* **153**, 223.
Quinton, P. M., Wright, E. M., and Tormey, J. McD. (1973). *J. Cell Biol.* **58**, 724.
Raimondi, A. J., Evans, J. P., and Mullan, S. (1962). *Acta Neuropathol.* **2**, 177.
Rapoport, S. I. (1970). *Am. J. Physiol.* **219**, 270.
Rapoport, S. I. (1976a). *Exp. Neurol.* **52**, 467.
Rapoport, S. I. (1976b). "Blood-Brain Barrier in Physiology and Medicine." Raven Press, New York.
Rapoport, S. I., Hori, M., and Klatzo, I. (1971). *Science* **173**, 1026.
Rapoport, S. I., Hori, M., and Klatzo, I. (1972). *Am. J. Physiol.* **223**, 323.
Rawson, R. A. (1943). *Am. J. Physiol.* **138**, 708.
Reese, T. S., and Karnovsky, M. J. (1967). *J. Cell Biol.* **34**, 207.
Reese, T. S., Feder, N., and Brightman, M. W. (1971). *J. Neuropathol. Exp. Neurol.* **30**, 137 (abstr.).
Renkin, E. M. (1978). *Microvasc. Res.* **15**, 123.
Reyners, H., Gianfelici de Reyners, E., Jadin, J. M., and Maisin, J. R. (1975). *Cell Tissue Res.* **157**, 93.
Richards, J. G. (1978). *J. Neurocytol.* **7**, 61.
Richardson, J., Bouchard, T., and Ferguson, C. C. (1976). *Lab. Invest.* **35**, 307.
Rodwald, R. (1973). *J. Cell Biol.* **58**, 189.
Rubin, B. T. (1977). *J. Theor. Biol.* **64**, 619.
Sandri, C., van Buren, J. M., and Akert, K. (1977). *Prog. Brain Res.* 46,
Santolaya, R. C., and Roderigues Echandia, E. L. (1968). *Z. Zellforsch. Mikrosk. Anat.* **92**, 43.
Sasaki, S., Ferszt, R., and Cervós-Navarro, J. (1977). *Acta Neuropathol.* **40**, 207.
Saunders, N. R. (1977). *Exp. Eye Res.* **25**, Suppl., 523.
Schneeberger, E. E., Walters, D. V., and Olver, R. E. (1978). *J. Cell Sci.* **32**, 307.
Schultz, R. L., and Karlsson, U. (1965). *J. Ultrastruct. Res.* **12**, 187.
Schultz, W. J., Brownfield, M. S., and Kozlowski, G. P. (1977). *Cell Tissue Res.* **178**, 129.
Seibel, W. (1974). *Am. J. Anat.* **140**, 213.
Shea, S. M., and Bossert, W. H. (1973). *Microvasc. Res.* **6**, 305.
Shea, S. M., and Karnovsky, M. J. (1966). *Nature (London)* **212**, 353.
Shea, S. M., Karnovsky, M. J., and Bossert, W. H. (1969). *J. Theor. Biol.* **24**, 30.
Simionescu, M., and Simionescu, N. (1978). *J. Cell Biol.* **79**, 381a (abstr.).
Simionescu, M., Simionescu, N., and Palade, G. E. (1974). *J. Cell Biol.* **60**, 128.
Simionescu, M., Simionescu, N., and Palade, G. E. (1975). *J. Cell Biol.* **67**, 863.
Simionescu, M., Simionescu, N., and Palade, G. E. (1976). *J. Cell Biol.* **68**, 705.
Simionescu, N., Simionescu, M., and Palade, G. E. (1973). *J. Cell Biol.* **57**, 424.
Simionescu, N., Simionescu, M., and Palade, G. E. (1975). *J. Cell Biol.* **64**, 586.

Simionescu, N., Simionescu, M., and Palade, G. E. (1978a). *Microvasc. Res.* **15**, 1.
Simionescu, N., Simionescu, M., and Palade, G. E. (1978b). *Microvasc. Res.* **15**, 17.
Simionescu, N., Simionescu, M., and Palade, G. E. (1978c). *J. Cell Biol.* **79**, 27.
Smith, D. E., Streicher, E., Milković, K., and Klatzo, I. (1964), *Acta Neuropathol.* **3**, 372.
Staehelin, L. A. (1973). *J. Cell Sci.* **13**, 763.
Staehelin, L. A. (1974). *Int. Rev. Cytol.* **39**, 191.
Staehelin, L. A. (1975). *J. Cell Sci.* **18**, 545.
Stehbens, W. E. (1965). *Nature (London)* **207**, 197.
Steinman, R. M., Silver, J. M., and Cohn, Z. A. (1974). *J. Cell Biol.* **63**, 949.
Stern, L., and Gautier, R. (1921), *Arch. Int. Physiol.* **17**, 138.
Straus, W. (1958). *J. Biophys. Biochem. Cytol.* **4**, 541.
Straus, W. (1977). *J. Histochem. Cytochem.* **25**, 215.
Sturrock, R. R. (1978). *Neuropathol. Appl. Neurobiol.* **4**, 307.
Tani, E., and Evans, J. P. (1965). *Acta Neuropathol.* **4**, 507.
Tani, E., Ibeda, K., Kudo, S., Yamagata, S., Higashi, N., and Fujihara, E. (1974). *J. Neurosurg.* **40**, 696.
Tani, E., Yamagata, S., and Ito, Y. (1977). *Cell Tissue Res.* **176**, 157.
Tennyson, V. M., and Pappas, G. D. (1961). *In* "Disorders of the Developing Nervous System" (W. S. Fields and M. M. Desmond, eds.), p. 267. Thomas, Springfield, Illinois.
Tennyson, V. M., and Pappas, G. D. (1964). *J. Comp. Neurol.* **123**, 379.
Thompson, A. M. (1970). *In* "Capillary Permeability" (C. Crone and N. A. Lassen, eds.), p. 459. Munksgaard, Copenhagen.
Tomlin, S. G. (1969). *Biochim. Biophys. Acta* **183**, 559.
Torack, R. M., and Barrnett, R. J. (1964). *J. Neuropathol. Exp. Neurol.* **23**, 46.
Torack, R. M., Besen, M., and Becker, N. H. (1961). *Neurology* **11**, 71.
Tschirgi, R. D. (1950). *Am. J. Physiol.* **163**, 756P.
Ussing, H. H., Erlij, D., and Lassen, U. (1974). *Annu. Rev. Physiol.* **36**, 17.
van Breemen, V. L., and Clemente, C. D. (1955). *J. Biophys. Biochem. Cytol.* **1**, 161.
van Deurs, B. (1976a). *J. Ultrastruct. Res.* **55**, 400.
van Deurs, B. (1976b). *J. Ultrastruct. Res.* **56**, 65.
van Deurs, B. (1977). *Brain Res.* **124**, 1.
van Deurs, B. (1978a). *J. Ultrastruct. Res.* **62**, 155.
van Deurs, B. (1978b). *J. Ultrastruct. Res.* **62**, 168.
van Deurs, B. (1978c). *Microvasc. Res.* **16**, 280.
van Deurs, B. (1978d). *Anat. Rec.* **190**, 569.
van Deurs, B. (1979). *Anat. Rec.* **195**, 73.
van Deurs, B., and Amtorp, O. (1978). *Neuroscience* **3**, 737.
van Deurs, B., and Koehler, J. K. (1978). *Electron Microsc., Proc. Int. Congr., 9th 1978* Vol. II, p. 336.
van Deurs, B., and Koehler, J. K. (1979). *J. Cell Biol.* **80**, 662.
van Deurs, B., and Luft, J. H. (1979). *J. Ultrastruct. Res.* **68**, 160.
van Deurs, B., Møller, M., and Amtorp, O. (1978). *Cell Tissue Res.* **187**, 215.
van Harreveld, A., and Khattab, F. I. (1968). *J. Cell Sci.* **3**, 579.
Várkonyi, T., and Joó, F. (1968). *Experientia* **24**, 452.
Vegge, T., and Haye, R. (1977). *Histochemistry* **53**, 217.
Vick, N. A., and Bigner, D. D. (1972). *J. Neurol. Sci.* **17**, 29.
Wade, J. B., and Karnovsky, M. J. (1974). *J. Cell Biol.* **60**, 168.
Wade, J. B., Revel, J. P., and DiScala, V. A. (1973). *Am. J. Physiol.* **224**, 407.
Wagner, H.-J., Pilgrim, C., and Brandt, J. (1974). *Acta Neuropathol.* **27**, 299.
Walker, W. A., Cornell, R., Davenport, L.M., and Isselbacher, K. J. (1972). *J. Cell Biol.* **54**, 195.

Walters, I. N., Teychenne, P. F., Claveria, L. E., and Calne, D. B. (1976). *Neurology* **26**, 1008.
Ware, R. A., Chang, L. W., and Burkholder, P. M. (1974). *Acta Neuropathol.* **30**, 211.
Weindl, A., and Joynt, R. J. (1973). *Arch. Neurol.* (Chicago) **29**, 16.
Weindl, A., and Sofroneiw, M. V. (1978). *In* "Treatment of Pituitary Adenomas" (R. Fahlbusch and K. v. Werder, eds.), p. 10. Thieme, Stuttgart.
Welch, K. (1962). *Proc. Soc. Exp. Biol. Med.* **109**, 953.
Welch, K., and Araki, H. (1975). *In* "Fluid Environment of the Brain" (H. F. Cserr, J. D. Fenstermacher, and V. Fencl. eds.), p. 157. Academic Press, New York.
Welch, K., and Sadler, K. (1966). *Am. J. Physiol.* **210**, 652.
Westergaard, E. (1975a). *Acta Neuropathol.* **32**, 27.
Westergaard, E. (1975b). *J. Ultrastruct. Res.* **50**, 383 (abstr.).
Westergaard, E. (1977). *Acta Neuropathol.* **39**, 181.
Westergaard, E. (1978). *J. Neural Transm. Suppl.* **14**, 9.
Westergaard, E., and Brightman, M. W. (1973). *J. Comp. Neurol.* **152**, 17.
Westergaard, E., Go, G. Klatzo, I., and Spatz, M. (1976). *Acta Neuropathol.* **35**, 307.
Westergaard, E., van Deurs, B., and Brøndsted, H. E. (1977). *Acta Neuropathol.* **37**, 141.
Williams, M. C., and Wissig, S. L. (1975). *J. Cell Biol.* **66**, 531.
Williams, S. K., and Wagner, R. C. (1978). *J. Cell Biol.* **79**, 382a (abstr.).
Williams, T. H., and Jew, J. Y. (1975). *Tissue & Cell* **7**, 407.
Winther, F. Ø. (1971). *Z. Zellforsch. Mikrosk. Anat.* **114**, 193.
Wislocki, G. B., and Ladman, A. J. (1958). *In* "Cerebrospinal Fluid. Production, Circulation and Regulation" (G. E. W. Wolstenholme and C. M. O'Connor, eds.), p. 55. Churchill, London.
Wislocki, G. B., and Leduc, E. H. (1952). *J. Comp. Neurol.* **96**, 371.
Wisse, E. (1972). *J. Ultrastruct. Res.* **38**, 528.
Wissig, S. L., and Williams, M. C. (1978). *J. Cell Biol.* **76**, 341.
Witzel, E. W., and Hunt, G. M. (1962). *J. Neuropathol. Exp. Neurol.* **21**, 250.
Wolff, J. (1963). *Z. Zellforsch. Mikrosk. Anat.* **60**, 409.
Wolfson, L. I., Katzman, R., and Escriva, A. (1974). *Neurology* **24**, 772.
Wright, E. M. (1972). *J. Physiol.* (London) **226**, 545.
Wright, E. M. (1975). *In* "Fluid Environment of the Brain" (H. F. Cserr, J. D. Fenstermacher, and V. Frencl, eds.), p. 139. Academic Press, New York.
Yamamoto, K., Fujimoto, S., and Takeshige, Y. (1976). *J. Ultrastruct. Res.* **54**, 22.
Yee, A. G., and Revel, J. P. (1975). *J. Cell Biol.* **66**, 200.

Note Added in Proof. Well-developed endothelial tight junctions (see p. 125) have recently been described in a freeze-fracture study of the blood–brain barrier in a reptile [Shivers, R. R. (1979). *Brain Res.* **169**, 221].

Experimental opening of tight junctions in the rat cerebral endothelium to HRP (see pp. 133–137) has been reported by Nagy and co-workers [Nagy, Z., Pappius, H. M., Mathieson, G., and Hüttner, I. (1979). *J. Comp. Neurol.* **185**, 569 and Nagy, Z., Mathieson, G., and Hüttner, I. (1979). *J. Comp. Neurol.* **185**, 579].

Reversible opening of the blood–brain barrier to enzymes by infusion of arabinose solutions (see pp. 133–137) has been discussed by Barranger *et al.* [Barranger, J. A., Rapoport, S. I., Fredericks, W. R., Pentchev, P. G., MacDermot, K. D., Steusing, J. K., and Brady, R. O. (1979). *Proc. Natl. Acad. Sci. U.S.A.* **76**, 481].

Immunochemistry of Cytoplasmic Contractile Proteins

UTE GRÖSCHEL-STEWART

*Institut für Zoologie der Technischen Hochschule, Darmstadt,
Federal Republic of Germany*

True philosophers, who are only eager for truth and knowledge, never regard themselves as already so thoroughly informed, but that they welcome further information from whomsoever and from wheresoever it may come; nor are they so narrow minded as to imagine any of the arts or sciences transmitted to us by the ancients in such a state of forwardness or completeness that nothing is left for the ingenuity and industry of others. On the contrary, very many maintain that all we know is still infinitely less than all that still remains unknown.

WILLIAM HARVEY, 1628.

193

I. Introduction

The pioneering work dealing with contractile phenomena in cells other than muscle dates back to the early 1950s, and we find that Loewy (1952) was the first to extract actomyosin-like protein from the slime mold *Physarum polycephalum*. At about the same time, Hoffmann-Berling (1953) convincingly demonstrated that glycerol-extracted fibroblasts and sarcoma cells contract after the addition of adenosine triphosphate (ATP). He later isolated and characterized the contractile proteins from these cells and found great similarities to their counterparts in true muscle cells (Hoffmann-Berling and Weber, 1955). The isolation of thrombosthenin, the actomyosin complex from blood platelets, by Bettex-Galland and Lüscher (1959), should also be considered an important event in the history of nonmuscle cell contractility. Since then, contractile proteins have been detected in many and various cells and tissues (even plant and prokaryotic cells), and several comprehensive reviews have been written describing their biochemical properties (Pollard and Weihing, 1974; Goldman *et al.*, 1976a; Clarke and Spudich, 1977; Korn, 1978). In 1973, Bray suggested that these proteins from nonmuscular tissue be called "cytoplasmic", a term still kept, but far too broad to describe contractile proteins found in protozoa, as well as in complex tissues of vertebrates. I will try to suggest a more appropriate subgrouping that will relate not only to origin but also to function.

A great advancement in the characterization of small amounts of contractile proteins (e.g., from single cells) was made when the analysis of proteins and protein subunits on polyacrylamide gels in the presence of sodium dodecyl sulfate (SDS–PAGE) was introduced in 1967 by Shapiro *et al.* This allowed molecular weight (MW) estimations, and thus, the comparison with the better defined analogous proteins from myofibrils. Subsequently, Ishikawa's ingenious method of decorating actin fibrils with heavy meromyosin (HMM) led to intracellular localization and even demonstrated the polarity of the filaments (Ishikawa *et al.*, 1969). The relationship of the other contractile proteins, such as myosin, tropomyosin, and α-actinin, to these actin filaments (microfilaments) was largely unknown until about five years later when immunological methods of localization were introduced, and though this technique was originally met with skepticism (e.g., Pollard and Weihing, 1974), it soon became an almost indispensible tool in cell biology studies (see Goldman *et al.*, 1976a). It is noteworthy to mention here that the first papers describing the immunogenicity of actin and myosin appeared in Hungary as early as 1949 (Kesztyüs *et al.*, 1949). Extensive immunological investigations on the distribution of striated muscle myosin and their relationship to myogenesis were made by Finck, Holtzer, and Pepe (for review, see Pepe, 1968). Later, Rubinstein *et al.* (1974) reported that antibodies to "adult-type" contractile proteins would not crossreact with embryonic cells (fibroblasts, chondroblasts, and presumptive myoblasts), even though myosin and actin could be shown to be present by other analytical methods. These findings clearly asked for antibodies directed to contractile proteins from nondifferentiated muscle (or even smooth muscle, as certain structural similarities are present); however, these studies lagged behind due to the difficulties encountered in the purification of these proteins.

In the following pages, I will begin with a short and mostly descriptive section on the biochemistry of those contractile proteins used as immunogens, and since there is a flowing transition in both morphology and function from smooth muscle to single contractile cells (Benninghoff, 1926), I will, in this and the subsequent sections on the immunology, of necessity, include smooth muscle proteins. In further sections, the effect of the antibodies on contractile proteins *in vitro* and *in vivo* will be discussed, as well as the localization of contractile elements in cells and tissues by immunohistochemical methods.

I have tried to accumulate all pertinent literature, but I will apologize in advance to all who feel they were unduly left out. It was unintentional.

II. Biochemistry of Cytoplasmic Contractile Proteins

A. ACTIN

Five to ten percent of the total cellular protein seems to be actin (Clarke and Spudich, 1977). The cytoplasmic actins from various sources closely resemble each other, as well as the actins from vertebrate smooth and striated muscle; this suggests that actin has been highly conserved during evolution. As a matter of interest, Ogievetskaya's (1977) calculations indicate an extremely low evolutionary rate; at most, three amino acid changes have accumulated per hundred residues in 100 million years. Actins are 42,000 dalton monomers (G-actin) built up of 374–375 amino acid residues. Being of identical size, all actins coelectrophorese in SDS–PAGE. Despite these similarities of actins, some chemical differences have recently been described. First, differences in the fingerprints of CNBr-fragmented (Booyse *et al.*, 1973) or tryptically digested (Gruenstein and Rich, 1975) actins of muscle or cytoplasmic origin were noted. Also, actins from various sources were shown to be distinguishable on the basis of their isoelectric points. The most acidic forms were isolated from *Physarum polycephalum* (Zechel and Weber, 1978) and *Dictyostelium discoideum* (Uyemura *et al.*, 1978). They migrate ahead of the α-form of vertebrate skeletal and cardiac actin. Two less acidic forms are found in smooth muscle and nonmuscle cells, named β and γ (Whalen *et al.*, 1976; Storti and Rich, 1976; Storti *et al.*, 1976; Garrels and Gibson, 1976; Rubenstein and Spudich, 1977; Landon *et al.*, 1977). β-Actin alone is found in human erythrocyte membranes (Pinder *et al.*, 1978a), in the muscles of some annelids, molluscs, and arthropods (Schachat *et al.*, 1977; de Couet *et al.*, 1980), and in the liver of *Torpedo marmorata* (Zechel and Weber, 1978). The proportions of β- and γ-actin in nonmuscle cells vary from an approximate β:γ = 6:1 in thrombocytes and fibroblasts to about 1:1 in nervous tissue such as brain, synaptosomes, and cultured neurons (Landon *et al.*, 1977; Choo and Bray, 1978; Marotta *et al.*, 1978). In smooth muscle of chicken gizzard, γ-actin is the predominant form (according to Zechel and Weber, 1978, it is a γ-like form). *Acanthamoeba* actin has the most alkaline isoelectric point; it is called δ (Gordon *et al.*, 1977). The minor and short-lived forms have not been considered here. Whereas isoelectric focusing is an interesting means of "typing" actins in a variety of tissues from one individual or from closely related species, amino acid sequence studies are more relevant when it comes to establishing phylogenetic relationships between actins (Vandekerckhove and Weber, 1978a,b). For example, the "near α-type" actin from *Physarum* and the δ-type actin from *Acanthamoeba* are, by the available sequence data (Vandekerckhove and Weber, 1978c; Elzinga, cited in Korn, 1978), rather similar; yet both differ significantly from the α-actin of rabbit skeletal muscle, and less from the mammalian β,γ-cytoplasmic actins. In addition, isoelectric focusing will neither distinguish between cardiac and skeletal actins, which differ in two amino acid residues (Collins and Elzinga, 1975; Lu and Elzinga, 1976), nor will it indicate the closer structural relationship of smooth and striated muscle actins as compared to smooth muscle and nonmuscle types of actin (Elzinga and Kolega, 1978). The information gained by isoelectric focusing, as well as by sequence analysis, indicates that actins in a higher mammal are coded by at least six different genes (Vandekerckhove and Weber, 1978d).

B. Actin-Associated Proteins

In contrast to muscle tissue, cytoplasmic actins are not always organized into filaments (F-actin). It is estimated that between 40–70% of the cellular actin can exist in the monomeric or oligomeric form, e.g., in platelets (Probst and Lüscher, 1972; Abramowitz *et al.*, 1975), in invertebrate sperm (Tilney, 1975), and in fibroblasts and brain (Bray and Thomas, 1976). One might assume that this nonfilamentous actin represents a storage form that would only polymerize into filaments when needed for physiological functions. Factors have indeed been isolated recently that will keep actin in a stable monomeric state, whereas others will influence its polymerization and aggregation. Possibly, there is a physiological turnover of filaments in polymerization and depolymerization cycles.

1. *Factors That Keep Actin in the Monomeric Form*

a. *DNase I.* It has been known for many years that this 31,000 dalton pancreatic enzyme can be inhibited by a protein factor present in many tissues. In

1974, this factor was identified as actin by Lazarides and Lindberg. G-actin and DNase I form a stable 1:1 complex, which inactivates the enzyme. DNase I also depolymerizes fibrillar actin. The physiological importance of this association (at least from the actin's point of view) is still unknown. Nevertheless, actin—DNase I complex formation has been useful in estimating the degree of actin polymerization in various cells (Blikstad *et al.*, 1978), for the isolation and purification of cytoplasmic actins (Vandekerckhove and Weber, 1978b) and for the removal of unwanted actin from myosin preparations (Burkl *et al.*, 1979).

b. *Profilin.* Profilin, a 16,000 dalton polypeptide, has been isolated from spleen and other tissue (Carlsson *et al.*, 1976), including platelets (Markey and Lindberg, 1978; Harris and Weeds, 1978); it is, however, not present in muscle or erythrocytes. Profilin forms a stable 1:1 crystalline complex with G-actin.

2. *Factors Promoting Actin Polymerization and Aggregation*

a. *α-Actinin.* α-Actinin was first isolated and characterized by Ebashi and co-workers (Ebashi and Ebashi, 1965; Maruyama and Ebashi, 1965). This protein, consisting of two 100,000 dalton polypeptide chains, has since been isolated in highly purified form from striated muscle (Suzuki *et al.*, 1973; Robson and Zeece, 1973) and from smooth muscle (Fujiwara *et al.*, 1978; Craig and Pardo, 1979; Geiger and Singer, 1979). It is also present in brain tissue (S. Puszkin *et al.*, 1976), in platelets, and other nonmuscle cells (Schollmeyer *et al.*, 1973, 1978). There is a close association of α-actinin to the Z-disks of striated muscle (Stromer and Goll, 1972) and to dense bodies and attachment plaques in smooth muscle and nonmuscle cells (Schollmeyer *et al.*, 1973). These findings and the early observations of Ebashi and co-workers that α-actinin, a rod-shaped molecule (Podlubnaya *et al.*, 1975), stimulates superprecipitation of actomyosin gels and the cross-linking of actin filaments *in vitro,* suggest that the physiological role of α-actinin in nonmuscular cells may be the promotion of microfilament formation and the attachment of such filaments to the membrane (Schollmeyer *et al.*, 1976).

b. *Actin-Binding Protein.* Actin-binding protein, a high-molecular-weight, membrane-associated protein (250,000 dalton subunit) has been isolated from macrophages (Hartwig and Stossel, 1975), polymorphonuclear leukocytes (Boxer and Stossel, 1976), and platelets (Schollmeyer *et al.*, 1978). This factor (possibly in collaboration with α-actinin) induces actin gelation. Under appropriate conditions, this can lead to the formation of a rigid three-dimensional network or of parallel bundles such as observed in platelet pseudopods. There is possibly a structural, but apparently not a functional, relationship of this factor (Brotschi *et al.*, 1978) to filamin.

c. *Filamin.* Filamin, a dimeric protein with a 220,000 dalton subunit, can be isolated from avian and mammalian smooth muscles (Wang *et al.*, 1975;

Shizuta *et al.*, 1976b; Wang, 1977; Wallach *et al.*, 1978). Its presence in certain nonmuscle cells has also been inferred by immunological studies (see Section V). Filamin will bind to F-actin *in vitro* and inhibit its ability to activate myosin Mg^{2+}-ATPase (Davies *et al.*, 1977). The aggregation and gelling of actin filaments by filamin seems to restrict their ability to interact with myosin (Wang and Singer, 1977).

d. *Gel-actins.* Gel-actins are four very potent low-molecular-weight gelation factors (dimers with MWs ranging from 23,000 to 78,000) purified from *Acanthamoeba castellani* (Maruta and Korn, 1977a).

3. *Factors Whose Interaction with Actin Is Still Controversial*

a. *β-Actinin.* β-Actinin, originally described as a monomeric 62,000 dalton protein in striated muscle (Maruyama, 1971), inhibited the interfilamental interaction of F-actin in solution. Heizmann and Häuptle (1977) have recently isolated and characterized a similar 65,000–67,000 dalton protein monomer from chicken striated muscle and, on the basis of immunological data, claimed that an identical factor is also present in nonmuscle tissue. Their preparation was shown to inhibit the polymerization of G-actin and to promote the depolymerization of preformed F-actin filaments *in vitro*. While this work was in press, Maruyama *et al.* (1977) reported, in contradiction to their earlier work, that β-actinin was composed of two subunits with chain weights of 37,000 and 34,000. While still inhibiting the network formation of F-actin, this factor retards depolymerization of F-actin and accelerates polymerization of G-actin, quite in contrast to Heizmann and Häuptle's result. To the best of my knowledge, this recent discrepancy has not yet been resolved.

b. *Spectrin.* Spectrin is a high-molecular-weight protein that appears to be confined to the erythrocyte membrane (Marchesi *et al.*, 1969). It is a dimeric protein consisting of two immunologically distinct subunits of approximately 220,000 MW (Kirkpatrick *et al.*, 1978). Spectrin can interact with erythrocyte actin, and this interaction, which is not fully understood, might influence shape and deformability of the red blood cell. It seems that in the membrane itself, spectrin and actin are present in a nonpolymerized form (Tilney and Detmers, 1975; Pinder *et al.*, 1978a), although *in vitro,* spectrin may cause polymerization of actin under conditions that usually depolymerize F-actin (Pinder *et al.*, 1975; S. Puszkin *et al.*, 1978). Pinder *et al.* (1978a) propose that, in the membrane, spectrin is present as a series of tetramers, linked to each other and to monomers of actin. The phosphorylation of spectrin possibly causes additional spectrin-actin cross-linkings (Pinder *et al.*, 1977).

c. *Others.* A high-molecular-weight protein doublet (250,000 and 230,000) has been isolated from *Thyone* sperm periacrosomal cup, which may complex with nonpolymerized actin, keeping it in a "profilamentous" form (Tilney, 1977). A low-molecular-weight protein may also be involved in this process.

C. MYOSIN

Myosin represents about 0.3-1.5% of the total protein of nonmuscle cells. With the exception of *Acanthamoeba* (see following), all presently known myosins from nonmuscle sources closely resemble smooth muscle myosin in their gross morphology (Elliot *et al.*, 1976) and their molecular architecture: a 450,000 to 500,000 dalton macromolecule composed of two heavy chains (MW around 200,000) and two pairs of light chains (MWs about 20,000 and 17,000 respectively). The presence of three light chain pairs is reported for myosins from brain and cultured sympathetic nerve cells (Burridge, 1976b) and for chicken brain myosin (Kuczmarski and Rosenbaum, 1979a). The light chains are associated with the globular head of the myosin molecule, where actin-binding sites and ATPase activity are localized. The rod portions of the heavy chains are involved in the assembly of myosin molecules into bipolar filaments. While it has generally been assumed that nonmuscle myosins assemble in much shorter filaments (about 300 nm long) than true muscle myosins (Burridge and Bray, 1975; Niederman and Pollard, 1975), it was recently shown that the nonmuscle myosins can also form large filaments (3-5 μm) under appropriate experimental conditions (Hinssen *et al.*, 1978).

Differences are expected in the amino acid composition of smooth muscle and cytoplasmic myosin heavy chains (Burridge, 1974, 1976b; Burridge and Bray, 1975), since both tryptic digestion, as well as chemical cleavage by cyanylation, will produce different peptide patterns. In contrast, the light chains from fibroblast and smooth muscle myosin are reported to be identical (Burridge, 1974, 1976b).

The ATPase activity of cytoplasmic myosins is rather low. As in smooth muscle myosin (Gröschel-Stewart, 1971), the K^+-EDTA-activated ATPase is often higher than the Ca^{2+}-stimulated activity. Pure myosins will not show any significant Mg^{2+}-stimulated ATPase activity after the addition of actin, unless the 20,000 dalton light chain is phosphorylated by Ca^{2+}-dependent myosin light chain kinase. This enzyme is found in smooth muscle (Aksoy *et al.*, 1976; Sobieszek and Small, 1977; Dabrowska *et al.*, 1978) and also in nonmuscle cells such as platelets (Daniel and Adelstein, 1976), astrocytes (Scordilis *et al.*, 1977), platelets and brain (Dabrowska *et al.*, 1978), synaptosomes (Schulman and Greengard, 1978), and baby hamster kidney (BHK 21) cells (Yerna *et al.*, 1979).

As mentioned before, *A. castellani* offers some unusual variations to the myosin pattern (Maruta and Korn, 1977b). Next to a "normal" two-headed myosin (II) with two 170,000 dalton heavy chains and two pairs of light chains (17,500 and 17,000), three single-headed subforms with one heavy chain (130,000 dalton) and two light chains ranging from 27,000 to 14,000 dalton have also been isolated (myosin IA,B,C). The relationship between myosin I and II, which can both be produced in the same cell, is still unknown; however, they seem to be products of different genes and are not derived from each other (Maruta *et al.*, 1978). *Acanthamoeba* (Maruta and Korn, 1977c) and *Dictyostelium* myosin (Rahmsdorf *et al.*, 1978) are phosphorylated on the heavy chains. Maruta *et al.* (1978) showed that the 27,000 and 14,000 dalton light chains could be removed from *Acanthamoeba* myosin IB without altering the specific ATPase activity found in the original enzyme.

D. TROPOMYOSIN

Tropomyosin has also been highly conserved during evolution. Nevertheless, there are distinct differences between the tropomyosins of striated muscle, smooth muscle, and cytoplasmic origin. The muscle proteins are rod-shaped molecules 41 nm in length (MW about 70,000), composed of two α-helical polypeptide chains twisted to form a coiled-coil structure (for review, see Mannherz and Goody, 1976). In skeletal muscle, the two chains (α and β) differ slightly in their amino acid composition and their molecular weight (α, 37,000; β, 33,000), and different muscles vary in their proportions of α and β (Cummins and Perry, 1974). Smooth muscle tropomyosin subunits have electrophoretic mobilities different from those of their striated muscle counterparts. Uterine tropomyosin migrates as a single band (Cummins and Perry, 1974), chicken gizzard tropomyosin migrates as a doublet, with apparent MW of 39,000 and 36,000 (Driska and Hartshorne, 1975) or 43,000 and 32,000 (Izant and Lazarides, 1977). Differences in the peptide maps of smooth and striated tropomyosins allow Fine and Blitz (1975) to assume that they are coded by different genes.

Cohen and Cohen (1972), working with human platelets, gave first evidence that cytoplasmic tropomyosins are smaller than those isolated from muscle. The rod-shaped molecules have a length of 35 nm and a subunit MW of 30,000. Fine and Blitz (1975) extended these results to tropomyosins from brain, pancreas, and fibroblasts. A "small" tropomyosin subunit of 29,000 dalton has also been isolated from sea urchin eggs (Ishimoda-Tagaki, 1978). Bretscher and Weber (1978a) report that tropomyosins from brain, thymus, and SV40-transformed 3T3 cells can also be separated into two chains of slightly different molecular weight.

In striated muscle, the tropomyosin rods lie in the grooves of the double-stranded actin-filaments, and together with the troponin complex, they regulate the ability of actin filaments to interact with myosin (Ebashi and Kodama, 1966). While the function of tropomyosin in smooth muscle and nonmuscle cells is still unclear, cytoplasmic tropomyosins can replace to some extent muscle tropomyosins in the striated muscle regulatory system (Fine *et al.*, 1973). It is questionable, however, if they can still bind the troponin complex (Côté *et al.*, 1978). Tropomyosin may have some regulatory role in nonmuscle systems, too, since it can interfere with the reaction of actin filaments and actin-associated proteins: at room temperature or above, tropomyosin can dislodge α-actinin from thin filaments (Drabikowski *et al.*, 1968; Stromer and Goll, 1972); it can inhibit the interaction of F-actin and filamin (Maruyama and Ohashi, 1978); and it can interfere with the spectrin–α-actinin enhanced interaction of actin and myosin (S. Puszkin *et al.*, 1978).

E. Regulatory Proteins

The troponin complex of striated muscle mentioned earlier is composed of one mole each troponin C (Tn-C; Ca^{2+}-binding protein), troponin I (Tn-I; inhibitory protein), and troponin T (Tn-T tropomyosin-binding protein). The presence of a similar trio in smooth muscle and nonmuscle cells is still under dispute (Hitchcock, 1977). A Tn-I-like protein has recently been isolated from erythrocyte membranes (Maimon and Puszkin, 1978). Evidence is accumulating that the Tn-C-like proteins previously isolated from smooth muscle and nonmuscle tissues are identical to the low-molecular-weight Ca^{2+}-binding protein (about 17,000 dalton) present in most eukaryotic cells (Drabikowski *et al.*, 1978); Barylko *et al.*, 1978). This intracellular Ca^{2+}-dependent regulator protein is a most potent activator of many biological systems such as $3',5'$-cyclic nucleotide phosphodiesterase and adenylate cyclase (for review, see Dedman *et al.*, 1977b). It is also present as the "regulatory subunit" in myosin light chain kinase; together with a "catalytic subunit" of about 100,000 MW, it regulates the contractile activity in smooth muscle and nonmuscle cells (Dabrowska *et al.*, 1979). There are structural and functional similarities between this regulator protein and two other Ca^{2+}-binding proteins, namely troponin-C and parvalbumin (Dedman *et al.*, 1977a, 1978a). A Ca^{2+}-dependent regulator protein with properties very similar to those described for the vertebrate protein has recently been isolated from an anthozoan coelenterate (Jones *et al.*, 1979), which adds another member to the list of proteins that have been highly conserved during evolution.

III. Antibodies to Contractile Proteins

A. Autoimmune Antibodies to Smooth Muscle

Holborow and his co-workers, using the immunofluorescent technique, were the first to describe antibodies to smooth muscle in patients with liver disease (Johnson *et al.*, 1965, 1966) and with malignant disease (Whitehouse and Holborow, 1971). Some of these antibodies were also shown to react with microfilament bundles of cultured cells and with hepatocyte membranes (Farrow *et al.*, 1971). A few of such autoimmune antibodies have a rather broad spectrum, reacting with smooth and striated muscle myosin (Fairfax and Gröschel-Stewart, 1977), others are specific for smooth muscle and can be absorbed with crude smooth muscle actomyosin. After making antisera to contractile proteins of smooth muscle in rabbits, Holborow and his group could reproduce with these antibodies the staining patterns seen with the autoimmune antibodies of patients (Trenchev *et al.*, 1974; Holborow *et al.*, 1975). About this time, Gabbiani *et al.* (1973) were able to absorb the smooth muscle activity in sera of patients with chronic active hepatitis with actin from human platelets, showing that these autoantibodies were directed to actin. These findings have since been confirmed by several other groups (Lidman *et al.*, 1976; Bottazzo *et al.*, 1976; Andersen *et al.*, 1976; Toh *et al.*, 1976). A very thorough study on this actin specificity was performed by Fagraeus and co-workers (Lidman *et al.*, 1976; Fagraeus *et al.*, 1978; Utter *et al.*, 1978), in which they analyzed the immunoprecipitates by SDS–PAGE and showed them to contain actin and the Ig heavy and light chains only, and that the Ig fraction released from such precipitates had retained its antiactin activity. Some of the autoimmune antisera (at a dilution of 1:10) still visibly precipitate rabbit skeletal actin at a concentration of 0.3 mg/ml; however, all authors mentioned here note that they need rather large amounts of actin to absorb the antiactin globulins from the patients' sera (6-30 mg/ml). As Utter *et al.* (1978) point out, F-actin may have few available antigenic determinants for antibodies of human origin. It was interesting to find out that the immunofluorescent reaction of such autoimmune sera with fibroblasts cannot be blocked by pretreatment with experimentally produced rabbit antiactins (Fagraeus *et al.*, 1978). Since both types of sera gave the same broad staining pattern (striated muscle, smooth muscle, and cytoplasmic actin), it seems likely that different epitopes are recognized in the production of these two types of antibodies. The discovery of autoimmune antibodies to actin terminated a not too well founded but firm prejudice that actin is not immunogenic due to its conservative structure. The mechanism by which humans and animals can form antibodies to their own actin is, however, still unknown. It is quite likely that virus infections play a role in their genesis; and one can speculate that virus-induced alterations of the plasma membrane expose membrane-associated actin in a stabilized and therefore more immunogenic form (Fagraeus *et al.*, 1978; Holborow, 1979).

B. Antibodies to Actin

1. *Preparation of the Immunogen*

Although actin antibodies are now available to several research groups, the immunogenicity of actin and the specificity of the obtained antisera is still rather problematic. According to Wilson and Finck (1971), one must assume (although one cannot prove) that all earlier actin preparations were contaminated with tropomyosin, thus producing tropomyosin-, rather than actin-specific antibodies. By monitoring the actin purification steps by SDS-PAGE, it is now relatively easy to prepare highly purified actins. For vertebrate actins, I prefer to prepare acetone-dried powders according to Carsten and Mommaerts (1963) and to extract actin following Spudich and Watt (1971); for invertebrate tissues, my co-workers separate actin from the actomyosin complex under actin-depolymerizing conditions, as recommended by Stossel and Pollard (1973). Unless only very small amounts of tissue are available, e.g., in the case of 3T3 cells (Lazarides and Weber, 1974), SDS-PAGE is not an obligatory step in obtaining "tropomyosin- and myosin-free" actin (Lazarides, 1975b). To enhance the immunogenicity of actin preparations, however, many authors still recommend pretreatment of the immunogen with SDS or SDS and 2-mercaptoethanol (Lazarides, 1975b; Weber *et al.*, 1976). In doing so, it must be kept in mind that nonspecific antibodies to SDS–protein complexes will be formed, as was shown by the complement fixation technique (Lompre *et al.*, 1979). Such antibodies may then cause nonspecific crossreactions, especially with membrane structures.

Tyrosylation, diazotizing and treatment with SH-blockers (*N*-ethylmaleimide) will not enhance the antigenicity of actin (U. Gröschel-Stewart, unpublished observation). Attachment of G-actin to inert carrier particles (Affigel 702, Biorad Labs) will increase the immunogenicity, but also restrict the specificity (Gröschel-Stewart *et al.*, 1977a). Jockusch *et al.* (1978) adsorb native and SDS-denatured actin to alum precipitates, this method has also been used by Owaribe and Hatano (1975) for *Physarum* actin. While Trenchev and Holborow (1976) recommend "aging" to enhance the immunogenicity of actin, Fagraeus *et al.* (1978) and our own group (Gröschel-Stewart et al., 1977a) use native actin without pretreatment. It seems noteworthy that the last four groups were not able to obtain antibodies to SDS-treated actin.

2. Immunization Procedure

Wilson and Finck (1971) already pointed out that the use of small doses of immunogen will be more satisfactory in the production of actin antibodies. This has been fully confirmed both by groups using SDS-denatured or nondenatured actin (generally in a total dose between 0.8–3 mg). Trenchev and Holborow (1976) required single doses of 1–2 mg of human uterine actin to produce antibodies in 14 out of 36 rabbits. Although we started out in 1972 using total doses of 13–16 mg per rabbit over several months, with varying success, we have now reduced the total amount to about 1 mg (4–5 injections in 21-day intervals) when chicken actin or invertebrate actins are used. Jockusch *et al.* (1978) consistently obtained precipitating antibodies after 40 days, injecting 2.3 mg immunogen per rabbit. Karsenti *et al.* (1978) recommend several footpad injections of 500 μg each in monthly intervals. It is also possible to isoimmunize rabbits by 3 to 5 injections of 1–5 mg rabbit actin per animal (Fagraeus *et al.*, 1978).

3. Specificity of Antisera

Specificity tests of a more general kind that pertain to all antigen–antibody systems discussed here will be dealt with in Section IV,A and C.

It is of interest to compare the species- and tissue-specificity of the antibodies obtained by the various methods. It appears that most antiactins from muscle and nonmuscle sources have the same broad spectrum as the autoimmune antibodies, especially in the immunofluorescence test. It is not surprising, however, that the very remote species, such as antibodies to *Physarum* actin and vertebrate actin, will not crossreact (Owaribe and Hatano, 1975). In contrast, however, antibodies to chicken smooth muscle actin are reported to react with *Dictyostelium* actin (Eckert and Lazarides, 1978). The antibodies we prepared to chicken smooth muscle actin are rather species nonspecific, but have a high degree of tissue specificity within vertebrates (Chamley *et al.*, 1977a,b; Chamley-Campbell *et al.*, 1978). We observed that these antibodies, which were raised to a smooth muscle γ-type actin will react with all muscle actins, including the more closely related (Elzinga and Kolega, 1978) α-actins from skeletal and cardiac muscle. They apparently do not recognize the β-form of actin very well; and therefore we find no reaction with erythrocyte membranes (all β) or with cells that have a high β:γ ratio such as platelets and fibroblasts. Cytoplasmic actins with almost equal ratios of β and γ, such as brain and glial cells, will still be recognized (Gröschel-Stewart *et al.*, 1977b). More experiments are needed to confirm this hypothesis.

The reason why antiactins, irrespective of their pretreatment and the mode of injection, differ in their specificity, cannot as yet be explained. It certainly cannot be assumed that SDS treatment invariably will cause the production of broad spectrum antibodies, since Fellini *et al.* (1978) report on an antibody specific to γ-actin of chicken smooth muscle, and the antigen used was gizzard actin eluted from preparative SDS gels.

C. Antibodies to Actin-Associated Proteins

1. Antibodies to DNase I

Injections of about 1 mg of DNase I into rabbits will lead to the formation of precipitating antibodies (Wang and Goldberg, 1978; Lindberg *et al.*, 1979; H. G. Mannherz and U. Gröschel-Stewart, unpublished result). They have been used by the first two groups to visualize actin filaments in cultured cells, either directly or after treatment with DNase I. For limitations, see Section V,D,5 and 6.

2. Antibodies to α-Actinin

a. *Preparation of the Immunogen.* Native and highly purified α-actinin from chicken smooth muscle (Schollmeyer *et al.*, 1976; Fujiwara *et al.*, 1978;

Craig and Pardo, 1979), from porcine skeletal muscle (Lazarides, 1976b; Jockusch *et al.*, 1977), from canine skeletal muscle (S. Puszkin *et al.*, 1977), or SDS-denatured chicken gizzard α-actinin (Geiger and Singer, 1979) were used.

b. *Immunization Procedure.* Fujiwara *et al.* (1978) gave two 250 μg subcutaneous injections at multiple sites. Lazarides (1976b) and Jockusch *et al.* (1977) used one subcutaneous injection (250 μg), followed by multiple intravenous injections of alum-precipitated immunogen (total about 2 mg). S. Puszkin *et al.* (1977) gave 6–8 weekly injections of 1 mg each. The SDS-denatured α-actinin (Geiger and Singer, 1979) was administered in three intradermal injections. Rabbits were used by all.

c. *Specificities of the Antisera.* The antibodies described seem to react with α-actinin of smooth, striated muscle and of nonmuscle origin. Fujiwara *et al.* (1978) find crossreaction with both native and denatured α-actinin and note a certain degree of species specificity. Their antibody will not react with the 95,000 dalton protein from brushborder (see also Bretscher and Weber, 1978c).

3. Antibodies to Actin-Binding Protein

Six to eight weekly injections of 20–40 μg of human leukocyte actin-binding protein elicited the production of precipitating antibodies in rabbits, which did not crossreact with erythrocyte spectrin or leukocyte myosin, and which specifically inhibited the gelation of warmed leukocyte extracts (Boxer and Stossel, 1976).

4. Antibodies to Filamin

Lymph node injections were used to obtain antibodies to chicken gizzard filamin (Wang *et al.*, 19756). They did not react with myosin, nor did they detect filamin in skeletal muscle. Filamin or a crossreacting antigen were found in kidney, liver, and brain. Wallach *et al.* (1978) prepared filamin from guinea pig vas deferens and immunized a goat, injecting 100 μg every 3 weeks for 9 weeks. The antibodies they obtained cross-reacted with chicken gizzard filamin, rabbit alveolar macrophage actin-binding protein (Hartwig and Stossel, 1975), and with platelet and fibroblast (but not brain) high-molecular-weight protein (filamin). No reaction was noted with spectrin, and only a very weak one with skeletal and cardiac muscle. Wehland *et al.* (1977) gave no details about their antifilamin serum.

5. Antibodies to ''β-Actinin''

In spite of the controversy about this protein, the antibodies obtained to this 66,000 dalton protein will be mentioned here, since they also localized the antigen in nonmuscular tissue such as brain, pancreas, kidney, and blood (Heizmann and Häuptle, 1977).

6. *Antibodies to Spectrin*

Spectrin antibodies were first obtained by Nicolson and Painter (1973) after injecting about 7 mg of purified human spectrin into rabbits. Kirkpatrick *et. al.* (1978) showed that injection of spectrin will induce preferential formation against the stronger immunogenic Component 1; and there is no crossreaction between Component 1 and 2 and their respective antibodies. Sheetz *et al.* (1976a,b) find a weak crossreaction between spectrin antibodies and human uterine myosin. It would have been interesting if this crossreaction had been checked with the sensitive radioimmunoassay for human spectrin, as published by Hiller and Weber (1977). This assay, sensitive to 5 ng/ml, was not able to detect spectrin in human tissue culture cells.

D. ANTIBODIES TO MYOSIN

1. *Preparation of the Immunogen*

Both actomyosin and myosin are better immunogens than actin, and so most of the antibodies described here were elicited to highly purified nondenatured proteins. E. G. Puszkin *et al.* (1977) enhanced the immunogenicity by adsorbing platelet myosin and myosin rod to polystyrene (Lytron) particles. Immunoprecipitates of myosin antibodies with myosins from human uterine muscle (Rukosuev and Chekina, 1973) or chicken gizzard (U. Gröschel-Stewart, unpublished result) have also been used. Miller *et al.* (1976) and Webster *et al.* (1978) extracted gizzard myosin heavy chains from SDS gels. Gizzard myosin light chains were isolated by chromatography in the presence of SDS (U. Gröschel-Stewart, unpublished result).

2. *Immunization Procedure*

Antibody formation to smooth muscle and cytoplasmic myosins can be induced by repeated intravenous injection without the help of adjuvants. The tendency to reduce the amount of immunogen is especially marked in this case. Whereas Finck (1965) started with 50–90 mg of gizzard myosin, we now use less than 1 mg total to obtain precipitating antibodies (Gröschel-Stewart *et al.*, 1976a, also unpublished result). In the case of human uterine myosin, we always required more immunogen (up to 20 mg, Burkl *et al.*, 1979). Only 2–4 mg were needed by E. G. Puszkin *et al.* (1977) when human platelet myosin rod was adsorbed to Lytron particles. In techniques where Freund's adjuvant is included and where the immunogens are applied via the subcutaneous, intradermal, or intramuscular route, much less protein is generally needed. To quote only a few examples, one injection of 1 mg of human uterine myosine sufficed to obtain precipitating antibodies (Trenchev *et al.*, 1974), and total doses of 0.5 mg elicited, in rabbits, antibodies to myosins from mouse L-cells (Willingham *et al.*, 1974), from human platelet myosin and myosin rod (Fujiwara and Pollard,

1976), from bovine adrenal medulla (Creutz, 1977), from human platelets (Moore *et al.*, 1977), from BHK 21 cells (Yerna *et al.*, 1978); and only 0.185 mg of *Physarum* myosin, injected into the footpads, were needed to produce precipitating antibodies. To obtain precipitating antibodies to chicken gizzard myosin light chains, 2 mg of the SDS-denatured immunogen were injected intravenously (U. Gröschel-Stewart, unpublished result).

3. Specificity of Antisera

Most of the antibodies to smooth muscle and cytoplasmic myosins seem to be directed to the heavy chains rather than the light chains, and also predominantly to the globular head portion (see Section IV,C). Only antibodies to platelet myosin were so far shown to be specific for the rod part of the molecule (Fujiwara and Pollard, 1976). Generally, there is no crossreaction between striated muscle myosin and smooth muscle/cytoplasmic myosins (Rukosuev, 1966; Aita *et al.*, 1968; Becker and Murphy, 1969; Gröschel-Stewart and Doniach, 1969), the exceptions being anti-granulocyte myosin (Stossel and Pollard, 1973) and anti-bovine striated muscle myosin (Trifarò *et al.*, 1978). Antisera to myosins of lower phyla (*Physarum,* starfish) do not crossreact with vertebrate myosins. However, Mabuchi and Okuno (1977) noticed that anti-starfish egg myosin will crossreact with bovine platelet and rabbit skeletal myosin heavy chain after immunoelectrophoretic separation in the presence of 0.05% SDS. The immunological relationship between smooth muscle and cytoplasmic myosins is far from clear. Some antibodies to cytoplasmic myosins will crossreact with smooth muscle (anti-platelet myosin: Becker and Nachman, 1973; Moore *et al.*, 1977; anti-bovine adrenal medulla myosin: Creutz, 1977); others will not (anti-platelet myosin: Booyse *et al.*, 1971; E. G. Puszkin *et al.*, 1977; anti-mouse L-cell myosin: Willingham et al, 1974; anti-chicken brain myosin: S. Puszkin *et al.*, 1972; anti-equine leukocyte myosin: Shibata *et al.*, 1975; anti-BHK 21 cell myosin: Yerna *et al.*, 1978; see, however, Section V,A ,2,b). Fujiwara and Pollard (1976) find that some, but not all antisera to platelet myosin crossreact with smooth muscle. Conversely, most antibodies to uterine myosins will crossreact with platelets and other nonmuscle cells such as fibroblasts and endothelium (Becker and Murphy, 1969; Gröschel-Stewart, 1971; Trenchev *et al.*, 1974; Gabbiani *et al.*, 1978; Burkl *et al.*, 1979). Pollard suggested in 1976 that there might be two types of myosin, muscular and cytoplasmic, in the uterus (Pollard *et al.*, 1976); and we have further supported this concept by selectively absorbing out the cytoplasmic component with platelet actomyosin (Burkl *et al.*, 1979). Some authors find antibodies to chicken gizzard myosin (Gordon, 1978) and to SDS-denatured gizzard myosin heavy chain (Miller *et al.*, 1976; Webster *et al.*, 1978) to react with both smooth muscle and cytoplasmic myosins. In our hands, antibodies to chicken gizzard myosin (Gröschel-Stewart *et al.*, 1976a), will not react with true fibroblasts, platelets, or endothelial cells. This has been a consistent finding. We did find, however, one example of a rather individual variation in two litter mates immunized under completely identical conditions; whereas one animal gave the expected response, the other rabbit produced antibodies with a 10- to 20-fold higher titer by immunodiffusion, and also gave a very strong and specific reaction with peripheral nerves (Unsicker et al., 1978a,b; Drenckhahn, 1979; Drenckhahn et al., 1979). It was rather surprising to note that antibodies to chicken gizzard light chains were not only tissue-type, but also species-specific, quite in contrast to antibodies to the whole myosin molecule (U. Gröschel-Stewart, unpublished result).

E. ANTIBODIES TO TROPOMYOSIN

Tropomyosins are quite immunogenic, and there are no special problems in raising antibodies to the native protein from chicken, bovine, and human muscle by injecting from 2–10 mg into rabbits, or, in the case of the rabbit immunogen, into guinea pigs. The specificities of the antibodies obtained by various groups, however, vary so much, that it is impossible to reconcile the different results. Cummins and Perry (1974) reported that vertebrate striated muscle α and β subunits are immunologically distinct from each other, and both distinct from smooth muscle tropomyosin (uterus, gizzard) and from invertebrate tropomyosin (Pecten). Antibodies to the β subunit were found to be species-specific; those to the α subunit were not. The latter also reacted with the α tropomyosin from heart. In a recent paper, Hayashi and Hirabayashi (1978) report that tropomyosins from various organs, although differing in their subunit structure, still have "common antigenic sites." In their hands, antisera to chicken skeletal tropomyosin will crossreact in immunodiffusion with tropomyosins from all chicken organs, including brain, also with tropomyosins from other vertebrates and from invertebrates, even with extracts from a protozoan (Tetrahymena). These findings are in agreement with the tissue and species nonspecificity of the anti-chicken skeletal tropomyosins used by Lazarides (1975a). Antibodies to smooth muscle tropomyosin from uterus (Trenchev et al., 1974) or gizzard (Masaki, 1975) show no crossreaction with the striated muscle protein. Schollmeyer et al., (1976) find a weak crossreaction with antibodies to SDS-denatured gizzard tropomyosin; we find that antibodies to native gizzard tropomyosin react species-nonspecifically with tropomyosin from smooth and skeletal muscle (Chamley-Campbell et al., 1977), but will hardly react with true fibroblasts. In contrast, an antibody to bovine cardiac tropomyosin was found to be specific for striated muscle (D. Drenckhahn and U. Gröschel-Stewart, unpublished result). The specificity of our antitropomyosins is best illustrated in Figs. 1a–d

---

FIG. 1.   Serial frozen sections of a pulmonary vein and the bronchiolar wall of the rat lung, stained by the indirect immunofluorescence technique with antibodies to (a) chicken gizzard myosin, (b) chicken gizzard tropomyosin, (c) chicken pectoral myosin, and (d) bovine cardiac tropomyosin. The bronchiolar smooth muscle (asteriks) and the vascular smooth muscle react strongly with anti-

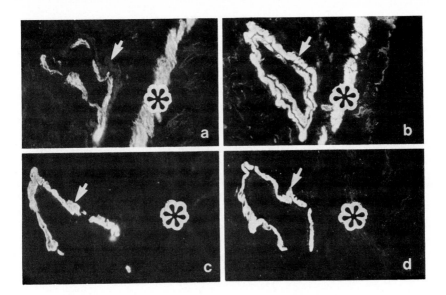

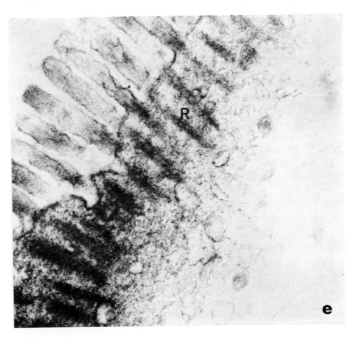

gizzard myosin (a) and anti-gizzard tropomyosin (b). Arrows point to a layer of cardiac muscle cells surrounding the pulmonary vein. They will not stain with anti-gizzard myosin (a), but with anti-pectoral myosin (c), and with both anti-gizzard tropomyosin (b) and anti-cardiac tropomyosin (d). ×140. (e) Electron micrograph of rat intestinal brush border, demonstrating the distribution of smooth muscle myosin-specific immunoreactivity (anti-chicken gizzard myosin, 2 μg/ml), using the unlabeled antibody enzyme technique. Note the intense labeling along the microvillar filament rootlets (R). ×37,500. (Courtesy D. Drenckhahn.)

and 6). Unfortunately, the specificity of the only antibody to cytoplasmic tropomyosin (bovine brain) is not described further (Webster *et al.*, 1978).

F. ANTIBODIES TO CALCIUM DEPENDENT REGULATOR (CDR) PROTEIN

There are presently only two reports on antibodies to this protein. Dedman *et al.* (1978b) immunized a goat with a total of 20 mg of rat testis CDR and obtained nonprecipitating antibodies, which inhibited the biological activity of the regulator protein. Andersen *et al.* (1978), using bovine cerebrum CDR both in its native and SDS-denatured form, obtained precipitating antibodies in rabbits, using the immunization schedule recommended by Lazarides (1976a) for α-actinin. Andersen's antibody is shown to give a straight line of precipitation with CDR, located almost in the middle between antigen and antiserum well, which actually suggests that CDR (17,000 dalton) and IgG (160,000 dalton) are close in molecular weight and diffusion coefficient.

IV. Interactions of Antibodies and Contractile Proteins

A. TESTING FOR THE PRESENCE OF ANTIBODIES *in Vitro*

1. *Precipitation Reactions and Immunoelectrophoresis*

The most simple test for the presence of precipitating antibodies is still the double diffusion test according to Ouchterlony on agar- or agarose-coated slides. For the myosin-antimyosin system, the gels are best prepared in buffers of an ionic strength $\mu \leqslant 0.3$. We found that the presence of glycerol (50%) or urea (2 M) will facilitate the diffusion of these large antigens into the gel without interfering with the immune reaction. SDS (0.05–0.1%) can also be included in the gels when whole antiserum is used in the experiment (Ig alone is sensitive to SDS) (Masaki, 1974). Many of the earlier actin antibodies may have been overlooked, since actin will not easily diffuse into the gels when the ionic strength of the buffer favors the formation of F-actin. Lidman *et al.* (1976) therefore recommend the use of low ionic strength buffers and the presence of dissociating agents such as 2-mercaptoethanol. An alternative is the use of actomyosin as the test antigen (Gröschel-Stewart *et al.*, 1977a). When antibody titers are still low, precipitation bands are visible only after drying and staining of the gels, preceded by extensive washing. Since several authors have pointed out the presence of spontaneously occurring antibodies to actin (and other cell constituents), preimmune controls should be included in these and all following tests (Karsenti *et al.*, 1977; Fagraeus *et al.*, 1978). Immunoelectrophoresis has been used to ascertain the specificity of the antibodies to *Physarum* actin (Owaribe and Hatano, 1975) and, in the presence of SDS, to test the activity of antisera to starfish egg myosin

(Mabuchi and Okuno, 1977). Fujiwara and Pollard (1976), in addition, recommend the very sensitive method of crossed immunoelectrophoresis in low ionic strength pyrophosphate buffers to detect the possible presence of small amounts of contaminating antibodies, which may appear after prolonged immunization. A modification of the classical immunoelectrophoresis is the localization of antigens on SDS gels. With this method, we were able to ascertain the presence of myosin in the rat corneal epithelium by staining myosin heavy chain with fluorescein-labeled antibodies to smooth muscle myosin (Drenckhahn and Gröschel-Stewart, 1977). Possibly even more sensitive is the localization of the antigen with radioactively labeled antibodies, as recommended for actin localization by Burridge (1976a). With this method, diffusible and bound actin has been identified in the nuclei of *Xenopus* oocytes (Clark and Merriam, 1977). The components of a thoroughly washed immunoprecipitate can also be identified by SDS-PAGE: ideally, only the polypeptide chains of the antigen and Ig heavy and light chain (MWs of about 50,000 and 25,000, respectively) should be seen. A fine example of this method is the analysis of an actin-antiactin complex by Utter *et al.* (1978).

2. Agglutination Reaction and Complement Fixation Test

Passive hemagglutination tests have apparently not been too successful in the analysis of actin antibodies (Fagraeus *et al.*, 1978). With the very sensitive immunoadherence method, we have been able to demonstrate immunological differences between smooth and striated muscle actomyosins, using 20 μg of antigen and antisera dilutions of 1/600–1/1600, and we also established the immunological relationship between placental (King and Gröschel-Stewart, 1965; Gröschel-Stewart and Gigli, 1968) and uterine actomyosin. Quantitative complement fixation tests determined the myosin content in migrating *Physarum* plasmodia to be 0.77% (Kessler *et al.*, 1976), and actin was estimated by densitometric analysis of SDS gels to comprise 15–25% of the total protein. Detectable crossreactions were only observed with the myosins of the order Physarales, whereas mycomycetes outside the order, or *Dictyostelium* (often referred to as a cellular slime mold) did not react. This sensitive method should indeed be recommended to study the relationship of cytoplasmic and muscular contractile proteins under standardized conditions; minor structural changes can easily be detected, as was successfully shown for cardiac and skeletal myosins by Schwartz *et al.*, (1977).

3. Immunoenzymatic Methods and Radioimmunoassay

Kurki (1978) introduced an enzyme-linked immunosorbent assay for actin antibodies (antiactin-ELISA). The principle of this assay is similar to the horseradish peroxidase labeling described later. In short, actin is linked to polystyrene tubes and the antiactins are allowed to react with the bound antibody. In a second step, an antibody to the first Ig, labeled with alkaline phosphatase, is added. After thorough washing, the complex is allowed to react with phosphatase substrate (e.g., Na-p-nitrophenylphosphate), the amount of enzyme substrate liberated can be quantitated and allows a calculation of the amount of actin antibody bound in the first step. Autoimmune and experimentally produced antibodies to actin behave differently in this assay, stressing again that they might be directed to quite different epitopes.

Radioimmunoassay would also allow the comparison of the immunological properties of muscle and cytoplasmic contractile proteins, but so far, only an assay for erythrocyte spectrin has been introduced (Hiller and Weber, 1977).

B. ISOLATION OF IMMUNOGLOBULINS AND PURIFICATION OF ANTIBODIES

While most of the preceding studies can be performed with the immune sera directly, further purification is mandatory for other experiments. If antibodies are to be used without affinity purification, I recommend that the immunoglobulins be separated from the immune and control sera, e.g., by salting out and ion-exchange chromatography according to Harboe and Ingild (1973). By SDS-PAGE analysis, such a preparation consists of Ig heavy and light chains only and are free of lipoproteins, which are often responsible for background staining in the immunohistochemical techniques. One should be aware, however, that even these highly purified globulins still may contain traces of plasmin-type proteolytic activity. For concentration and purification of antibodies by affiniy chromatography, crude 50% ammonium sulfate precipitates can be used. In this method, immunoadsorbents are prepared by rendering the antigen insoluble, e.g., by binding it to glutaraldehyde-activated polyacrylamide beads (Ternynck and Avrameas, 1972) or to CNBr-activated Sepharose. The crude immunoglobulin fraction is then passed through the adsorbent, unbound protein is washed out, and the specific antibody molecules bound to the antigen are then released by buffers of low pH or by concentrated salt solutions (for a review of techniques, see, e.g., Fuchs and Sela, 1978). The specificity of antibodies can also be enhanced by affinity chromatography, provided the bound antigen is really of higher purity than the original immunogen. There are, however, some drawbacks to this method. (1) Some of the eluted specific antibody may be denatured, especially when low pH buffers are used for elution (Jockusch et al., 1977). (2) The antibodies may lose their ability to precipitate (Karsenti et al., 1978). (3) Certain specificities, such as the staining of the mitotic spindle with anti-platelet myosin, may get lost (Fujiwara and Pollard, 1976). (4) The unwanted reactivities are not always removed: one of our antibodies to chicken pectoral myosin showed a very unusual nonspecific staining of rat thymus epithelial cell, and this was not abolished, but rather intensified by affinity chromatography with a myosin batch different than the one used for immunization (D. Drenckhahn and U. Gröschel-Stewart, unpublished result).

C. The Effect of Antibodies on Biological Activities of Contractile Proteins *in Vitro* and *in Vivo*

The experiments reported here can be taken as further tests for the specificity of antibodies, provided proper controls are included. In 1967, Nachman and co-workers specifically inhibited platelet myosin ATPase activity with antimyosin, but not with antifibrinogen. We were able to show in 1971 that antibodies to human uterine actomyosin (being myosin-specific) inhibit superprecipitation of actomyosin and the Ca^{2+}-activated ATPase of uterine actomyosin, myosin, and HMM, whereas antibodies to pectoral myosin did not (Gröschel-Stewart, 1971). Antibodies to platelet actomyosin inhibit platelet actomyosin ATPase considerably (Becker and Nachman, 1973). Enzyme activity and superprecipitation are also inhibited in the *Physarum* myosin-antimyosin system (Nachmias and Kessler, 1976). All these examples indicate that these myosin antibodies are directed to the head and/or hinge region of the molecule. In contrast, antibodies to platelet myosin rod (E. G. Puszkin *et al.*, 1977), although abolishing superprecipitation, rather enhance ATPase activity. Actin antibodies were shown to interfere with G–F-actin transformation in smooth muscle and *Physarum* (Trenchev and Holborow, 1976; Owaribe and Hatano, 1975); and anti-gizzard actins are potent inhibitors of actin-activated Mg^{2+}-ATPase in skeletal muscle actomyosin (Gröschel-Stewart *et al.*, 1977a). Isolated smooth muscle cells with damaged membranes were shown to specifically bind anti-smooth muscle myosin, and this significantly inhibited their capacity to contract upon the addition of ATP (Kominz and Gröschel-Stewart, 1973). Mabuchi's and Okuno's (1977) antibody to starfish egg myosin did not affect actin-activated ATPase; however, microinjection of more than 0.3 ng of antimyosin Ig (to give a ratio of Ig: egg myosin = 10:1) into blastomeres at the two-cell stage at interphase inhibited the subsequent cleavage (see also Section V,C). Antibodies to CDR inhibit regulator-dependent phosphodiesterase activity in both rat testis and bovine brain. The homologous Ca^{2+}-binding proteins, troponin C (from rabbit muscle), and parvalbumin (from carp muscle) will not interfere with antibody-regulator binding. *In vivo* experiments with anti-CDR, such as microinjection into marine eggs, were under way at the time this article was written.

D. Immunohistochemical Methods

Most information on the presence and distribution of cytoplasmic contractile proteins has been obtained from localization studies *in situ*. The actual presentation of the resutls of such studies shall be preceded by a discussion of the most important methods used in this field.

1. General Remarks on the Methodology

a. *Immunoglobulins.* As has been pointed out, it is best to isolated pure globulin fractions from the immune and control sera. When dealing with cells known to have surface receptors for the Fc portion of immunoglobulins (such as monocytes, mast cells, lymphocytes), it is advisable to remove this fragment by digestion with either pepsin (to yield the divalent F(ab')$_2$ fragment) or papain (to yield the monovalent Fab fragments). The monovalent antibodies should also be preferred when steric hindrance of the antibody–antigen reaction is expected.

b. *Fixation Methods.* For immunocytochemical staining of contractile proteins, mild fixation methods are usually recommended, even though they often do not guarantee optimal fixation of the antigens nor optimal preservation of morphology.

i. *Sections.* Four to six micron air-dried cryostat sections of freshly frozen tissue are very suitable for the localization of contractile proteins by immunohistochemistry. If the tissue is rich in lipids, the sections have a tendency to "float away," and so I fix them with 2% paraformaldehyde in phosphate-buffered saline, pH 7.3, for 5–10 minutes at room temperature prior to staining. Fixation with 95% ethanol at 4°C is also recommended (Creutz, 1977). Paraffin-embedded sections, although not suitable for immunofluorescent staining, do permit localization by the immunoperoxidase methods (Nakane and Pierce, 1966; Sternberger *et al.*, 1970).

ii. *Cell cultures, cell smears.* For surface staining, cells may be fixed with buffered formalin only, but to visualize contractile elements within the cells, the membranes must also be ruptured. Most authors treat with organic solvents such as methanol and/or acetone at temperatures below 0°C, either following or preceding the fixation with formalin (Lazarides and Weber, 1974; Weber *et al.*, 1976; Fujiwara and Pollard, 1978). Lysis of cells has also been accomplished with buffers containing nonionic detergents such as Triton X-100 (0.04%), followed by fixation with formalin and treatment with acetone–water (Cande *et al.*, 1977; Jockusch *et al.*, 1978). Karsenti *et al.* (1978) recommend fixation in 4% paraformaldehyde, 10% glycerol and 10% dimethylsulfoxide.

iii. *Fixation artifacts.* In both sections and cultures, fixation with glutaraldehyde causes a strong yellowish autofluorescence and nonspecific staining that can be overcome by reduction of the free aldehyde groups with NaBH$_4$ (Schachner *et al.*, 1977; Weber *et al.*, 1978). It is then possible to examine tissue culture cells in parallel by immunofluorescence and low power electron microscopy (Weber *et al.*, 1978). Osmium tetroxide is also unsuitable as a fixative, since it destroys the antigenicity of most proteins. Furthermore, Szamier *et al.* (1975)

and Maupin-Szamier and Pollard (1978) showed that osmium tetroxide treatment will fragment and disarrange pure actin filaments, and only those with bound tropomyosin will be preserved. Thus, for electron microscopy, the immunoreaction after mild fixation is the mandatory first step, followed by osmication and the necessary further processing.

c. *Staining Methods. i. Immunofluorescence.* Two stains are most commonly used to form fluorescent antibody conjugates in the studies presented here: (1) fluorescein isothiocyanate, FITC (excitation wavelength 495 nm, emission maximum 520–525 nm; fluorescence apple green) and (2) tetramethylrhodamine isothiocyanate = TRITC (excitation wavelength 545–555 nm, emission maximum 580 nm; fluorescence orange red). In the case of FITC, the molar ratio of dye:protein should not exceed 1.5, in order to minimize nonspecific staining and antibody denaturation (Hudson and Hay, 1976; Nairn, 1976; Peters and Coons, 1976). The dye:protein ratio is apparently not as stringent in the case of TRITC. Successful and specific staining was obtained by Cebra and Goldstein (1976) with conjugates having a 280 nm: 515 nm ratio varying from 2.1–9.4; this corresponds to a molar dye:protein ratio of approximately 0.7–3. Both FITC- and TRITC-conjugates may be fractionated further by DEAE-cellulose chromatography, and the optimally conjugated fractions can then be selected. A useful alternative to the better controlled but time consuming chromatographic separation is the absorption of the conjugates with tissue homogenates or tissue powders, usually from liver. This procedure removes nonspecific staining and also molecules with a high negative charge (=overconjugated Igs). I recommend the use of tissue from the same species to be examined with the absorbed conjugate. I have also successfully used pellets of the noncrossreacting actomyosin (e.g., striated muscle versus anti-smooth muscle). All absorbed conjugates should be filtered (e.g., using a 0.45 μm Millipore membrane filter) and should be used immediately. Immunofluorescence is most commonly performed at room temperature, and incubation times of 20 minutes to 2 hours are recommended. To prevent "capping" and endocytosis, intact living cells are often stained at 0–4°C. Frequently, immunofluorescent staining of tissue culture cells is performed at 37°C, and though this intensifies the staining, I personally feel that it also increases the degree of nonspecific staining, especially in the dense perinuclear region. There are two ways of performing immunofluorecence: the *direct* method, where the antibody to the cellular antigen is labeled; and the *indirect* method, where the first antibody is unconjugated and a labeled antibody specific for the first Ig is used to demonstrate the sites of the antigen–antibody reaction. The advantage of the direct method is that it more easily allows the simultaneous localization of two antigens with the two antibodies bearing different fluorescent stains. The indirect method, however, is more sensitive since there is a geometrical increase in the amount of second antibody bound. It is often considered to be less specific than the direct method, but since the concentration of the first antibody can be significantly decreased, the problem of nonspecific staining is also decreased.

ii. *Immunoenzyme technique.* An indirect method, where the second antibody is labeled with the enzyme horseradish peroxidase, is often used (Nakane and Pierce, 1966). The more widely applied method at the present time, however, seems to be the three-step "unlabeled antibody enzyme method" technique of Sternberger *et al.* (1970). This method, which is not as easily performed as immunofluorescence, allows the localization of antigenic sites both by light microscopy, and, after appropriate processing, also by electron microscopy. The reaction is performed as follows: (1) a specific antibody to the antigen in the section is applied (e.g., rabbit anti-chicken myosin); (2) an antibody to the Ig of the first step follows (e.g., goat anti-rabbit Ig); (3) an antibody to horseradish peroxidase, which has peroxidase bound to it in a controlled manner and which must be raised in the same animal as the first antibody (e.g., rabbit antiperoxidase), is then allowed to react with the free valences of the second layer antibody. The original antigenic sites are then localized by applying peroxidase substrate (3,3'-diaminobenzidine and H_2O_2). This triple layering has a great potentiating effect and therefore very low concentrations of the first antibody are needed (for antimyosin, as little as 1 μg/ml is sufficient; D. Drenckhahn, unpublished result).

iii. *Immunoferritin technique.* This method has primarily been designed for the ultrastructural localization of antigens, since the electron-dense ferritin complex with its iron core can easily be located in the electron microscope (Rifkind, 1976). Immunoferritin labeling is very efficient for localizing surface antigens, but there are the usual technical difficulties in making intracellular antigens accessible to the large immunoglobulin–tracer complex. The problem of steric hindrance, as described by Utter *et al.* (1978) could, at least in part, be overcome by the use of monovalent antibodies. Nonspecific ferritin deposits are often a problem here, which makes the results hard to evaluate.

d. *Controls.* Stringent controls are necessary to ascertain the specificity of the stainings observed. In the direct immunofluorecent method, pretreatment of the tissues with nonlabeled antibody should completely abolish all staining with the labeled antibody. In all indirect immunohistochemical methods, the omission of the first (tissue-specific antibody) should abolish the reaction of all further steps. Preimmune sera are often used as controls, and while they assure that no spontaneous antibody was already present in the animals to be immunized, they will not allow any conclusions as to antibodies (other than to the immunogen) that may have formed during the immunization procedure, such as antibodies to the mycobacteria present in Freund's adjuvant, or nonspecific antibodies to SDS–protein complexes. It is therefore better to use another immune serum, raised in an identical manner to a heterologous antigen, as control. As mentioned before, the noncrossreactivity of smooth and striated muscle myosins provides such an ideal "internal control." Single or repeated absorptions of the antibodies with the immunogens should also reduce all staining reaction to zero. Another good control globulin is found in the nonretained fractions from affinity columns (Fujiwara *et al.*, 1978), which raises the question as to what is actually removed by this process.

V. Localization of Cytoplasmic Contractile Proteins

A. CULTURED CELLS: AN OVERVIEW

1. *Continuous Cell Lines*

a. *Normal Cells.* When freshly trypsinized cells of established lines are replated, the round cells exhibit a diffuse and uniform fluorescence when treated with FITC-labeled antibodies to actin, α-actinin, or tropomyosin (Lazarides, 1975b, 1976a). As the cells (e.g., rat embryo fibroblastoid cells, human skin fibroblasts, 3T3 cells) begin to spread and to attach to the substratum, some of them develop a very regular network, which seems to precede the formation of straight filament bundles. Actin and α-actinin (Lazarides, 1976a; Lazarides and Burridge, 1975) can be located in the vertices of these nets, and actin fibers can be seen to radiate from them towards the edge of the cells and into the areas of membrane ruffling (also see Rathke *et al.*, 1977). Lazarides' interpretation is that α-actinin is involved in the organization of actin from a diffuse form into microfilament bundles. In this outgrowth of filaments, tropomyosin seems to lag behind and rarely associates with newly formed fibers, but preferentially binds to already assembled bundles (Lazarides, 1976b). When the cells are fully spread, parallel running microfilament bundles either span the cell in its entire length or converge at so-called "focal points." These fibrous structures can be seen in living 3T3 cells by phase contrast, Nomarski, and polarized light optics. Electron microscopy shows that these submembranous bundles are preferentially located on the attachment side of the cell (Goldman *et al.*, 1975). After fixation and acetone treatment to open the cell membrane, the microfilament bundles stain uniformly in their entire length with antibodies to actin, using the indirect immunofluorescent technique (Lazarides and Weber, 1974) or indirect immunoperoxidase staining (Karsenti *et al.*, 1978). When cultured cells are extracted with nonionic detergents, which keeps the microfilament system largely intact (Osborn and Weber, 1977; Small and Celis, 1978), single microfilaments (5–7 nm) and very thin bundles, invisible to light microscopy, can be seen in the electron microscope. Small and Celis also demonstrated that the peripheral web of fine fibers, which interconnects radiating linear filament bundles ("microspikes"), stains as a brightly fluorescent band with fluorescent spikes radiating from it after application of autoimmune antibodies to actin. When the filaments of the extracted cells are decorated with HMM, the arrowheads show uniform polarity, pointing into the cell interior in those filaments attached to the membrane (e.g., Small and Celis, 1978). The arrowhead decoration can be dramatically enhanced when 0.2% tannic acid is included in the fixation medium (Begg et al., 1978), and in such preparations it can be noted that the filaments in the cytoplasm show an antiparallel polarity in both myoblasts and CHO cells. These findings strongly support the concept of a sliding filament mechanism in cell motility. The uniform fluorescent staining of filament bundles with antiactin alone would not have allowed any such conclusions. Similar to antiactin, antibodies to filamin (a protein postulated to be involved in the cross-linking of actin) will stain the microfilament bundles the same way actin antibodies do, also reaching into the cell periphery and the membrane ruffles (Wang et al., 1975; Webster et al., 1978). One of the antibodies raised to CDR will uniformly stain the microfilament bundles of a variety of cells, next to showing diffuse staining of the cytoplasm (Dedman et al., 1978b). Tropomyosin antibodies will bind to microfilaments involved in structural support in a discontinuous pattern (Lazarides, 1975a), showing fluorescent segments of about 1.2 μm length and nonfluorescent spacings of about 0.4 μm. Antibodies to α-actinin (Lazarides and Burridge, 1975; Gordon, 1978) also stain the filaments in an interrupted pattern with spacings inverse to those of the tropomyosin antibody; the fluorescent segments being 0.25–0.5 μm long, the nonfluorescent spacings 1–1.5 μm. When these two antibodies are applied in sequence, a "complementary" nonstriated fluorescent pattern is noted. Antibodies to myosin also stain in a discontinuous pattern, as was first shown by Weber and Gröschel-Stewart (1974) in 3T3 cells and human fibroblasts, using an antibody to chicken smooth muscle myosin. Similar staining patterns were also found by Fujiwara and Pollard (1976) in HeLa and Enson cells with antibodies to human platelet myosin and myosin rod; by Ash et al. (1976) in normal rat kidney cells with an antibody to human uterine myosin, and by Yerna et al. (1978) in BHK 21 cells with an antibody to BHK 21 myosin. The latter group found the striated staining pattern not only in fixed cells, but also in unfixed glycerinated cells, proving that "interruptions" are not artifacts due to the fixation. As a matter of interest, isolated smooth muscle cells from chicken gizzard occasionally also display regular banding patterns with a periodicity of about 1.5 μm when treated with antibodies to gizzard myosin (Bagby and Pepe, 1978). In cultured gerbil fibroma cells, antimyosin-stained segments are about 0.4–0.7 μm wide; the nonfluorescent spacings are 1–1.5 μm and are complementary to the α-actinin pattern (Gordon, 1978). In these cells, Gordon occasionally observed fibers with an overall periodicity of 1 μm when stained with antibodies to actin. Although this orderly arrangement of some of the contractile proteins on or near the microfilament bundles is very impressive and indeed does suggest a sarcomere-like structure, it still has to be shown that the fluorescent images really reflect the distribution *in vitro*. It must be remembered that fluorescent stainings only correspond to points of light emission, which makes exact measurements difficult. Also, the binding of antibodies, especially in the indirect fluorescent method, where two layers of Ig molecules are applied, "blows" up the original structure enormously. As Webster *et al.* (1978) estimate, a single 6 nm actin fiber can, after decoration with 2–4 IgG molecules (diameter of one IgG is about 9 nm) appear as a 30–60 nm structure. Furthermore, steric hindrance may not allow the antibodies to reach all antigenic sites, just as antiactins will not stain the region of thick and thin filament overlap in the isolated striated myofibril. The spatial relationship of the contractile proteins to each other can obviously only be evaluated on an ultrastructural level. Two interesting and promising approaches to this problem have recently been published. Webster *et al.* (1978) exposed the cytoskeleton of detergent-extracted rat kangaroo (PtK2) cells to various fluorescent antibodies to contractile proteins, and the cells were then viewed both by immunofluorescence and low power electron microscopy. The images obtained closely resemble each other and possibly also reflect the *in vivo* state. However, higher resolution electron microscopy, after application of antibodies with electron-dense markers is still the method of choice. In the second paper, Goldman *et al.* (1979) report on the first clue they obtained as to the supramolecular organization of myosin in microfilament bundles. When BHK 21 cells are processed for electronmicroscopy with glutaraldehyde fixative containing 0.2% tannic acid, striated patterns are frequently observed that consist of regularly spaced electron-dense and electron-lucid bands, and comparative measurements indicate that this pattern is rather similar to the striations seen in indirect immunofluorescence with BHK 21 myosin antibodies. A combination of both methods, antibody treatment followed by tannic acid fixation, would have been most instructive and was surely attempted by the authors, but was probably hindered by the technical difficulties we all face in this particular field. In this context, it would also be interesting to find out if the small "node-like" structures with fine tails, attached to the 4–12 nm filaments in fibroblasts and supposed to represent myosin monomers or oligomers, would decorate with antibodies to myosin (Buckley *et al.*, 1978).

    b. *Transformed Cells.* When tissue culture cells are transformed, particularly by tumor viruses, they show abnormal growth and are, in addition, recognizable by their rounded shape, disordered alignment, and low adhesion to the substratum (Pastan and Willingham, 1978). Although the growth properties per se are unchanged (Willingham *et al.*, 1977), there seems to be a defect in adhesion that is responsible for the altered morphology observed in transformation. Pollack and Rifkin (1975) postulated that proteolytic enzymes might be responsible for this loss of anchorage, since there is a correlation between this phenomenon and the production of plasminogen activator in transformed cells. On an ultrastructural level, McNutt *et al.* (1973) noted that the microfilament bundles usually seen in normal cells were disrupted and that the same results were obtained with fluorescent antibodies to actin in SV40-transformed 3T3 cells (Pollack *et al.*, 1975; Weber *et al.*, 1975; Tucker *et al.*, 1978) and in transformed 3T3 cells and fibroblasts (Edelman and Yahara, 1976). The disruption of filaments is expressed as a shift from membrane-bound to "soluble" actin upon transformation (Wickus *et al.*, 1975; Fine and Taylor, 1976). Also, the synthesis of actin in transformed cells was found to decrease by about 20% (Fine and Taylor, 1976). *a*-Actinin also gets either lost or dispersed in transformed cells (Schollmeyer *et al.*, 1976). The fine striational pattern usually seen in normal cells with antibodies to myosin also becomes greatly disordered (Pollack *et al.*, 1975; Ash *et al.*, 1976). Ash and co-workers showed that the normal myosin pattern can be regained either when the Rous sarcoma virus-infected cells are grown at temperatures nonpermissive for the virus, or when protein synthesis inhibitors are added. This indicates that transformation, but not myosin distribution, is a protein synthesis dependent process. Myosin synthesis is, however, also reduced in transformed cells, and again it concerns the membrane-associated part (Ostlund *et al.*, 1974; Shizuta *et al.*, 1976a). Goldman *et al.* (1976b), in an attempt to correlate ultrastructural changes and fluorescent images in normal and transformed cells, warns that there is not always a true correspondence in the results of the two techniques, and that this makes the relationship between microfilament structure and transformation even harder to interpret.

c. *Cells Treated with Drugs.* i. *Cytochalasins.* These fungal metabolites have a still enigmatic effect on isolated culture cells, and it seems to vary considerably from one cell to another (Goldman, 1972). Presumably, both the membrane and the microfilament system are affected by this group of drugs. The different cytochalasins (A, B, and D) differ in their potencies, cytochalasin A and D being more powerful than B in provoking visible changes in cells (Miranda *et al.*, 1974a,b). Studies with fluorescent antibodies to actin have not really been helpful in elucidating the complex mechanism of action of the cytochalasins but have added some interesting information. Cultured lymphoid cells and platelets, when treated with cytochalasin B, will lose their surface projections and their microfilament bundles (Norberg *et al.*, 1975b). When such cultures are stained with antibodies to actin, diffuse staining of the cells is noted and pads of fluorescent material are left behind on the coverslip where the cells have retracted; this material may possibly be bits of membranes with microfilaments attached to them. Pollack and Rifkin (1976), using subcultures of rat embryo cells and an antibody to actin, find this noncoordinated retraction and detachment of cells and the loss of microfilament bundles to be a process dependent on dosage and exposure to cytochalasin B. The rapid disorganization of microfilament bundles in 3T3 cells into star-like patches, and later, the contraction into arborized structures not only involves actin, but also tropomyosin (Weber *et al.*, 1976) and presumably the other contractile proteins, too. Cytochalasin D (Miranda *et al.*,

1974a,b) and also cytochalasin B only act *in vivo* and will not influence the ATP-induced contraction in glycerinated rat mammary cells (Weber *et al.*, 1976), suggesting that some hypothetical receptor is removed by glycerination. The modification of microfilament structures seen in immunofluorescence microscopy with antibodies to actin were of better reproducibility when cytochalasin A was used (Rathke *et al.*, 1977). At a concentration of 5 μg/ml, the cells began to contract. When they were fixed and stained after 30 minutes of drug action, three main structural changes were noted in the cells: a dense central aggregate; patches as seen after cytochalasin B treatment; and very regularly sized rod-shaped elements of about 5 μm length, which were especially frequent in the flattened cell regions. The authors discuss the attractive hypothesis that the regular length of these actin-containing rods reflects the periodicity with which the submembranous microfilament bundles attach to the surface membrane. It would be desirable also to locate the other contractile proteins in such altered cells, preferably on an ultrastructural level.

ii. *Phalloidin.* Wieland and his co-workers (Lengsfeld *et al.*, 1974) found that phalloidin, the toxin from *Amanita phalloides,* will react *in vitro* stoichiometrically with actin, promoting actin polymerization and stabilizing such actin polymers. The pathological action of phalloidin on the liver, which also involves actin filaments, will be discussed in Section V,D,7. A concentration-dependent effect of this toxin on tissue cultured cells (3T3 and PtK2) was shown, using antibodies to actin and filamin (Wehland *et al.*, 1977). When low doses of phalloidin (0.2 m*M*) are injected into cells, which are fixed and stained several hours thereafter, stress fibers are still visible, but, in addition, numerous islands of actin and filamin are noted. When concentrations of 1 m*M* are injected, the cells will contract, leading to an aberrant morphology. Locomotion and cell growth can also be retarded in a concentration-dependent way. Since we found that myosin is affected similarly to actin by chronic phalloidin intoxication (Heine *et al.*, 1976), it would have been interesting to see the effect of phalloidin on the myosin distribution in cultured cells.

While it cannot be denied that continuous and subcloned cell lines have been most useful and instructive in studies of cell growth and the distribution of cytoplasmic contractile proteins in normal and transformed cells, it must be kept in mind that the origin of these cells, which are often prepared from whole embryos or whole organs, cannot be traced back to a particular cell type. So while many cells do indeed have the appearance and morphology of a mature fibroblasts, they may be modulated cells of nonfibroblastic origin (Pastan and Willingham, 1978). Since we now know, for instance, that at least six genes may be responsible for the coding of actins within one individual (Vandekerckhove and Weber, 1978d), one must include cells of defined origin in studies of the immunochemistry of cytoplasmic contractile proteins. Primary cell cultures will fulfil this need.

2. Primary Cell Cultures

a. *Fibroblasts.* A first attempt to study the distribution and ultrastructure of microfilament bundles in nonmuscle cells with the help of actin antibodies was made by Perdue (1973). After having isolated and characterized an actin-like protein from chick embryo fibroblasts (Yang and Perdue, 1972), Perdue attempted to raise antibodies to this protein. He did not obtain precipitating antibodies, but he used the serum for ultrastructural localization. Using an indirect technique and a hybrid antibody to rat IgG and ferritin as the second layer, he observed a quite variable binding of the antibodies to microfilaments and to amorphous regions in the cell. Some staining was also found at the tips of intact fibroblasts and along the ventral surface, which the author attributed to external microfilaments that have been disposed of by the retreating cell. Such tracks will also stain with antibodies to actin, tropomyosin (Lazarides, 1975b), and myosin (Gröschel-Stewart et al., 1975b).

b. *Smooth Muscle Cells.* Although the following studies were primarily designed to study smooth muscle cells, we also obtained information on the antigenicity of contractile proteins of fibroblasts and endothelial cells often present in these cultures. Isolated smooth muscle cells from organs of newborn vertebrates, as well as fibroblasts, have been well characterized in their morphology and their behavior (Campbell et al., 1971, 1974; Chamley et al., 1973; Chamley and Campbell, 1974). When such cultures were fixed and stained in the indirect immunofluorescence technique with antibodies to chicken smooth muscle myosin (Gröschel-Stewart et al., 1975a, 1976a), the contractile-state smooth muscle cells stained brightly in the described interrupted pattern, when the incubation with the first and second antibody was done at room temperature for 30 minutes each. In early cultures (1 day), the stained fibrils were often in register and gave the cells a striated appearance. In later cultures (2–3 days), coarser fluorescent fiber bundles, which often appeared noninterrupted, were seen in addition. At this stage, the pattern resembled the one described for 3T3 cells, where we had stained at 37°C for 60 minutes each with the first and second antibody (Weber and Gröschel-Stewart, 1974). In these primary cultures, however, the noncontractile fibroblasts stained only very weakly with the myosin antibody when the reaction was performed at room temperature. The staining could be considerably enhanced by using the incubation conditions described for 3T3 cells, although it was never as strong as in the contractile smooth muscle cell. The same antibody also reacted very weakly with the so-called "modulated" smooth muscle cell (Gröschel-Stewart et al., 1975b; Chamley et al., 1977b; Chamley-Campbell et al., 1979). These are cells that reversibly modulate over several days from a phenotypic state in which they are contractile, contain thick myofilaments, and do not undergo migration or mitosis, to another state in which they become capable of mitosis and migration, lose thick filaments and contractility, and show an increase of organelles associated with synthesis.

Another smooth muscle antibody, however, directed to human uterine myosin (Burkl *et al.*, 1979), stained not only the contractile-state muscle cell, but also the modulated smooth muscle cell and the fibroblasts under the standard conditions (30 minutes at room temperature). The possible presence of two immunologically distinct myosins in the human uterus (see Section III,D,3) is further supported by these findings. The only antibody to cytoplasmic myosin (BHK 21 cells) that was tested so far on these primary cultures was not specific to cytoplasmic myosin; it stained contractile-state muscle cells as well (R. D. Goldman and J. H. Chamley-Campbell, oral communication).

Antibodies to native chicken gizzard actin (Gröschel-Stewart *et al.*, 1977a) stained both contractile-state and modulated smooth muscle cells and striated muscle cells. They did, however, not react with fibroblasts, prefusion myoblasts, or endothelial cells, irrespective of the length or temperature of incubation. As we mentioned before, our antibodies do not seem to recognize the β-form of actin. Only Fellini *et al.* (1978) found a similar specificity.

The antibodies we raised to chicken gizzard tropomyosin also have different specificities than the ones described by others. They will stain the I-band in striated muscle cells and in contractile-state or modulated smooth muscle cells in the described interrupted pattern. Fibroblasts and endothelial cells stain so weakly at both 20 and 37°C that the fine pattern of striation can only be seen at high magnification. 3T3 cells, however, do show a moderate reaction with this antibody, which raises again the question as to the derivation of these cells.

All in all, I feel that the antibodies we raised to chicken gizzard contractile proteins almost seem to discriminate against mesenchymal cells in primary cultures (see also Section V,D–G); and whereas they are not helpful in the localization of cytoplasmic contractile proteins, they are very useful for the distinction between fibroblasts and modulated smooth muscle cells—a problem that pertains to questions in wound healing and the genesis of atherosclerotic plaques.

B. Cytoplasmic Contractile Proteins and the Plasma Membrane

1. *Are They Present on the Cell Surface?*

Since I have personally added to the controversy over this problem, I feel I should defend my results. In all our early experiments on the localization of contractile proteins at the surface of embryonic chick cells with antibodies to smooth muscle actomyosin, freshly trypsinized cells were used, and the antibodies were added immediately after the removal of the proteolytic enzyme (Gröschel-Stewart *et al.*, 1970; ap Gwynn *et al.*, 1974). Although these cells are still viable by the dye exclusion test, one can assume that some parts of the cell exterior have been removed in the trypsinization process. Chamley *et al.* (1973), for instance, showed that trypsinized smooth muscle cells have no basal lamina during the first few days in culture, which, however, does generally reconstitute by day 5. Moreover, enzyme-treated adult smooth muscle cells, whose membrane damage was shown by their lack of response to serotonin, bound excessive amounts of antibodies to smooth muscle actomyosin, and thereby lost their ATP-induced myogenic contractile response (Kominz and Gröschel-Stewart, 1973). In further experiments with primary cell cultures, we allowed cells to attach and flatten in the presence of myosin antibodies (Gröschel-Stewart et al., 1976b), and we never observed any binding of antibody to the surface of healthy cells. However, damaged cells, which could not attach and flatten, often showed a nonspecific uptake of FITC-labeled IgG in the form of a halo, as we had described earlier (Garnett et al., 1973). Furthermore, intact mouse peritoneal macrophages, allowed to settle and attach in the presence of FITC-labeled anti-smooth muscle myosin, did not show any surface binding of antibody; there was only nonspecific uptake of all labeled globulins offered by microendocytosis (Gröschel-Stewart and Gröschel, 1974). I feel that these experiments conclusively show that the binding of antibodies was due to the treatment with proteolytic enzymes and that these results do not allow any conclusions as to the presence of contractile proteins on the surface of *intact* cells. The problem of surface localization in intact cells has not been settled. Neither Fagraeus et al. (1974) nor Fujiwara and Pollard (1976) observe binding of antiactin or antimyosin to the surface of unfixed cells. Painter et al. (1975), using immunoferritin staining, find that antibodies to human uterine myosin bind to the cytoplasmic surface of the plasma membrane of human fibroblasts. Ferritin-labeled antibodies to myosin from polymorphonuclear leukocytes also bind only to the cytoplasmic side of the cell membrane (Rikihisa and Mizuno, 1977). In contrast, E. G. Puszkin et al. (1977) find a faint fluorescence when unfixed platelets are stained by indirect immunofluorescence with antibody to platelet myosin rod, and Willingham and co-workers (1974) showed that antisera to purified mouse L-cell myosin agglutinate fibroblasts and that myosin can be localized by immunofluorescence, peroxidase immunocytology, and lactoperoxidase-catalyzed iodination on the surface of intact cells (Willingham et al., 1974; Olden et al., 1976). Such conflicting results should be double-checked by using both antigens and antibodies from a related crossreacting system. In continuation of the surface localization problem, Olden et al. (1976) found no actin (but myosin) on the outer surface of the plasma membrane, whereas Owen et al. (1978), using lactoperoxidase-catalyzed iodination and antibodies to porcine myosin and actin, found no myosin, but considerable amounts of actin on the surface of B (and to a lesser extent of T) lymphocytes. The latter authors also postulate that the surface staining of Ig-bearing cells (but not of non-Ig bearing cells) with FITC-labeled antibodies to actin and myosin solely reflects the presence of antibodies that crossreact with Ig, since it can be depleted by adsorption to Ig-Sepharose; a fact that should be well remembered in all such experiments.

2. Proteins That Mediate the Binding of Microfilaments to Cell Membranes

While the argument as to whether microfilaments attach to the inside of the membrane or whether they traverse it is still in flux, immunological studies have indicated that microfilament bundles of smooth muscle cells (Schollmeyer et al., 1976) and cultured fibroblasts (Lazarides and Burridge, 1975) attach to membranes near the sites where α-actinin is localized. An almost classic example for the insertion of microfilament bundles into a dense matrix is the microvillus of the intestinal brush border (Mooseker and Tilney, 1975). Next to actin (the major protein of the isolated microvili), a 95,000 dalton polypeptide was also found in SDS-PAGE (Tilney and Mooseker, 1971; Mooseker, 1976); and since it coelectrophoresed with α-actinin from muscle, it was designated as the attachment and bridging protein of the microvillus. Other authors have confirmed that the isolated microfilament cores of microvilli only contain actin and the 95,000 dalton polypeptide in a 10:1 ratio (Bretscher and Weber, 1978b,c); but it has recently been shown by immunological methods that α-actinin and the 95,000 dalton protein are not identical (Mooseker et al., cited in Fujiwara et al., 1978; Bretscher and Weber, 1978c). With antibodies to the 95,000 dalton protein becoming available (Bretscher and Weber, 1978c), the role of this protein in the microvillus, and possibly in other cells, should soon be elucidated (see also Section V,D,1).

3. Transmembrane Relationships

Patching and Capping. When certain polyvalent ligands (e.g., antibodies, lectins) are allowed to react with surface structures on cell membranes at low temperatures to prevent pinocytosis, the freely floating membrane structures are agglutinated by the ligands and appear as "patches" (this process does not require energy). If the cells are then allowed to warm up, the patches coalesce and form a "cap" over one pole of the cell (an energy-requiring process) and the complex is eventually endocytosed. Both patching and capping seem to involve the action of cytoplasmic microfilaments. In 1977, Bourguignon and Singer suggested a molecular mechanism for the patching–capping phenomenon: membrane-associated actin is bound to a hypothetical protein X present in the membrane of eukaryotic cells. When any receptor of the membrane is aggregated by its specific multivalent ligand, the aggregate binds to X and so gets connected to actin and myosin on the cytoplasmic surface. The contractile proteins then collect the aggregates into a cap, and this movement requires part of the energy needed in capping. The relationship between external ligands and the intracellular components is best studied by double fluorescent staining. First, the ligand, labeled with one stain (e.g., FITC), is allowed to react with the unfixed cell in suspension; after careful fixation and opening of the membrane, the cytoplasmic components are then visualized with a specific anti-body labeled with another dye (e.g., TRITC). Using a thin sectioning technique (sections < 2 μm) and antibodies to human uterine myosin and chick brain tubulin, and an HMM–biotin–rhodamine-conjugated–avidin complex to visualize actin (Heggeness and Ash, 1977), Bourguignon et al. (1978) showed that lectin- and antibody-induced caps in mouse B and T lymphocytes and in other cells are always accompanied by concentrations of actin and myosin directly under the cap. These "subcaps" are shown to be quite massive, reaching about 0.1–0.3 μm from the membrane into the cytoplasm. Not only capping, but also the preceeding patching pattern are accompanied by actin and myosin accumulations. Very recently, Geiger and Singer (1979) showed that α-actinin is also involved in these processes. Interestingly enough, the membrane-bound α-actinin will not be internalized during endocytosis, suggesting that it may dissociate again from the patches and caps. I would feel more confident in these results if the authors had not used SDS-denatured α-actinin as the immunogen (see Section III,B,1). In contrast to the contractile proteins, tubulin does not show any redistribution when lectin-induced caps were observed in these studies. While Singer and co-workers (Bourguignon and Singer, 1977; Bourguignon et al., 1978) feel that all capping phenomena are mediated by the described mechanism, Schreiner and Unanue (1976) postulate two forms of capping: one represented by surface IgG, which actively involves cytoplasmic filaments, and another, represented by histocompatibility antigens (e.g., H-2 in the mouse) and lectins, which results simply from cross-links and is independent of cytoplasmic motility (also see Schreiner et al., 1977). To explain this difference, Braun et al. (1978) included a kinetic analysis of the capping process in lymphocytes, using a cell smear technique and an antibody to human platelet myosin rod. They agree with Singer's group in the type of capping induced by surface Ig, Fc receptor, and thymus leukemia antigen, which they describe as a rapid process involving cytoplasmic myosin aggregates. They disagree when it comes to caps induced by histocompatibility molecules (H-2 and Theta antigen), which they find to form in a slow process, not connected with myosin accumulations. Lectin-induced capping (here, Concanavalin A) showed an intermediate response. At first, the results of the two groups seem difficult to reconcile. In their latest report, Geiger and Singer (1979) point out, however, that the difference may lie in the technique used to study the "subcaps". Using the cell smear technique in their present experiments, they had more difficulties in locating the intracellular accumulations of contractile proteins than they did in their previous work using the thin sectioning technique.

The relationship between microfilaments and capping have also been studied using autoimmune antibodies to actin (Sundqvist and Ehrnst, 1976; Toh and Hard, 1977; Gabbiani et al., 1977) and the results were comparable; except that Gabbiani finds tubulin to cocap with the IgG receptors. Other studies showed that cytochalasin D can induce cocapping of leukemia viral proteins with actin in infected cells (Mousa *et al.*, 1978) and that cytochalasin B can enhance the polarization of membrane components and actin, as well as capping in transformed (contrast to normal) cells (Sundqvist *et al.*, 1978). One wonders if such mechanisms are related to the postulated *in vivo* "presentation" of actin on the cell surface of virus-infected cells, which may ultimately lead to the formation of autoimmune antibodies to actin.

Biochemical evidence now begins to support the immunological findings on transmembrane relationships. In a murine mastocytoma cell line (P815), large amounts of plasma membrane-derived material is shed by the cell. These "exfoliates" were shown to consist mainly of detergent-stable complexes of actin and the histocompatibility antigen H-2 (Koch and Smith, 1978). Direct and specific cross-linkings between actin and surface IgG in patches (P3 myeloma cells) and caps (lymphocytes) were also shown by biochemical methods after gentle rupture of these cells (Flanagan and Koch, 1978).

In human fibroblasts, transmembrane linkages were also described by double fluorescent techniques for the membrane integral proteins β_2-microglobulin (part of the HLA-complex), and the enzymes aminopeptidase and Na^+,K^+-ATPase, with intracellular actin and myosin (Ash *et al.*, 1977). When the membrane proteins were clustered by their specific antibodies, they apparently became aligned with the actin–myosin filaments underneath the membrane. It seems that there is a very aimed transduction of external signals to the intracellular molecular machinery to produce specific endocytosis of the ligand–receptor complex, a prerequisite to their biological action. Such a mechanism may have more applications than is currently realized.

Another example of transmembrane linkage was presented by Nicolson and Painter (1973) in erythrocyte membranes. The linkage was demonstrated between the inner surface protein spectrin and the integral membrane component glycophorin, using antibodies to spectrin and colloidal iron to label the outer membrane. Lastly, fibronectin, one of the major surface proteins of fibroblasts, and actin were shown to distribute coincidently during cell spreading (Heggeness *et al.*, 1978; Hynes and Destree, 1978). Since actin filaments will appear before the surface streaks of fibronectin, microfilaments may be involved directly in determining the surface distribution of fibronectin, which may then mediate processes such as attachment to or detachment from the substratum.

C. Mitosis and Cytokinesis

1. *Interphase*

The question whether actin and myosin are present in the vertebrate interphase nucleus has been a matter of much dispute and will not be discussed here. Although nuclear staining is rarely observed with specific and purified antibodies to contractile proteins, these findings do not really exclude their presence in amounts too small to be detected by immunofluorescence. During cell division, however, this picture changes drastically and both studies with antibodies and with FITC-labeled HMM have shown a rather characteristic distribution of contractile proteins, even if some of the details are still somewhat contradictory. Aronson (1965), introducing FITC-labeled HMM, was the first to show that actin participates in cell division, and this was later confirmed by Behnke et al. (1971a) and Gawadi (1971). In the following, I will try to summarize the more recent findings. For sake of clarity, I will give detailed references in the first step (prophase); in the following steps, I will quote the authors only when there are disagreements.

2. Prophase

The fluorescent staining of cytoplasmic fibers generally seen in cultured cells at interphase with either FITC-labeled HMM (rat kangaroo cell, Sanger, 1975; HeLa and PtK2 cells, Herman and Pollard, 1978) or with antibodies to actin (PtK1 cells, Cande et al., 1977), myosin (HeLa cells, Fujiwara and Pollard, 1976, 1978), α-actinin (chick embryo cells, Fujiwara et al., 1978), and CDR protein (PtK1 cells, Welsh et al., 1978) becomes more or less diffuse throughout the cells. Diffuse cytoplasmic staining with anti-CDR both in interphase and prophase was described in HeLa cells by Andersen et al. (1978), who also noted a strong fluorescence of the two centriolar poles at late prophase. Nuclear or chromosomal staining was not seen by any author.

3. Metaphase

Actin-antibodies are shown to concentrate in the region between chromosomes and poles, sometimes spindle fibers can be seen extending from the pole toward the chromosomes. Myosin-specific staining, although often diffuse throughout the cytoplasm, delineates the spindle and clearly excludes the chromosomes. There is also some cortical staining with antimyosin. Antibodies to α-actinin weakly and diffusely stain the entire cytoplasm including the spindle, and there are beads of staining noted in the cortex. Anti-CDR staining is restricted to the pole-related part of the spindle, reaching only one-third of the pole–kinetochore distance according to Andersen et al. (1978) or all the way to the chromosomes (Welsh et al., 1978).

4. Anaphase

Sanger (1975) and Cande et al. (1977) still find actin concentrated between chromosomes and poles and, in addition, a faint staining of the spindle interzone where the cleavage furrow will form. In contrast, Herman and Pollard (1978) not only find intense fluorescence between the poles and chromosomes, but also in the entire interzone separating the chromosomes, rather than in the cleavage furrow. The distribution of myosin and α-actinin is now rather similar: diffuse staining of the spindle and a brightly fluorescent ring around the equator between the separated chromosome. CDR-specific staining is still restricted to the poles and not present in the interzone. Welsh *et al.* (1978) observe occasional fluorescent patches associated with the chromosomes.

5. *Telophase*

Actin is still concentrated in the cleavage furrow according to Sanger (1975) and Cande *et al.* (1977), whereas Herman and Pollard (1978) still find actin throughout the entire interzone. As cell division progresses, antimyosin and anti-α-actinin staining intensifies in the concave portion of the furrow. When the midbody forms, however, the staining pattern becomes different. Antimyosin staining decreases and vanishes completely by late telophase, whereas anti-α-actinin staining will persist. A very individual pattern is seen in the distribution of CDR protein: it will rather abruptly appear in the interzone region, while the polar fluorescence diminishes. In late telophase, very intensely fluorescent patches on either side of the bridge are noted, leaving the central part unstained.

The pattern of the multifunctional regulator protein does not necessarily reflect any association with the contractile events during cell division. As a matter of fact, Marcum *et al.* (1978), finding that CDR can both inhibit and reverse *in vitro* microtubule assembly in a Ca^{2+}-dependent manner, claim that it regulates microtubular disassembly during mitosis. The distribution of actin, myosin, and α-actinin in the dividing cell support the electron microscopic evidence that a contractile mechanism is involved in *cleavage* (Schroeder, 1976), and a possible sequence of events is proposed by Fujiwara *et al.* (1978). The available data, however, are still insufficient to explain an active participation of contractile elements in the movement of chromosomes. The possibility that myosin may not be involved in chromosome separation is actually suggested by the experiments of Mabuchi and Okuno (1977). For whereas the injection of small amounts of antimyosin will inhibit cytokinesis and more or less disturb the formation of the mitotic apparatus, the antibody had no effect on an already formed mitotic apparatus and did not prevent chromosomal separation and formation of daughter nuclei. In addition, chromosomal movement *in vitro* is supposedly not inhibited by antibodies to myosin, but by antibodies to dynein, which puts it all back to the microtubular system (Sakai *et al.*, cited in Mabuchi and Okuno, 1977).

While these first sections have dealt mainly with the immunofluorecent localization of cytoplasmic contractile proteins in isolated cells, the following sections will also include results obtained with sectioned tissue.

D. EPITHELIAL CELLS

1. *Intestinal Epithelial Cells*

Columnar epithelial cells of the intestine have a refractile border with very delicate vertical striations. Such a striated or brush border consists of numerous cylindrical cell processes called microvilli. Each microvillus contains a core of actin filaments that is anchored beneath the striated border in a transverse zone of filamentous cytoplasm called the terminal web. It has been postulated by Mooseker (1976) that the microvillar movement is mediated by interaction between the basal ends of the actin cores and myosin present in the terminal web. Using antibodies to platelet myosin rod, Mooseker *et al.* (1978) could indeed localize the antigen exclusively in the terminal web. Shortly thereafter, Bretscher and Weber (1978c) showed that antibodies to actin decorated, as expected, the microvilli, whereas antibodies to myosin heavy chain, to tropomyosin, filamin, and α-actinin only stained the seemingly less-ordered microfilaments of the terminal web. The fluorescence was localized preferentially around the edge of the cells, presumably the regions of the zonulae adherentes. These authors found some regular arrangement or "striations" in their staining patterns, as previously described for tissue culture cells. On an ultrastructural level, we (Drenckhahn and Gröschel-Stewart, 1979) were able to demonstrate antimyosin binding along the rootlets of the microvillar filament bundles and also, though less intensely, within the terminal web (Fig. 1e). Craig and Pardo (1979) also find a preferential localization of anti-α-actinin in a polygonal pattern corresponding to zonulae occludentes (tight junctions) and/or zonulae adherentes (belt desmosomes). An arrangement where actin is localized in surface extensions and myosin more at the cell base may not be exclusive for the intestinal brush border. Chen (1977) finds less myosin in the leading and retracting edges of moving fibroblasts, and Pollard *et al.* (1977) finds myosin to be concentrated in the cell body of activated platelets (see also Section V,G). The described arrangement of contractile proteins in the brush border could explain how surface extensions can be withdrawn or wiggled from side to side.

2. *Ciliated Epithelia*

Actin was shown to be present, both by immunofluorescent staining with autoimmune antibodies to actin, as well as by HMM decoration, in tracheobronchial epithelial cells of rat and man (Reverdin *et al.*, 1975). The microfilaments were shown to be localized around the basal bodies, and this may be coincident with the centers of ATPase activity in the basal bodies of the oviduct, as reported by Anderson (1977). Anti-smooth muscle myosin will also stain the basal bodies of the ciliary epithelium in the human oviduct (U. Gröschel-Stewart and D. Lehmann, unpublished result) and in *Paramecium caudatum* (Gröschel-Stewart and Reder, 1975). However, we found that preimmune sera and nonrelated immune sera give a similar, although less intense staining (e.g., the apical zone in ependymal cells of rat diencephalon; Gröschel-Stewart *et al.*, 1977b). Apparently many nonimmune sera give a strong and "specific" staining of centriole and basal bodies (Connolly and Kalnins, 1978). Maybe this staining could also be overcome by adsorption of the antibodies on fixed epithelial cells, as Bretscher and Weber (1978c) suggested in their work with brush border.

3. Epithelia of Renal Tubules

There are differing reports on the presence of contractile elements in these structures. While Becker (1972) finds no immunoreactivity in kidney epithelial cells with anti-uterus actomyosin, Holborow *et al.* (1975), using immunoperoxidase staining and antibodies to several smooth muscle contractile proteins, report the presence of contractile fibers in the apical portion of kidney tubules, and they consider this finding characteristic for sites of endocytotic activity. We report similar findings, and we have also found strong myosin- and tropomyosin-specific reactivity in the base of the epithelium of proximal and distal tubules (D. Drenckhahn and U. Gröschel-Stewart, unpublished result).

4. Ocular Epithelia (Cornea, Lens, Pigment Epithelium)

Corneal epithelial cells are known for their striking migratory potency when it is necessary to cover tissue defects. We were able to localize smooth muscle-type actin and myosin in these cells and were also able to identify myosin heavy chain in pure corneal abrasions (Drenckhahn and Gröschel-Stewart, 1977). Anterior lens epithelium displayed also actin- and myosin-specific immunoreactivity, which was mainly confined to the apical portion of those epithelial cells that are of ectodermal origin. Another interesting result of these studies has been the localization of myosin and actin in the apical portion of retinal pigment epithelial cells in both rat and frog.

5. Exocrine Glands

Antibodies to actomyosin from normal human colon (Archer and Kao, 1968) and to myosin from human uterus (Rukosuev, 1973) were shown to brilliantly stain myoepithelial cells present in human salivary, sweat, and mammary glands. In a more detailed study (Drenckhahn *et al.*, 1977), we found that smooth muscle-type actin and myosin can be localized in various salivary glands and in exocrine pancreas in three major sites: (1) in myoepithelial cells (see Fig. 2); (2) underneath the cell membrane bordering the acinar lumen of secretory glands (except in the Harderian and mucous lingual glands); and (3) in the epithelial cells of the various secretory ducts of all glands, in a similar distribution to the acinar cells. While the contractile machinery in the ectodermal myoepithelium is similar to that of smooth muscle and the function of these cells can therefore be explained, the presence of microfilaments beneath the acinar lumen (see Fig. 3) is not so easily understood. The function of the filaments here may be similar to that in endocrine cells (vide infra). Another interesting aspect is the association of actin with enzymes (Bray, 1975), especially with the potent pancreatic DNase I, where it acts as a natural inhibitor.

6. Endocrine Glands

Actin has been identified with autoimmune antibodies in the microfilament web at the periphery of pancreatic islet cells (Gabbiani *et al.*, 1974), and an

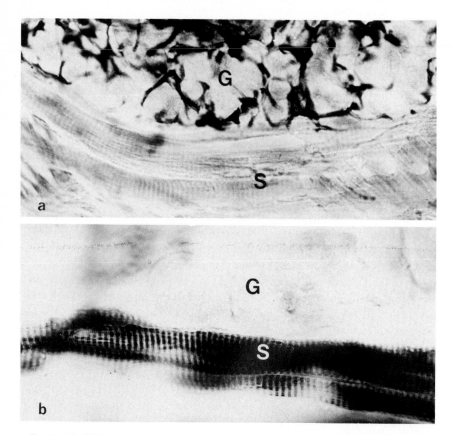

FIG. 2.   Serial frozen sections of rat tongue, stained before embedding by the three layer immunoperoxidase technique with antibodies to chicken gizzard myosin (a) and to chicken pectoral myosin (b). The myoepithelial cells of the lingual glands (G) react with the gizzard myosin antibody; the skeletal muscle fibers (S) react with the pectoral muscle antibody only. Note the A-band staining. ×560. (Courtesy D. Drenckhahn.)

interesting model about its role in the secretion of insulin has been proposed by Malaisse and Orci (1979). According to this theory, the secretory granules are kept away from the plasma membrane by the microfilament web when the cell is not stimulated. During the first phase of secretory release, granules in the close vicinity of the cell web are extruded. In the later phase, granules of the cell interior are transported along oriented microtubular pathways and incorporated into the labile peripheral pool. A similar function may be ascribed to the contractile proteins found near the acinar lumina of exocrine glands (vide supra). Since it has recently been suggested that antibodies to DNase I can be used to localize actin filaments (Wang and Goldberg, 1978), I wish to mention that we were unable to stain, with antibodies to DNase I, the microfilament web of the islet

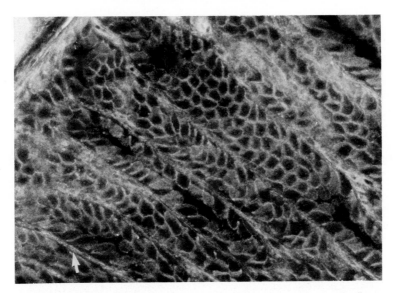

FIG. 3.   Frozen section of the chicken proventricle, stained by the indirect immunofluorescence technique with antibodies to chicken gizzard actin. In this oblique section, note the staining of the cytoplasm beneath the cell membrane of the acinar cells, and of the smooth muscle of the muscularis mucosae (arrow). ×255.

cells, whereas the exocrine cells showed the same staining pattern as reported above for antiactin and antimyosin (H. G. Mannherz and U. Gröschel-Stewart, unpublished result).

The immunofluorescent localization of myosin in exocrine and endocrine glands has now been supported by the isolation and characterization of myosin from secretory tissue (Ostlund *et al.*, 1978), and these authors could further show that *in vitro* salivary gland myosin will crossreact with antibodies to smooth muscle and to fibroblast myosin.

7. *Hepatocytes*

As early as 1966, Vassiletz and Zubzhitsky showed by immunofluorescence that the myosin-like protein from the external membrane of liver cells is distinct from myofibrillar protein, assigning the difference to the changing function of the liver protein from contraction to membrane transport. Liver cells were also among the first nonmuscle cells shown to react with smooth muscle autoimmune antibodies (Johnson *et al.*, 1966). The staining patterns were either continuous (polygonal) or double and interrupted (bile-canalicular), and I have never found out if they actually represent related structures. With experimentally produced antibodies to smooth muscle actin, myosin, and tropomyosin, one generally observes fluorescence in double lines or dots in the cytoplasm around the bile canaliculi (Heine *et al.*, 1976; Miller *et al.*, 1976; Creutz, 1977; D. Drenckhahn, unpublished result). It seems interesting that bile canalicular-like structures can also be seen when isolated hepatocytes are allowed to reaggregate into epithelial cell sheets (Miettinen *et al.*, 1978). These structures and other functional complexes were shown to stain intensely with autoimmune antibodies to actin.

Phalloidin was shown to preferentially bind to the plasma membrane of liver cells (Govindan *et al.*, 1972). Chronic administration of low doses of the toxin to rats will cause a most striking increase of microfilament structures in the hepatocytes, especially around the bile canalicular system. This was shown by immunofluorescent studies with autoimmune antibodies to actin (Gabbiani *et al.*, 1975). Since the toxin is known to irreversibly stabilize F-actin filaments *in vitro*, this increase was attributed to a disturbance of the physiological turnover of microfilamentous actin in the cell. Myosin immunoreactivity was shown by us to increase as dramatically as actin (Heine *et al.*, 1976), and it would be worthwhile to test whether other contractile proteins are involved in this disturbance, too.

8. *Thymic Epithelial Cells*

The reticular epithelial cells of this gland form a loose three-dimensional network, which presumably has a supportive and skeleton-like function. This network and the Hassal's bodies arising from it will stain strongly with antibodies to smooth muscle actin and myosin in rat, guinea pig, cat, chicken, and man (Drenckhahn *et al.*, 1978b). Reticular epithelial cells of chicken bursa and thymus were shown to react strongly with autoimmune antibodies to actin (Boyd *et al.*, 1977). The physiological function of such a contractile network is still unclear; Boyd and co-workers postulate that it may assist in the expulsion of the mature lymphocytes. In their immunoreactivity, the epithelial cells contrast with the myoid cells also found in this gland, which will only stain with antibodies to striated muscle myosin (Hayward, 1972; Drenckhahn *et al.*, 1978b) and the autoimmune antibodies of patients with myasthenia gravis (e.g., Van der Geld and Strauss, 1966).

In summarizing the findings on epithelial cells, Drenckhahn, in a personal communication, drew my attention to the fact that only those cells having a functional polarity (acinar and duct cells of exocrine glands, intestinal epithelium, liver cells, and pigment epithelium) will have the contractile proteins concentrated beneath the luminal border, whereas the other epithelia display a rather diffuse distributional pattern.

E. Blood Vessels

1. *Endothelium*

Endothelial contractility may play a role in the regulation of vascular permeability, inflammatory processes, and the initiation of thrombus formation. The

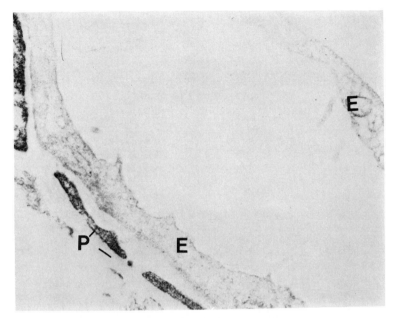

FIG. 4.   Electron micrograph of a capillary in rat submandibular gland, showing the distribution of smooth muscle myosin-specific immunoreactivity (anti-chicken gizzard myosin), using the unlabeled antibody enzyme technique. The endothelial cells (E) of the capillary do not react, only the surrounding processes of the pericytes (P). No heavy metal counterstain. ×19,000. (Courtesy D. Drenckhahn.)

cells have been shown to contain contractile proteins antigenically similar to platelet and uterine actomyosin (Becker and Nachman, 1973; Jaffe *et al.*, 1973; Burkl *et al.*, 1979) and antigenically dissimilar to actin and myosin from chicken smooth muscle (Chamley *et al.*, 1977a). The staining of capillary walls that we have occasionally noted with the latter antibodies, as in adrenal medulla or brain (Unsicker and Gröschel-Stewart, 1978; Owman *et al.*, 1978), has now been attributed in ultrastructural studies to the presence of perivascular cells such as pericytes and astrocytes (D. Drenckhahn, unpublished results), as shown in Fig. 4. An attempt to raise antibodies to endothelial myosin has so far not been successful (Moore *et al.*, 1977).

2. Perivascular Cells

By their morphology, pericytes, which are contractile cells closely associated with the endothelium of capillaries, are often grouped in between fibroblasts and smooth muscle cells. They resemble fibroblasts in that they react strongly with antibodies to uterine myosin (with a muscle and a cytoplasmic component). They differ from fibroblasts in that they also react with antibodies to gizzard myosin (which has only a muscle component) (D. Drenckhahn, unpublished result). A similar immunological behavior is shown by the mesangial cells of the glomerulum, which react with antibodies to uterine actomyosin (Becker, 1972; Scheinman *et al.*, 1978; as in contrast to Rukosuev and Nanaev, 1975). The cell body and the processes of podocytes and the capsular epithelium of rat corpuscles give a sharp immunofluorescent staining with antibodies to gizzard myosin and tropomyosin (D. Drenckhahn, unpublished results). Both mesangial cells and podocytes were shown to react with ferritin-labeled autoimmune antibodies to actin (Accinni *et al.*, 1975).

F. CONNECTIVE TISSUE CELLS

1. *Fixed Cells*

The most common representatives of these cells are the fibroblasts, synthesizing the precursors of extracellular fibrous and amorphous components of the connective tissue proper. They have been extensively dealt with in Section V,A, and shall not be discussed further. A very interesting relative of the fibroblast is the "contractile interstitial cell" in the intraalveolar septa of the lung, shown to react with antibodies to actin (autoimmune) and to myosin (Kapanci *et al.*, 1979), as well as to tropomyosin (D. Drenckhahn, unpublished result). These cells, which have microfilament bundles organized into intracytoplasmic "muscles" are supposed to regulate the ventilation/perfusion ratio of the lung at an alveolar level.

It should also be mentioned here that all embryonic cells, such as myoblasts, chondroblasts, and osteoblasts do not give any reaction with our antibodies to chick smooth muscle actin and myosin (Chamley *et al.*, 1977a; D. Drenckhahn, unpublished result).

2. *Mobile Cells*

Cells capable of ameboid movement and, in part, of phagocytosis, emigrate from the blood into the connective tissue. These cells, which are closely connected with the organism's defense mechanism, comprise histiocytes, monocytes, lymphocytes, and granulocytes, and they all were shown to have the components of a contractile system. Thymus-derived and peripheral lymphocytes (Fagraeus *et al.*, 1973, 1974) react in their periphery with autoimmune antibodies to actin, provided the cell membrane had been damaged by acetone fixation; and these findings have been fully confirmed by Gabbiani (1973). When cells from lymphoid cell lines are suspended in citrate buffer or treated with other chelating agents, they form microvillar protrusions that stain very strongly with autoantibodies to actin, as well as with antibodies to rabbit actin (Fragraeus *et al.*, 1978). At the same time, the staining of the cell body is greatly reduced, and so these findings stress again the importance of membrane protrusions for cellular motility. Both mature B and T cells were shown to react with autoantibodies to actin (Boyd *et al.*, 1977), although these authors feel that cell-to-cell contact was needed for its actual expression. Antibodies to leukocyte myosin have not been very efficient in localizing myosin in leukocytes by the immunoperoxidase method (Shibata *et al.*, 1975). This was probably due to technical difficulties, since the same group (Senda *et al.*, 1979), using HMM decoration, find that the immobile state leukocyte has microfilaments located mainly in the cytoplasm beneath the cell membrane, whereas the moving cell is stained predominantly in the pseudopod and the anterior part of the granuloplasm. Actin-binding protein has also been localized on the cytoplasmic side of polymorphonuclear leukocytes by immunofluorescent studies on "inside-out" membrane preparations (Boxer *et al.*, 1976), but it also seems to be present (by immunoperoxidase labeling) on the plasma membrane itself. Our own attempts to localize actin and myosin in mouse peritoneal macrophages are not too convincing due to improper fixation (Gröschel-Stewart and Gröschel, 1974), although the studies were helpful in solving the problem of myosin surface localization. In unfixed tissue sections, none of the mobile cells of the connective tissue were found to react with antibodies to gizzard contractile proteins (D. Drenckhahn, unpublished result).

An impressive study of Berlin and Oliver(1978) describe the distribution of actin and tubulin in phagocytizing rabbit peritoneal macrophages by immuno-fluorescence. In resting cells, they find a well-oriented system of fibers and fluorescent patches (attachment points) with an autoantibody to actin. Cells actively engaged in phagocytosis of opsonized erythrocytes show a marked concentration of antiactin fluorescence at the sites where the particle to be engulfed is attached, whereas such areas are quite devoid of antitubulin staining. The surface structure changes seen here are reminiscent of those described for capping.

G. THROMBOCYTES

The presence of contractile proteins in platelets is well established by biochemical methods. The question as to whether they are present on the cell surface is, however, still unsolved. It has yet to be ruled out (as in all other surface localization studies, too) that there was no contamination of the original immunogen with membrane constituents; also, in the case of the very labile platelets, membrane damage and actomyosin spilling have to be ruled out as a possible source of coating on the surface of intact cells. Platelets have been shown to react without surface staining with autoantibodies to actin (Gabbiani *et al.*, 1972; Norberg *et al.*, 1975a), and with antibodies to platelet myosin rod (Pollard *et al.*, 1977). Surface staining was noted with antibodies to platelet myosin (Booyse *et al.*, 1971), to platelet myosin rod (E. G. Puszkin *et al.*, 1977), and to platelet and uterine actomyosin (Becker and Nachman, 1973). Using mostly fixed platelets, we did not check for surface staining. We found platelets to react with antibodies to human uterine actomyosin and myosin (Gröschel-Stewart, 1971; Burkl *et al.*, 1979) and to platelet actin (U. Gröschel-Stewart and I. Thiele, unpublished results), but not with antibodies to the gizzard contractile proteins. The precise subcellular localization of microfilaments by immunohistochemical methods is still somewhat unclear. Pollard *et al.* (1977) report that ADP-activated platelets, known to form numerous filopodia that consist predominantly of microfilament bundles (Zucker-Franklin *et al.*, 1967; White, 1968; Behnke *et al.*, 1971b) of the same polarity (Asch *et al.*, 1975), will have most of their myosin localized in the cell body and only little will extend for a short distance into the filopod. This arrangement very much resembles the morphology described for the brush border, and a contractile mechanism, helping in clot retraction, seems quite feasible.

In summarizing the findings on connective tissue cells, I am still tempted to claim that our antibodies to avian smooth muscle contractile proteins give no or only minimal reaction with these cells; the exceptions being the rather mobile vascular pericytes, the glomerular podocytes, and possibly also the peritoneal macrophages.

H. Smooth Muscle-Like Cells (Myoid Cells, "Myofibroblasts")

In the reproductive system of adult male vertebrates, epithelioid cells with cytological and contractile properties of smooth muscle cells are found. Due to their atypical shape and their organization, they cannot be called true muscle cells and have hence been named "myoid". The myoid or peritubular cells of the lamina propria surrounding the seminiferous tubules, presumably participating in sperm transport, can be decorated with HMM (Toyoma, 1977) and will stain in immunofluorescence with antibodies to uterine myosin (Rukosuev, 1976) and to gizzard smooth muscle myosin, actin (Gröschel-Stewart and Unsicker, 1977), and tropomyosin (D. Drenckhahn, unpublished result). The latter noticed that striational patterns can be observed when low antibody concentrations are applied. In sections of guinea pig testes, we have not been able to localize contractile proteins in Sertoli cells, although Franke *et al.* (1978) were able to show the presence of actin- and α-actinin-immunoreactivity in the junctions between spermatid heads and the remnants of Sertoli cells. The presence of actin in the sperm head is still under debate. Talbot and Kleve (1978), using antibodies to shrimp tail actin, find positive staining in the concave margin and the equatorial segment in the acrosomal region, in the connecting piece, and in the principal piece of hamster sperm. Autoimmune antibodies to actin are reported to stain the postacrosomal region of mammalian spermatozoa (Clarke and Yanagimachi, 1978). Amsterdam and myself found staining of rat sperm heads not only with antibodies to actin and myosin, but also with nonspecific antibodies (A. Amsterdam and U. Gröschel-Stewart, unpublished result), and this is in agreement with the findings of Franke *et al.* (1978).

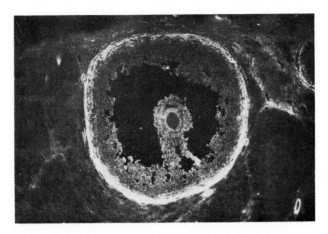

FIG. 5.   Frozen section of a Graafian follicle of a mature rabbit ovary, stained in the indirect immunofluorescence technique with an antibody to chicken gizzard myosin. Note the intense staining of a coherent band of smooth muscle-cells in the theca externa, the thin fluorescent band beneath the zona pellucida of the oocyte, and the peripheral staining of the granulosa cells. ×50.

The theca externa of rodent ovaries also has immunoreactive myosin-containing cells (Kapinus and Rukosuev, 1975). With antibodies to gizzard actin and myosin, Amsterdam et al. (1977) showed that the growth and maturation of the rodent ovary is attended by the development of a smooth muscle layer in the theca externa, suggesting that its contractility may play a role in the extrusion of the oocyte. A close association of these contractile cells with adrenergic nerve fibers (Walles et al., 1978) shows that these cells are under nervous control (in contrast to the myogenic activity proposed for the testicular peritubular cells). The large oocytes in mature ovaries were shown to also have a very narrow band of actin- and myosin-specific fluorescence located just beneath the zona pellucida (see Fig. 5); their function seemed unclear to us (Amsterdam et al., 1977). Franke et al. (1976), who identified considerable amounts of immunoreactive actin in the cortex of amphibian oocytes, assume that it may, at least in part, represent storage material that becomes functional during early embryo development.

I. NEURONAL CELLS AND TISSUES

1. Autonomic Nervous System

The autonomic nervous system mediates activity by two motor neurons placed in series, the first lying in the central nervous system, the second in a peripheral ganglion. Chromaffin cells of adrenal medulla represent a modified second autonomic neuron. By origin, the catecholamine containing granules of chromaffin cells are homologous to the synaptic vesicles of sympathetic nerve cells. They are released from the adrenal medulla by exocytosis, a process probably requiring the presence of contractile proteins. Microfilaments were shown to be present in a discontinuous network subtending the plasma membrane, and they could be decorated by HMM (Cooke and Poisner, 1976). Actomyosin-like protein had also been isolated from bovine adrenal medulla (Poisner, 1970), and this was followed by the isolation of pure actin (Phillips and Slater, 1975) and myosin (Creutz, 1977; Johnson et al., 1977). The light chain pattern of medullar myosin resembled that of smooth muscle and cytoplasmic myosins (20,000 and 17,000 dalton); contrast this to the three light chain pattern found in brain and sympathetic nerves (see Section II,C). One must assume, however, that the preparations of medullar contractile proteins consist of mixtures of neuronal and vascular proteins. Immunofluorescent studies with antibodies to medullar myosin (Creutz, 1977) showed no crossreaction with striated muscle, but a strong reaction with platelets, leukocytes, and large and small blood vessels. Surprisingly, even in sections of brain, no other structures but blood vessels and endothelia were seen to fluoresce. In the adrenal medulla itself, a strong reaction was noted in the cells located between the chords of chromaffin cells, which form a peripheral layer around chromaffin cell groups. These smooth muscle-like interstitial cells also display a strong immunoreactivity with antibodies to gizzard actin, myosin, and tropomyosin (Unsicker and Gröschel-Stewart, 1978; D. Drenckhahn, unpublished result). The chromaffin cells themselves stained rather weakly with the antibody to adrenal medullar myosin (Creutz, 1977), and no specific fluorescence was noted by us in guinea pig or rat chromaffin cells (Unsicker and Gröschel-Stewart, 1978). Jokusch et al. (1977), showed a rather specific association of α-actinin fluorescence with isolated chromaffin granules. It was rather a surprise that Trifaró et al. (1978) found intense myosin-specific fluorescence in fixed and unfixed chromaffin cells with an antibody to bovine striated muscle myosin. At the present time, the immunological classification of chromaffin cell contractile proteins is far from being clear.

2. Peripheral Nervous System

Within axons and dendrites of neurons, organelles and macromolecules are transported from the cell body (the main site of synthesis) to the synaptic terminal (anterograde) or from the terminals backward to the cell body (retrograde). The rate of movement of material is different, and fast and slow axonal transport are differentiated. It has been suggested that the slow transport may be mediated by an interaction between myosin attached to neurofilaments and actin anchored in the axolemma (Lasek and Hoffman, 1976). One of our antibodies to chicken gizzard myosin (see Section III,D,3) gives a strong and specific fluorescence with the axoplasm of various peripheral nerves (sciatic, cranial nerves V,VII,X) and the dorsal and ventral spinal roots, as well as with Schwann cells (Unsicker et al., 1978a,b). The white and grey matter of the spinal cord had only a few fibers reacting with the antibody, and these were mainly confined to the dorsal column. On an ultrastructural level, after staining according to Sternberger's unlabeled antibody enzyme technique (Sternberger *et al.*, 1970), the following antibody binding sites were seen: the axolemma, the membranes of the agranular endoplasmic reticulum, and the neurofilaments. Here, an interrupted irregular pattern was frequently noted (D. Drenckhahn, J. R. Wolff, and U. Gröschel-Stewart, unpublished results). A similar staining pattern was observed when Ca^{2+}, Mg^{2+}-activated ATPase was tested on sequential sections. These findings, together with the observation that actin may be integrally associated with neuro-filament preparations (Schlaepfer and Freeman, 1978), may lend additional support to the theory of Lasek and Hoffman. Hopefully, we can substantiate our cytochemical findings with biochemical data in the near future.

3. *Central Nervous System*

a. *Brain.*   The criticism that the early preparations of actomyosin-like pro-teins from whole brain (Poglazov, 1961; Berl and Puszkin, 1970; S. Puszkin and Berl, 1972) were mainly derived from blood vessels, endothelium, and other cells, rather than from the neuronal tissue itself, was soon reconciled by the selective use of synaptosomal fractions of brain for the extraction procedure (S. Puszkin *et al.*, 1972; Blitz and Fine, 1974). In addition, Puszkin and co-workers showed that antibodies to brain actomyosin did not crossreact with actomyosin from aorta. Unfortunately, immunofluorescent localization was not attempted with these early antibodies. Very recently, highly purified myosin and actin were prepared and characterized from (whole !) brains from one-day-old chickens (Kuczmarski and Rosenbaum, 1979a,b). Antibodies were prepared to the myosin, and they did not crossreact in the immunodiffusion test with myosin from smooth and striated muscle. In direct immunofluorescence, anti-brain myosin was shown to stain cultured nerve cells of the chicken dorsal root gan-glion. Diffuse staining was seen in the growth cones, in neurites, and in the cell body, excluding the nucleus. Satellite cells present in the culture, and cardiac myoblasts were also stained; the stress fibers of fibroblasts showing the known regular periodicities. From these data, the authors conclude that either large amounts of brain myosin come from Schwann cells, glial cells, and astrocytes; or that the cytoplasmic myosins from nerve and satellite cells are immunologically similar. It seemed surprising that the crossreaction with smooth muscle cells and endothelium was not tested in the more sensitive immunofluorescent test, too, since whole brains would certainly contain these cells as additional sources of immunogen. It would have been most interesting to test this possible crossreac-tion in view of the observations of Creutz (1977) and also in view of our own data:antibodies to chicken gizzard actin and myosin, shown to react minimally or not at all with fibroblasts, endothelia, and most other mesenchymal cells, strongly and specifically stain astrocytes and ependymal cells of the rat and

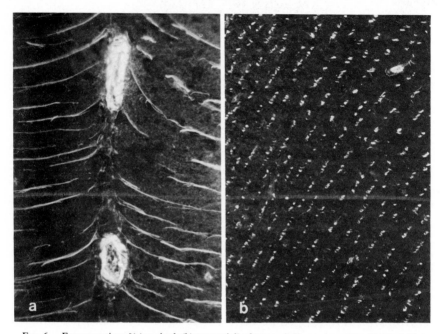

Fɪɢ. 6.   Frozen sections [(a) sagittal; (b) tangential] of rat cerebellar cortex, stained in the indirect immunofluorescent technique with antibodies to chicken gizzard tropomyosin. Note the intense reaction of the neuroglial cells (Golgi epithelial cell, Bergmann fibers) and of the blood vessel walls. ×560. (Courtesy D. Drenckhahn.)

human central nervous system (Gröschel-Stewart *et al.*, 1977b; Braak *et al.*, 1978; Drenckhahn *et al.*, 1978a). The staining of neuroglial fibers of rat and human brain was even crisper with antibodies to tropomyosin, as seen in Fig. 6 (D. Drenckhahn, unpublished result), and also with antibodies to glial fibrillary acidic protein (Braak *et al.*, 1978), than with antibodies to myosin.

b. *Retina.*   A strong actin- and myosin-specific immunofluorescence was localized in rod inner segments (Drenckhahn and Gröschel-Stewart, 1977) and in the external plexiform layer of rat retina (Drenckhahn, 1979). Electron microscopy revealed a specific immunoreactivity (peroxidase label) both in the myoid and ellipsoid portion of rod inner segments, and in the photoreceptor synapses located in the external plexiform layer. In the cytoplasm of rod inner segments, immunoperoxidase label showed a diffuse distribution, whereas a clear association with the membranes of synaptic vesicles was seen in the photoreceptor synapses (Drenckhahn, 1979). These immunological results have recently been supported by the isolation of myosin from bovine retina (Hesketh *et al.*, 1978).

4. *Tissue Cultures of Neural Cells*

Considering all the possible contaminants with nonneuronal cells in tissue extracts, cultured cells should be an ideal source for the extraction of specific actin and myosin, provided that these cultures do not undergo "modulation," as has been described for smooth muscle cells (see Section V,A,2,b). Myosin has been isolated in quantities sufficient for immunizations from clonal rat glial cells (Ash, 1975) and from cloned rat glioma and mouse neuroblastoma cells (Miller and Kuehl, 1976). Cultured neuroblastoma cells have also been used for the immunofluorescent localization of actin (Lessard *et al.*, 1976; Isenberg *et al.*, 1977). The authors noted that flattened cells usually displayed a rather unoriented network of actin filaments, with concentrations in the growth cones, microspikes, and membrane ruffles.

K. INVERTEBRATE TISSUE

Unfortunately, little immunofluorescence work has been performed with invertebrate cytoplasmic contractile proteins. This may be largely due to the lack of crossreactivity with antibodies to vertebrate proteins, to the lack of specific antibodies to invertebrate contractile systems, to nonspecific fluorescence (which I found to be quite a problem), and, lastly, to problems of fixation, especially in protozoa (Pollard and Ito, 1970). There is encouragement from three recent reports; one by Eckert and Lazarides (1978), who studied the distribution of actin in feeding and migrating *Dictyostelium discoideum* with antibodies to avian smooth muscle actin. The other report (Erlandsen *et al.*, 1978) describes the localization of actin- and α-actinin-specific immunofluorescence in the rim of an adhesive disc in an intestinal protozoon, *Giardia muris;* the antibodies used were directed to avian smooth muscle proteins. Finally, Pinder and co-workers (1978b) raised an antibody to human spectrin, and found by immunofluorescence that it not only stains the membranes of human, mouse, chicken, frog, and newt erythrocytes, but also the membrane of the large and complex erythrocytes of the most primitive animal to enclose its respiratory pigment in cells, the polychaete *Terebella lapidaria*. The crossreacting antigen was found to be a 270,000 dalton polypeptide. (The list of highly conserved structural proteins keeps growing.) And lastly, in my own group, antibodies to highly purified insect muscle myosin and actin are successfully being applied to the localization of epitheliomuscular cells in invertebrates.

VI. Conclusions and Outlook

The recent progress in the immunological aspects of cytoplasmic contractile protein research is really impressive, as more than one-third of the references quoted here have appeared in print within the last three years. Since 1974, we have gone far beyond the point of merely identifying "stress-fibers" by immunofluorescent staining. Had Isenberg and collaborators (1976) not shown that these structures are capable of contraction, we might still wonder about their function, since they react with so many and various antibodies.

The main emphasis of the present research has changed from topographical localization to physiological function. While most of us are still biased by the concept of the sarcomere as the basic contractile unit, fascinating new concepts of contractility are emerging such as the studies on brush border and thrombocyte filopodia.

Transmembrane linkages, either directly or via a mediator, bring up additional aspects about "remote control" interactions of external stimuli and the internal contractile system such as in capping and phagocytosia. Ca^{2+} ions are important mediators of all these functions, and Durham's (1974) impressive treatise on a unified theory of the control of actin and myosin in nonmuscle movements has gained additional weight by the recent findings that a Ca^{2+}-dependent protein acts as the modulator of myosin kinases, which regulate the interaction of actin and myosin in nonmuscle cells.

The quibblings about whose antibodies are specific and whose are not have quieted down somewhat. Basically, we have all more or less reached the same conclusions as to the localization of cytoplasmic contractile proteins. When it comes to the problem of the phylogeny and ontogeny of these proteins, however, it seems important that we should all use antibodies of comparable specificities. Some agreement to have reference standards, as are available from NIH and MRC for endocrinological work, would surely be desirable.

Drenckhahn and I have attempted in the last section of this chapter to group cytoplasmic contractile elements according to function and origin, and although this suggestion is in part based on the results with our own antibodies, we hope that it may be acceptable to others.

One decade ago, I was often teased about my enthusiasm for immunofluorescence by my colleagues in more abstract fields. In spite of this, I have found it to be the most esthetic aspect of my work with contractile proteins. However, the future of this research, I feel, will lie in ultrastructural localization. In this field (if the kind reader will please turn to Harvey's philosophizing at the beginning), there is much that still remains unknown.

ACKNOWLEDGMENTS

My contribution to the research on contractile proteins has been supported by the Deutsche Forschungsgemeinschaft since 1964. I wish to thank James C. Stewart for the emotional support and assistance in translation, and especially, I want to thank Detlev Drenckhahn, Anatomisches Institut, Kiel, for his inspiration and scientific assistance.

REFERENCES

Abramowitz, J. W., Stracher, A., and Detwiler, T. C. (1975). *Arch. Biochem. Biophys.* **167**, 230–237.

Accinni, L., Natali, P. G., Vassallo, L., Hsu, K. S., and Martino, C. (1975). *Cell Tissue Res.* **162**, 297–312.

Aita, M., Conti, G., Laszt, L., and Mandi, B. (1968). *Angiologica* **5**, 322–332.

Aksoy, M. O., Williams, D., Sharkey, E. M., and Hartshorne, D. J. (1976). *Biochem. Biophys. Res. Commun.* **69**, 35–41.

Amsterdam, A., Lindner, H. R., and Gröschel-Stewart, U. (1977). *Anat. Rec.* **187**, 311–328.

Andersen, B., Osborn, M., and Weber, K. (1978). *Cytobiologie* **17**, 354–364.

Andersen, P., Small, J. V., and Sobieszek, A. (1976). *Clin. Exp. Immunol.* **26**, 57–66.

Anderson, R. G. W. (1977). *J. Cell Biol.* **74**, 547–560.

ap, Gwynn, I. Kemp, R. B., Jones, B. M., and Gröschel-Stewart, U. (1974). *J. Cell Sci.* **15**, 279–289.

Archer, F. L., and Kao, V. C. Y. (1968). *Lab. Invest.* **18**, 669–674.

Aronson, J. F. (1965). *J. Cell Biol.* **26**, 293–298.

Asch, A., Elgart, E. S., and Nachmias, V. T. (1975). *J. Cell Biol.* **67**, 12a.

Ash, J. F. (1975). *J. Biol. Chem.* **250**, 3560–3566.

Ash, J. F., Vogt, P. K., and Singer, S. J. (1976). *Proc. Natl. Acad. Sci. U.S.A.* **73**, 3603–3607.

Ash, J. F., Louvard, D., and Singer, S. J. (1977). *Proc. Natl. Acad. Sci. U.S.A.* **74**, 5584–5588.

Bagby, R. M., and Pepe, F. A. (1978). *Histochemistry* **58**, 219–235.

Barylko, B., Kuźnicki, J., and Drabikowski, W. (1978). *FEBS Lett.* **90**, 301–304.

Becker, C. G. (1972). *Am. J. Pathol.* **66**, 97–110.

Becker, C. G., and Murphy, G. E. (1969). *Am. J. Pathol.* **55**, 1–37.

Becker, C. G., and Nachman, R. L. (1973). *Am. J. Pathol.* **71**, 1–22.

Begg, D. A., Rodewald, R., and Rebhun, L. I. (1978). *J. Cell Biol.* **79**, 846–852.

Behnke, O., Forer, A., and Emmersen, J. (1971a). *Nature (London)* **234**, 408–410.

Behnke, O., Kristensen, B. I., and Nielsen, L. E. (1971b). *J. Ultrastruct. Res.* **37**, 351–369.

Benninghoff, A. (1926). *Z. Zellforsch. Mikrosk. Anat.* **4**, 125–170.

Berl, S., and Puszkin, S. (1970). *Biochemistry* **9**, 2058–2067.

Berlin, R. D., and Oliver, J. M. (1978). *J. Cell Biol.* **77**, 789–804.

Bettex-Galland, M., and Lüscher, E. F. (1959). *Nature (London)* **184**, 276–277.

Blikstad, I., Markey, F., Carlsson, L., Persson, T., and Lindberg, U. (1978). *Cell* **15**, 935–943.

Blitz, A. L., and Fine, R. E. (1974). *Proc. Natl. Acad. Sci. U.S.A.* **71**, 4472–4476.

Booyse, F. M., Hoveke, T. P., Zschocke, D., and Rafelson, M. E., Jr. (1971). *J. Biol. Chem.* **246**, 4291–4297.

Booyse, F. M., Hoveke, T. P., and Rafelson, M. E., Jr. (1973). *J. Biol. Chem.* **248**, 4083–4091.

Bottazzo, G. F., Christensen, A. F., Fairfax, A., Granville, S., Doniach, D., and Gröschel-Stewart, U. (1976). *J. Clin. Pathol.* **29**, 403–410.

Bourguignon, L. Y. W., and Singer, S. J. (1977). *Proc. Natl. Acad. Sci. U.S.A.* **74**, 5031–5035.

Bourguignon, L. Y. W., Tokuyasu, K. T., and Singer, S. J. (1978). *J. Cell Physiol.* **95**, 239–258.

Boxer, L. A., and Stossel, T. P. (1976). *J. Clin. Invest.* **57**, 964–976.

Boxer, L. A., Richardson, S., and Floyd, A. (1976). *Nature (London)* **263**, 249–251.

Boyd, R. L., Toh, B. H., Muller, H. K., and Ward, H. A. (1977). *Int. Arch. Allergy Appl. Immunol.* **55**, 283–292.

Braak, E., Drenckhahn, D., Unsicker, K., Gröschel-Stewart, U., and Dahl, D. (1978). *Cell Tissue Res.* **191**, 493–499.

Braun, J., Fujiwara, K., Pollard, T. D., and Unanue, E. R. (1978). *J. Cell Biol.* **79**, 409–418.

Bray, D. (1973). *Cold Spring Harbor Symp. Quant. Biol.* **37**, 567–571.

Bray, D. (1975). *Nature (London)* **256**, 616.

Bray, D., and Thomas, C. (1976). *J. Mol. Biol.* **105**, 527–544.

Bretscher, A., and Weber, K. (1978a). *FEBS Lett.* **85**, 145–148.

Bretscher, A., and Weber, K. (1978b). *Exp. Cell Res.* **116**, 397–407.

Bretscher, A., and Weber, K. (1978c). *J. Cell Biol.* **79**, 839–845.

Brotschi, E. A., Hartwig, J. H., and Stossel, T. P. (1978). *J. Biol. Chem.* **253**, 8988–8993.

Buckley, I. K., Raju, T. R., Stewart, M., and Irving, D. O. (1978). *Congr. Int. Union Pure Appl. Biophys. 5th, 1978* Abstract, p. 305.

Burkl, B., Mahlmeister, C., Gröschel-Stewart, U., Chamley-Campbell, J., and Campbell, G. R. (1979). *Histochemistry* **60**, 135–143.

Burridge, K. (1974). *FEBS Lett.* **45**, 14–17.

Burridge, K. (1976a). *Proc. Natl. Acad. Sci. U.S.A.* **73**, 4457–4461.

Burridge, K. (1976b). *In* "Cell Motility" (R. Goldman, T. Pollard, and J. Rosenbaum, eds.), pp. 739–747. Cold Spring Harbor Lab., Cold Spring Harbor, New York.

Burridge, K., and Bray, D. (1975). *J. Mol. Biol.* **99**, 1–14.

Campbell, G. R., Uehara, Y., Mark, G., and Burnstock, G. (1971). *J. Cell Biol.* **49**, 21–34.

Campbell, G. R., Chamley, J. H., and Burnstock, G. (1974). *J. Anat.* **117**, 295–312.

Cande, W. Z., Lazarides, E., and McIntosh, J. R. (1977). *J. Cell Biol.* **72**, 552–567.

Carlsson, L., Myström, L. E., Lindberg, U., Kanaan, K. K., Cid-Dresdner, H., Lövgren, S., and Jörnvall, H. (1976). *J. Mol. Biol.* **105**, 353–366.

Carsten, M. E., and Mommaerts, W. F. H. M. (1963). *Biochemistry* **2**, 28–32.

Cebra, J. J., and Goldstein, G. (1976). *Methods Immunol. Immunochem.* **5**, 444–447.

Chamley, J. H., and Campbell, G. R. (1974). *Exp. Cell Res.* **84**, 105–109.

Chamley, J. H., Campbell, G. R., and Burnstock, G. (1973). *Dev. Biol.* **33**, 344–361.

Chamley, J. H., Gröschel-Stewart, U., Campbell, G. R., and Burnstock, G. (1977a). *Cell Tissue Res.* **177**, 445–457.

Chamley, J. H., Campbell, G. R., McConnell, J. D., and Gröschel-Stewart, U. (1977b). *Cell Tissue Res.* **177**, 503–522.

Chamley-Campbell, J. H., Campbell, G. R., Gröschel-Stewart, U., and Burnstock, G. (1977). *Cell Tissue Res.* **183**, 153–166.

Chamley-Campbell, J. H., Campbell, G. R., Gröschel-Stewart, U., and Vesely, P. (1978). *Folia Biol. (Prague)* **24**, 300–303.

Chamley-Campbell, J. H., Campbell, G. R., and Ross, R. (1979). *Physiol. Rev.* **59**, 1–61.

Chen, W. T. (1977). *J. Cell Biol.* **75**, 411a.

Choo, Q. L., and Bray, D. (1978). *J. Neurochem.* **31**, 217–224.

Clark, T. G., and Merriam, R. W. (1977). *Cell* **12**, 883–891.

Clarke, G. N., and Yanagimachi, R. (1978). *J. Exp. Zool.* **205**, 125–132.

Clarke, M., and Spudich, J. A. (1977). *Annu. Rev. Biochem.* **46**, 797–822.

Cohen, I., and Cohen, C. (1972). *J. Mol. Biol.* **68**, 383–387.

Collins, J. H., and Elzinga, M. (1975). *J. Biol. Chem.* **250**, 5915–5920.

Connolly, J. A., and Kalnins, V. I. (1978). *J. Cell Biol.* **79**, 526–532.

Cooke, P., and Poisner, A. M. (1976). *Cytobiologie* **13**, 442–450.

Côté, G. F., Lewis, W. G., Pato, M. D., and Smillie, L. B. (1978). *FEBS Lett.* **94**, 131–135.

Craig, S. W., and Pardo, J. V. (1979). *J. Cell Biol.* **80,**203–210.

Creutz, C. E. (1977). *Cell Tissue Res.* **178**, 17–38.

Cummins, P., and Perry, S. V. (1974). *Biochem. J.* **141**, 43–49.

Dabrowska, R., Sherry, J. M. F., and Hartshorne, D. J. (1979). *In* "Motility in Cell Function" (F. A. Pepe, ed.), 147–160. Academic Press, New York.

Dabrowska, R., Sherry, J. M. F., Aromatorio, D. K., and Hartshorne, D. J. (1978). *Biochemistry* **17**, 253–258.

Daniel, J. L., and Adelstein, R. S. (1976). *Biochemistry* **15**, 2370-2377.

Davies, P., Bechtel, P., and Pastan, I. (1977). *FEBS Lett.* **77**, 228-232.

de Couet, H. G., Mazander, K. D., and Gröschel-Stewart, U. (1980). *Experientia* (in press).

Dedman, J. R., Potter, J. D., and Means, A. R. (1977a). *J. Biol. Chem.* **252**, 2437-2440.

Dedman, J. R., Potter, J. D., Jackson, R. L., Johnson, J. D., and Means, A. R. (1977b). *J. Biol. Chem.* **252**, 8415-8422.

Dedman, J. R., Jackson, R. L., Schrieber, W. E., and Means, A. R. (1978a). *J. Biol. Chem.* **253**, 343-346.

Dedman, J. R., Welsh, M. J., and Means, A. R. (1978b). *J. Biol. Chem.* **253**, 7515-7521.

Drabikowski, W., Kuźnicki, J., and Grabarek, Z. (1978). *Comp. Biochem. Physiol. C* **60**, 1-6.

Drenckhahn, D. (1979). *Verh. Anat. Ges., Versamml., 73rd, 1978,* 1053-1054.

Drenckhahn, D., and Gröschel-Stewart, U. (1977). *Cell Tissue Res.* **181**, 493-503.

Drenckhahn, D., and Gröschel-Stewart, U. (1979). *Z. Physiol. Chem.* **360**, 1370.

Drenckhahn, D., Gröschel-Stewart, U., and Unsicker, K. (1977). *Cell Tissue Res.* **183**, 273-279.

Drenckhahn, D., Unsicker, K., and Gröschel-Stewart, U. (1978a). *Cytobiologie* **18**, 192.

Drenckhahn, D., Unsicker, K., Griesser, G. H., Schumacher, U., and Gröschel-Stewart, U. (1978b). *Cell Tissue Res.* **187**, 97-103.

Driska, S. P., and Hartshorne, D. J. (1975). *Arch. Biochem. Biophys.* **167**, 203-212.

Durham, A. C. H. (1974). *Cell* **2**, 123-136.

Ebashi, S., and Ebashi, F. (1965). *J. Biochem. (Tokyo)* **58**, 7-12.

Ebashi, S., and Kodama, A. (1966). *J. Biochem. (Tokyo)* **60**, 733-734.

Eckert, B. S., and Lazarides, E. (1978). *J. Cell Biol.* **77**, 714-721.

Edelman, G. M., and Yahara, I. (1976). *Proc. Natl. Acad. Sci. U.S.A.* **73**, 2047-2051.

Elliot, A., Offer, G., and Burridge, K. (1976). *Proc. R. Soc. London, Ser. B* **193**, 45-53.

Elzinga, M., and Kolega, J. (1978). *Fed. Proc., Fed. Am. Soc. Exp. Biol.* **37**, 1694.

Erlandsen, S. L., Schollmeyer, J. V., Feely, D. E., and Chase, D. G. (1978). *J. Cell Biol.* **79**, 264a.

Fagraeus, A., The, H., and Biberfeld, G. (1973). *Nature (London), New Biol.* **246**, 113-115.

Fagraeus, A., Lidman, K., and Biberfeld, G. (1974). *Nature (London)* **252**, 246-247.

Fagraeus, A., Norberg, R., and Biberfeld, G. (1978). *Ann. Immunol. (Paris)* **129c**, 245-254.

Fairfax, A. J., and Gröschel-Stewart, U. (1977). *Clin. Exp. Immunol.* **28**, 27-34.

Farrow, L. J., Holborow, E. J., and Brighton, W. D. (1971). *Nature (London), New Biol.* **232**, 186-187.

Fellini, S. A., Bennett, G. S., and Holtzer, H. (1978). *J. Cell Biol.* **79**, 266a

Finck, H. (1965). *Biochim. Biophys. Acta* **111**, 231-238.

Fine, R. E., and Blitz, A. L. (1975). *J. Mol. Biol.* **95**, 447-454.

Fine, R. E., and Taylor, L. (1976). *Exp. Cell Res.* **102**, 162-168.

Fine, R. E., Blitz, A. L., Hitchcock, S. E., and Kaminer, B. (1973). *Nature (London), New Biol.* **245**, 182-185.

Flanagan, J., and Koch, G. L. E. (1978). *Nature (London)* **273**, 278-280.

Franke, W. W., Rathke, P. C., Seib, E., Trendelenburg, M. F., Osborn, M., and Weber, K. (1976). *Cytobiologie* **14**, 111-130.

Franke, W. W., Grund, C., Fink, A., Weber, K., Jockusch, B. M., Zentgraf, H., and Osborn, M. (1978). *Biol. Cell.* **31**, 7-14.

Fuchs, S., and Sela, M. (1978). *In* "Handbook of Experimental Immunology" (D. M. Weir, ed.), 3rd ed., Vol. 1, Chapter 10. Blackwell, Oxford.

Fujiwara, K., and Pollard, T. D. (1976). *J. Cell Biol.* **71**, 848-875.

Fujiwara, K., and Pollard, T. D. (1978). *J. Cell Biol.* **77**, 182-195.

Fujiwara, K., Porter, M. E., and Pollard, T. D. (1978). *J. Cell Biol.* **79**, 268-275.

Gabbiani, G., Ryan, B., Badonnel, M.-C., and Majno, G. (1972). *Pathol. Biol.* **20**, Suppl., 6-8.

Gabbiani, G., Ryan, G. B., Lamelin, J. P., Vassali, P., Majno, G., Bouvier, C. A., Cruchaud, A., and Lüscher, E. F. (1973). *Am. J. Pathol.* **27**, 473-488.

Gabbiani, G., Malaisse-Lagae, F., Blondel, B., and Orci, L. (1974). *Endocrinology* **95**, 1630–1635.
Gabbiani, G., Montesano, R., Tuchweber, B., Salas, M., and Orci, L. (1975). *Lab. Invest.* **33**, 562–569.
Gabbiani, G., Chaponnier, C., Zumbe, A., and Vassali, P. (1977). *Nature (London)* **269**, 697–698.
Gabbiani, G., Chaponnier, C., and Hüttner, J. (1978). *J. Cell Biol.* **76**, 561–568.
Garnett, H., Gröschel-Stewart, U., Jones, B. M., and Kemp, R. B. (1973). *Cytobios* **7**, 163–169.
Garrels, J. I., and Gibson, W. (1976). *Cell* **9**, 793–805.
Gawadi, N. (1971). *Nature (London)* **234**, 410.
Geiger, B., and Singer, S. J. (1979). *Cell* **16**, 213–222.
Goldman, R. (1972). *J. Cell Biol.* **52**, 246–254.
Goldman, R. D., Lazarides, E., Pollack, R., and Weber, K. (1975). *Exp. Cell Res.* **90**, 333–344.
Goldman, R. D., Pollard, T., and Rosenbaum, J., eds. (1976a). "Cell Motility." Cold Spring Harbor Lab., Cold Spring Harbor, New York.
Goldman, R. D., Yerna, M. J., and Schloss, J. A. (1976b). *J. Supramol. Struct.* **5**, 155(107)-183(135).
Goldman, R. D., Chojnacki, B., and Yerna, M. J. (1979). *J. Cell Biol.* **80**, 759–766.
Gordon, D. J., Boyer, J. L., and Korn, E. D. (1977). *J. Biol. Chem.* **252**, 8300–8309.
Gordon, W. E., III (1978). *Exp. Cell Res.* **117**, 253–260.
Govindan, V. M., Faulstich, H., Wieland, T., Agostini, B., and Hasselbach, W. (1972). *Naturwissenschaften* **11**, 521–522.
Gröschel-Stewart, U. (1971). *Biochim. Biophys. Acta* **229**, 322–334.
Gröschel-Stewart, U., and Doniach, D. (1969). *Immunology* **17**, 991–994.
Gröschel-Stewart, U., and Gigli, I. (1968). *Experientia* **24**, 65–66.
Gröschel-Stewart, U., and Gröschel, D. (1974). *Experientia* **30**, 1152–1153.
Gröschel-Stewart, U., and Reder, R. (1975). In "Proceedings of the International Workshop on Hormones and Peptides" (T. A. Bewley, L. Ma., and J. Ramachandran, eds.), pp. 57–64. Chinese University, Hongkong.
Gröschel-Stewart, U., and Unsicker, K. (1977). *Histochemistry* **51**, 315–319.
Gröschel-Stewart, U., Jones, B. M., and Kemp, R. B. (1970). *Nature (London)* **227**, 280.
Gröschel-Stewart, U., Chamley, J. H., McConnell, J. D., and Burnstock, G. (1975a). *Histochemistry* **43**, 215–224.
Gröschel-Stewart, U., Chamley, J. H., Campbell, G. R., and Burnstock, G. (1975b). *Cell Tissue Res.* **165**, 13–22.
Gröschel-Stewart, U., Schreiber, J., Mahlmeister, C., and Weber, K. (1976a). *Histochemistry* **46**, 229–236.
Gröschel-Stewart, U., Chamley, J. H., McConnell, J. D., and Burnstock, G. (1976b). *Histochemistry* **47**, 285–289.
Gröschel-Stewart, U., Ceurremans, S., Lehr, I., Mahlmeister, C., and Paar, E. (1977a). *Histochemistry* **50**, 271–279.
Gröschel-Stewart, U., Unsicker, K., and Leonhardt, H. (1977b). *Cell Tissue Res.* **180**, 133–137.
Gruenstein, E., and Rich, A. (1975). *Biochem. Biophys. Res. Commun.* **64**, 472–477.
Harboe, N., and Ingild, A. (1973). *Scand. J. Immunol.* **2**, Suppl. 1, 161–164.
Harris, H. E., and Weeds, A. G. (1978). *FEBS Lett.* **90**, 84–88.
Hartwig, J. H., and Stossel, T. P. (1975). *J. Biol. Chem.* **250**, 5696–5705.
Hayashi, J. I., and Hirabayashi, T. (1978). *Biochim. Biophys. Acta* **533**, 362–370.
Hayward, A. D. (1972). *J. Pathol.* **106**, 45–48.
Heggeness, M. H., and Ash, J. F. (1977). *J. Cell Biol.* **73**, 783–788.
Heggeness, M. H., Ash, J. F., and Singer, S. J. (1978). *Ann. N.Y. Acad. Sci.* **312**, 414–417.
Heine, W. D., Altmann, H. W., and Gröschel-Stewart, U. (1976). *Verh. Dtsch. Ges. Pathol.* **60**, 321.

Heizmann, C. W., and Häuptle, M. T. (1977). *Eur. J. Biochem.* **80**, 443-451.

Herman, I. M., and Pollard, T. D. (1978). *Exp. Cell Res.* **114**, 15-25.

Hesketh, J. E., Virmaux, N., and Mandel, P. (1978). *Biochim. Biophys. Acta* **542**, 39-46.

Hiller, G., and Weber, K. (1977). *Nature (London)* **266**, 181-183.

Hinssen, H., D'Haese, J., Small, J. V., and Sobieszek, A. (1978). *J. Ultrastruct. Res.* **64**, 282-302.

Hitchcock, S. E. (1977). *J. Cell Biol.* **74**, 1-15.

Hoffmann-Berling, H. (1953). *Biochim. Biophys. Acta* **10**, 628.

Hoffmann Berling, H., and Weber, H. H. (1955). *Naturwissenschaften* **42**, 608-609.

Holborow, J. E. (1979). *Methods Achiev. Exp. Pathol.* **9**, 244-260 (Karger, Basel).

Holborow, E. J., Trenchev, P. S., Dorling, J., and Webb, J. (1975). *Ann. N.Y. Acad. Sci.* **254**, 489-504.

Hudson, L., and Hay, F. C. (1976). "Practical Immunology." Blackwell, Oxford.

Hynes, R. O., and Destree, A. T. (1978). *Cell* **15**, 875-886.

Isenberg, G., Rathke, P. C., Hülsmann, N., Franke, W. W., and Wohlfarth-Bottermann, K. E. (1976). *Cell Tissue Res.* **166**, 427-443.

Isenberg, G. Rieske, E., and Kreutzberg, G. W. (1977). *Cytobiologie* **15**, 382-389.

Ishikawa, H., Bischoff, R., and Holtzer, H. (1969). *J. Cell Biol.* **43**, 312-328.

Ishimoda-Takagi, T. (1978). *J. Biochem. (Tokyo)* **83**, 1757-1762.

Izant, J. G., and Lazarides, E. (1977). *Proc. Natl. Acad. Sci. U.S.A.* **74**, 1450-1454.

Jaffe, E. A., Nachman, R. L., Becker, C. G., and Minick, C. R. (1973). *J. Clin. Invest.* **52**, 2745-2756.

Jockusch, B. M., Burger, M. M., Richards, J. G., Chaponnier, C., and Gabbiani, G. (1977). *Nature (London)* **270**, 628-629.

Jockusch, B. M., Kelley, K. H., Meyer, R. K., and Burger, M. M. (1978). *Histochemistry* **55**, 177-184.

Johnson, D. H., McCubbin, W. D., and Kay, C. M. (1977). *FEBS Lett.* **77**, 69-74.

Johnson, G. D., Holborow, E. J., and Glynn, L. E. (1965). *Lancet* **2**, 878.

Johnson, G. D., Holborow, E. J., and Glynn, L. E. (1966). *Lancet* **2**, 416-418.

Jones, H. P., Matthews, J. C., and Cormier, M. J. (1979). *Biochemistry* **18**, 55-60.

Kapanci, Y., Mo Costabella, P., Cerutti, P., and Assimacopoulos, A. (1979). *Methods Achiev. Exp. Pathol.* **9**, 147-168.

Kapinus, L. N., and Rukosuev, V. S. (1975). *Bull. Eksp. Biol.* **9**, 95-97.

Karsenti, E., Guilbert, B., Bornens, M., and Avrameas, S. (1977). *Ann. Immunol. (Paris)* **128c**, 195-200.

Karsenti, E., Guilbert, B., Bornens, M., Avrameas, S., Whalen, R., and Pantaloni, D. (1978). *J. Histochem. Cytochem.* **26**, 934-947.

Kessler, D., Nachmias, V. T., and Loewy, A. G. (1976). *J. Cell Biol.* **69**, 393-406.

Kesztyüs, L., Nikodemusz, S., and Szilágyi, T. (1949). *Nature (London)* **163**, 136.

King, T. M., and Gröschel-Stewart, U. (1965). *Am. J. Obstet. Gynecol.* **93**, 253-258.

Kirkpatrick, F. H., Rose, D. J., and LaCelle, P. (1978). *Arch. Biochem. Biophys.* **186**, 1-8.

Koch, G. L. E., and Smith, M. J. (1978). *Nature (London)* **273**, 274-277.

Kominz, D. R., and Gröschel-Stewart, U. (1973). *J. Mechanochem. Cell Motil.* **2**, 181-191.

Korn, E. D. (1978). *Proc. Natl. Acad. Sci. U.S.A.* **75**, 588-599.

Kuczmarski, E. R., and Rosenbaum, J. L. (1979a). *J. Cell Biol.* **80**, 341-355.

Kuczmarski, E. R., and Rosenbaum, J. L. (1979b). *J. Cell Biol.* **80**, 356-371.

Kurki, P. (1978). *Clin. Immunol. Immunopathol.* **11**, 328-338.

Landon, F., Huc, C., Thomé, F., Oriol, C., and Olomucki, A. (1977). *Eur. J. Biochem.* **81**, 571-577.

Lasek, R. J., and Hoffman, P. N. (1976). *In* "Cell Motility" (R. Goldman, T. Pollard, and J. Rosenbaum, eds.), pp. 1021-1049. Cold Spring Harbor Lab., Cold Spring Harbor, New York.

Lazarides, E. (1975a). *J. Cell Biol.* **65,** 549–561.

Lazarides, E. (1975b). *J. Histochem. Cytochem.* **23,** 507–528.

Lazarides, E. (1976a). *J. Cell Biol.* **68,** 202–219.

Lazarides, E. (1976b). *J. Supramol. Struct.* **5,** 531(383)-563(415).

Lazarides, E., and Burridge, K. (1975). *Cell* **6,** 289–298.

Lazarides, E., and Lindberg, U. (1974). *Proc. Natl. Acad. Sci. U.S.A.* **71,** 4742–4746.

Lazarides, E., and Weber, K. (1974). *Proc. Natl. Acad. Sci. U.S.A.* **71,** 2268–2272.

Lengsfeld, A. M., Löw, I., Wieland, T., Dancker, P., and Hasselbach, W. (1974). *Proc. Natl. Acad. Sci. U.S.A.* **71,** 2803–2807.

Lessard, J. L., Goldblatt, D., Rein, D., and Carlton, D. (1976). *J. Cell Biol.* **70,** 150a.

Lidman, K., Biberfeld, G., Fagraeus, A., Norberg, A., Torstensson, R., Utter, G., Carlsson, L., Luca, J., and Lindberg, U., (1976). *Clin. Exp. Immunol.* **24,** 266–272.

Lindberg, U., Carlsson, L., Markey, F., and Nyström, L. E. (1979). *Methods Achiev. Exp. Pathol.* **8,** 143–170 (Karger, Basel).

Loewy, A. G. (1952). *J. Cell. Comp. Physiol.* **40,** 127–156.

Lompre, A. M., Bouveret, P., Leger, J., and Schwartz, K. (1979). *J. Immunol. Methods* **28,** 143–148.

Lu, R., and Elzinga, M. (1976). *In* "Cell Motility" (R. Goldman, T. Pollard, and J. Rosenbaum, eds.), pp. 487–942. Cold Spring Harbor Lab., Cold Spring Harbor, New York.

Mabuchi, I., and Okuno, M. (1977). *J. Cell Biol.* **74,** 251–263.

McNutt, N. S., Culp, L. A., and Black, P. H. (1973). *J. Cell Biol.* **56,** 412–428.

Maimon, J., and Puszkin, S. (1978). *J. Supramol. Struct.* **9,** 131–141.

Malaisse, W. J., and Orci, L. (1979). *Methods Achiev. Exp. Pathol.* **9,** 112–136.

Mannherz, H. G., and Goody, R. S. (1976). *Annu. Rev. Biochem.* **45,** 427–465.

Marchesi, S. L., Steers, E., Marchesi, V. T., and Tillack, T. W. (1969). *Biochemistry* **9,** 50–57.

Marcum, J. M., Dedman, J. R., Brinkley, B. R., and Means, A. R. (1978). *Proc. Natl. Acad. Sci. U.S.A.* **75,** 3771–3775.

Markey, F., and Lindberg, U. (1978). *FEBS Lett.* **88,** 75–79.

Marotta, C. A., Strocchi, P., and Gilbert, J. M. (1978). *J. Neurochem.* **30,** 1441–1451.

Maruta, H., and Korn, E. D. (1977a). *J. Biol. Chem.* **252,** 399–402.

Maruta, H., and Korn, E. D. (1977b). *J. Biol. Chem.* **252,** 6501–6509.

Maruta, H., and Korn, E. D. (1977c). *J. Biol. Chem.* **252,** 8329–8332.

Maruta, H., Gadasi, H., Collins, J. H., and Korn, E. D. (1978). *J. Biol. Chem.* **253,** 6297–6300.

Maruyama, K. (1971). *J. Biochem. (Tokyo)* **69,** 369–386.

Maruyama, K., and Ebashi, S. (1965). *J. Biochem. (Tokyo)* **58,** 13–19.

Maruyama, K., and Ohashi, K. (1978). *J. Biochem. (Tokyo)* **84,** 1017–1019.

Maruyama, K., Kimura, S., Ishii, T., Kuroda, M., Ohashi, K., and Muramatsu, S. (1977). *J. Biochem. (Tokyo)* **81,** 215–232.

Masaki, T. (1974). *J. Biochem. (Tokyo)* **76,** 441–449.

Masaki, T. (1975). *J. Biochem. (Tokyo)* **77,** 901–904.

Maupin-Szamier, P., and Pollard, T. D. (1978). *J. Cell Biol.* **77,** 837–852.

Miettinen, A., Virtanen, I., and Linder, E. (1978). *J. Cell Sci.* **31,** 341–353.

Miller, C., and Kuehl, W. M. (1976). *Brain Res.* **108,** 115–124.

Miller, F., Lazarides, E., and Elias, J. (1976). *Clin. Immunol. Immunopathol.* **5,** 416–428.

Miranda, A. F., Godman, G. C., Deitch, A. D., and Tanenbaum, S. W. (1974a). *J. Cell Biol.* **61,** 481–500.

Miranda, A. F., Godman, G. C., and Tanenbaum, S. W. (1974b). *J. Cell Biol.* **62,** 406–423.

Moore, A., Jaffe, E. A., Becker, C. G., and Nachman, R. L. (1977). *Br. J. Haematol.* **35,** 71–79.

Mooseker, M. (1976). *J. Cell Biol.* **71,** 417–433.

Mooseker, M., and Tilney, L. G. (1975). *J. Cell Biol.* **67,** 725–743.

Mooseker, M. S., Pollard, T. D., and Fujiwara, K. (1978). *J. Cell Biol.* **79,** 444–453.

Mousa, G. Y., Trevithick, J. R., Bechberger, J., and Blair, D. G. (1978). *Nature (London)* **274**, 808–809.

Nachman, R. L., Marcus, A. J., and Safir, L. B. (1967). *J. Clin. Invest.* **46**, 1380–1389.

Nachmias, V. T., and Kessler, D. (1976). *Immunology* **30**, 419–424.

Nairn R. C. (1976). "Fluorescent Protein Tracing." Churchill-Livingstone, Edinburgh and London.

Nakane, P. K., and Pierce, G. B., Jr. (1966). *J. Histochem. Cytochem.* **14**, 929–931.

Nicolson, G. L., and Painter, R. G. (1973). *J. Cell Biol.* **59**, 395–406.

Niederman, R., and Pollard, T. D. (1975). *J. Cell Biol.* **67**, 72–92.

Norberg, R., Fagraeus, A., and Lidman, K. (1975a). *Clin. Exp. Immunol.* **21**, 284–288.

Norberg, R., Lidman, K., and Fagraeus, A. (1975b). *Cell* **6**, 507–512.

Ogievetskaya, M. M. (1977). *Origins Life* **8**, 145–154.

Olden, K., Willingham, M., and Pastan, I. (1976). *Cell* **8**, 383–390.

Osborn, M., and Weber, K. (1977). *Exp. Cell Res.* **106**, 339–349.

Ostlund, R. E., Pastan, I., and Adelstein, R. S. (1974). *J. Biol. Chem.* **249**, 3903–3907.

Ostlund, R. E., Jr., Leung, J. T., and Kipnis, D. M. (1978). *J. Cell Biol.* **77**, 827–836.

Owaribe, K., and Hatano, S. (1975). *Biochemistry* **14**, 3024–3029.

Owen, M. J., Auger, J., Barber, B. H., Edwards, A. J., Walsh, F. S., and Crumpton, M. J. (1978). *Proc. Natl. Acad. Sci. U.S.A.* **75**, 4484–4488.

Owman, C., Edvinsson, L., Hardebo, J. E., Gröschel-Stewart, U., Unsicker, K., and Walles, B. (1978). *Adv. Neurol.* **20**, 35–37.

Painter, R. G., Sheetz, M., and Singer, S. J. (1975). *Proc. Natl. Acad. Sci. U.S.A.* **72**, 1359–1363.

Pastan, I., and Willingham, M. (1978). *Nature (London)* **274**, 645–650.

Pepe, F. A. (1968). *Int. Rev. Cytol.* **24**, 193–231.

Perdue, J. F. (1973). *J. Cell Biol.* **58**, 265–283.

Peters, J. H., and Coons, A. H. (1976). *Methods Immunol. Immunochem.* **5**, 424–444.

Phillips, J. H., and Slater, A. (1975). *FEBS Lett.* **56**, 327–331.

Pinder, J. C., Bray, D., and Gratzer, W. B. (1975). *Nature (London)* **258**, 765–766.

Pinder, J. C., Bray, D., and Gratzer, W. B. (1977). *Nature (London)* **270**, 752–754.

Pinder, J. C., Ungewickell, E., Bray, D., and Gratzer, W. B. (1978a). *J. Supramol. Struct.* **8**, 439–446.

Pinder, J. C., Phethean, J., and Gratzer, W. B. (1978b). *FEBS Lett.* **92**, 278–282.

Podlubnaya, Z. A., Tskhovrebova, L. A., Zaalishvili, M. M., and Stefanenko, G. A. (1975). *J. Mol. Biol.* **92**, 357–359.

Poglazov, B. F. (1961). *Bull. Eksp. Biol. Med.* **9**, 56–59.

Poisner, A. M. (1970). *Fed. Proc., Fed. Am. Soc. Exp. Biol.* **29**, 545.

Pollack, R., and Rifkin, D. B. (1975). *Cell* **6**, 495–506.

Pollack, R., and Rifkin, D. B. (1976). *In* "Cell Motility" (R. Goldman, T. Pollard, and J. Rosenbaum, eds.), pp. 389–401. Cold Spring Harbor Lab., Cold Spring Harbor, New York.

Pollack, R., Osborn, M., and Weber, K. (1975). *Proc. Natl. Acad. Sci. U.S.A.* **72**, 994–998.

Pollard, T. D., and Ito, S. (1970). *J. Cell Biol.* **46**, 267–289.

Pollard, T. D., and Weihing, R. R. (1974). *Crit. Rev. Biochem.* **2**, 1–65.

Pollard, T. D., Fujiwara, K., Niederman, R., and Maupin-Szamier, P. (1976). *In* "Cell Motility" (R. Goldman, T. Plllard, and J. Rosenbaum, eds.), pp. 689–724. Cold Spring Harbor Lab., Cold Spring Harbor, New York.

Pollard, T. D., Fujiwara, K., Handin, R., and Weiss, G. (1977). *Ann. N.Y. Acad. Sci.* **283**, 218–236.

Probst, E., and Lüscher, F. (1972). *Biochim. Biophys. Acta* **278**, 577–584.

Puszkin, E. G., Maldonado, R., Spaet, T. H., and Zucker, M. B. (1977). *J. Biol. Chem.* **252**, 4371–4378.

Puszkin, S., and Berl, S. (1972). *Biochim. Biophys. Acta* **256**, 695–709.

Puszkin, S., Nicklas, W. J., and Berl, S. (1972). *J. Neurochem.* **19**, 1319-1333.

Puszkin, S., Schook, W., Puszkin, E., Rouault, C., Ores, C., Schlossberg, J., Kochwa, S., and Rosenfield, R. E. (1976). *In* "Contractile Systems in Non-muscle Tissues" (S. V. Perry, A. Margreth, and R. S. Adelstein, eds.), pp. 67-80. Elsevier, Amsterdam.

Puszkin, S., Puszkin, E., Maimon, J., Rouault, C., Schook, W., Ores, C., Kochwa, S., and Rosenfield, R. (1977). *J. Biol. Chem* **252**, 5529-5537.

Puszkin, S., Maimon, J., and Puszkin, E. (1978). *Biochim. Biophys. Acta* **513**, 205-220.

Rahmsdorf, H. J., Malchow, D., and Gerisch, G. (1978). *FEBS Lett.* **88**, 322-326.

Rathke, P. C., Seib, E., Weber, K., Osborn, M., and Franke, W. W. (1977). *Exp. Cell Res.* **105**, 253-262.

Reverdin, N., Gabbiani, G., and Kapanci, Y. (1975). *Experientia* **31**, 1348-1350.

Rifkind, R. A. (1976). *Methods Immunol. Immunochem.* **5**, 458-463.

Rikihisa, Y., and Mizuno, D. (1977). *Exp. Cell Res.* **110**, 87-92.

Robson, M., and Zeece, M. G. (1973). *Biochim. Biophys. Acta* **295**, 208-224.

Rubenstein, P. A., and Spudich, J. A. (1977). *Proc. Natl. Acad. Sci. U.S.A.* **74**, 120-123.

Rubinstein, N. A., Chi, J. C., and Holtzer, H. (1974). *Biochem. Biophys. Res. Commun.* **57**, 438-446.

Rukosuev, V. S. (1966). *Arch. Pathol.* **2**, 55-60.

Rukosuev, V. S. (1973). *Biull. Eksp. Biol.* **9**, 116-118.

Rukosuev, V. S. (1976). *Biull. Eksp. Biol. Med.* **82**, 1499-1501.

Rukosuev, V. S., and Chekina, I. A. (1973). *Biull. Eksp. Biol. Med.* **76**, 123-125.

Rukosuev, V. S., and Nanaev, A. K. (1975). *Biull. Eksp. Biol.* **3**, 115-117.

Sanger, J. W. (1975). *Proc. Natl. Acad. Sci. U.S.A.* **72**, 2451-2455.

Schachat, F. H., Harris, H. E., and Epstein, H. F. (1977). *Biochim. Biophys. Acta* **493**, 304-309.

Schachner, M., Hedley-Whyte, E. T., Hsu, D. W., Schoonmaker, G., and Bignami, A. (1977). *J. Cell Biol.* **75**, 67-73.

Scheinman, J. I., Fish, A. J., Matas, A. J., and Michael, A. F. (1978). *Am. J. Pathol.* **90**, 71-88.

Schlaepfer, W. W., and Freeman, L. A. (1978). *J. Cell Biol.* **78**, 653-662.

Schollmeyer, J. E., Furcht, L. T., Goll, D. E., Robson, R. M., Stromer, M. H. (1976). *In* "Cell Motility" (R. Goldman, T. Pollard, and J. Rosenbaum, eds.), pp. 361-388. Cold Spring Harbor Lab., Cold Spring Harbor, New York.

Schollmeyer, J. S., Goll, D. E., Robson, R. M., and Stromer, M. H. (1973). *J. Cell Biol.* **59**, 306a.

Schollmeyer, J. V., Rao, G. H. R., and White, J. G. (1978). *Am. J. Pathol.* **93**, 433-447.

Schreiner, G. F., and Unanue, E. R. (1976). *Adv. Immunol.* **24**, 37-165.

Schreiner, G. F., Fujiwara, K., Pollard, T. D., and Unanue, E. R. (1977). *J. Exp. Med.* **145**, 1393-1398.

Schroeder, T. E. (1976). *In* "Cell Motility" (R. Goldman, T. Pollard, and J. Rosenbaum, eds.), pp. 265-277. Cold Spring Harbor Lab., Cold Spring Harbor, New York.

Schulman, H., and Greengard, P. (1978). *Nature (London)* **271**, 478-479.

Schwartz, K., Bouveret, P., Sebag, C., Leger, J., and Swynghedauw, B. (1977). *Biochim. Biophys. Acta* **425**, 24-36.

Scordilis, S. P., Anderson, J. L., Pollack, R., and Adelstein, R. S. (1977). *J. Cell Biol.* **74**, 940-949.

Senda, N., Shibata, N., Tamura, H., and Yoshitake, J. (1979). *Methods Achiev. Exp. Pathol.* **9**, 169-186.

Shapiro, A. L., Vinuela, E., and Maizel, J. V. (1967). *Biochem. Biophys. Res. Commun.* **28**, 815-820.

Sheetz, M. P., Painter, R. G., and Singer, S. J. (1976a). *In* "Cell Motility" (R. Goldman, T. Pollard, and J. Rosenbaum, eds.), pp. 651-664. Cold Spring Harbor Lab., Cold Spring Harbor, New York.

Sheetz, M. P., Painter, R. G., and Singer, S. J. (1976b). *Biochemistry* **15**, 4486-4492.

Shibata, N., Tatsumi, N., Tanaka, K., Okamura, Y., and Senda, N. (1975). *Biochim. Biophys. Acta* **400**, 222–243.

Shizuta, Y., Davies, P. J., Olden, K., and Pastan, I. (1976a). *Nature (London)* **261**, 414–415.

Shizuta, Y., Shizuta, H., Gallo, M., Davies, P., and Pastan, I. (1976b). *J. Biol. Chem.* **251**, 6562–6567.

Small, J. V., and Celis, J. E. (1978). *Cytobiologie* **16**, 308–325.

Sobieszek, A., and Small, J. V. (1977). *J. Mol. Biol.* **112**, 559–576.

Spudich, J. A., and Watt S. (1971). *J. Biol. Chem.* **246**, 4866–4871.

Sternberger, L. A., Hardy, P. H., Jr., Coculis, J. J., and Meyer, H. G. (1970). *J. Histochem. Cytochem.* **18**, 315–333.

Storti, R. V., and Rich, A. (1976). *Proc. Natl. Acad. Sci. U.S.A.* **73**, 2346–2350.

Storti, R. V., Coen, D. M., and Rich, A. (1976). *Cell* **8**, 521–527.

Stossel, T. P., and Pollard, T. D. (1973). *J. Biol. Chem.* **248**, 8288–8294.

Stromer, M. H., and Goll, D. E. (1972). *J. Mol. Biol.* **67**, 489–494.

Sundqvist, K. G., and Ehrnst, A. (1976). *Nature (London)* **264**, 226–231.

Sundqvist, K. G., Otteskog, P., and Ege, T. (1978). *Nature (London)* **274**, 915–917.

Suzuki, A., Goll, D. E., Stromer, M. H., Singh, I., and Temple, J. (1973). *Biochim. Biophys. Acta* **295**, 188–207.

Szamier, P. M., Pollard, T. D., and Fujiwara, K. (1975). *J. Cell Biol.* **67**, 424a.

Talbot, P., and Kleve, M. G. (1978). *J. Exp. Zool.* **204**, 131–136.

Ternynck, T., and Avrameas, S. (1972). *FEBS Lett.* **23**, 24–28.

Tilney, L. G. (1975). *J. Cell Biol.* **64**, 289–310.

Tilney, L. G. (1977). *In* "International Cell Biology" (B. R. Brinkley and K. E. Porter, eds.), pp. 388–402. Rockefeller Univ. Press, New York.

Tilney, L. G., and Detmers, P. (1975). *J. Cell Biol.* **66**, 508–520.

Tilney, L. G., and Mooseker, M. (1971). *Proc. Natl. Acad. Sci. U.S.A.* **68**, 2611–2615.

Toh, B. H., and Harrd, C. C. (1977). *Nature (London)* **269**, 695–697.

Toh, B. H., Gallichio, H. A., Jeffrey, P. L., Livett, B. G., Muller, H. K., Cauchi, M. N., and Clarke, F. M. (1976). *Nature (London)* **264**, 648–650.

Toyama, Y. (1977). *Cell Tissue Res.* **177**, 221–226.

Trenchev, P., and Holborow, E. J. (1976). *Immunology* **31**, 509–517.

Trenchev, P., Sneyd, P., and Holborow, E. J. (1974). *Clin. Exp. Immunol.* **16**, 125–135.

Trifaró, J. M., Ulpian, C., and Preiksaitis, H. (1978). *Experientia* **34**, 1568–1571.

Tucker, R. W., Sanford, K. K., and Frankel, F. R. (1978). *Cell* **13**, 629–642.

Unsicker, K., and Gröschel-Stewart, U. (1978). *Experientia* **34**, 102–105.

Unsicker, K., Drenckhahn, D., and Gröschel-Stewart, U. (1978a). *Cell Tissue Res.* **188**, 341–344.

Unsicker, K., Drenckhahn, D., Gröschel-Stewart, U., Schumacher, U., and Griesser, G. H. (1978b). *Neuroscience* **3**, 301–306.

Utter, G., Biberfeld, P., Norberg, R., Thorstensson, R., and Fagraeus, A. (1978). *Exp. Cell Res.* **114**, 127–133.

Uyemura, D. G., Brown, S. S., and Spudich, J. A. (1978). *J. Biol. Chem.* **253**, 9088–9096.

Vandekerckhove, J., and Weber, K. (1978a). *Proc. Natl. Acad. Sci. U.S.A.* **75**, 1106–1110.

Vandekerckhove, J., and Weber, K. (1978b). *Eur. J. Biochem.* **90**, 451–462.

Vandekerckhove, J., and Weber, K. (1978c). *Nature (London)* **276**, 720–721.

Vandekerckhove, J., and Weber, K. (1978d). *J. Mol. Biol.* **126**, 783–802.

Van der Geld, H. W. R., and Strauss, A. J. L. (1966). *Lancet* **1**, 57–60.

Vassiletz, I. M., and Zubzhitsky, Y. N. (1966). *Biokhimiya* **31**, 453–457.

Wallach, D., Davies, P. J., and Pastan, I. (1978). *J. Biol. Chem.* **253**, 3328–3335.

Walles, B., Gröschel-Stewart, U., Owman, C., Sjöberg, N. O., and Unsicker, K. (1978). *J. Reprod. Fertil.* **52**, 175–178.

Wang, E., and Goldberg, A. R. (1978). *J. Histochem. Cytochem.* **26**, 745–749.

Wang, K. (1977). *Biochemistry* **16**, 1857–1865.

Wang, K., and Singer, S. J. (1977). *Proc. Natl. Acad. Sci. U.S.A.* **74**, 2021–2025.

Wang, K., Ash, J. F., and Singer, S. J. (1975). *Proc. Natl. Acad. Sci. U.S.A.* **72**, 4483–4486.

Weber, K., and Gröschel-Stewart, U. (1974). *Proc. Natl. Acad. Sci. U.S.A.* **71**, 4561–4564.

Weber, K., Lazarides, E., Goldman, R. D., Vogel, A., and Pollack, R. (1975). *Cold Spring Harbor Symp. Quant. Biol.* **39**, Pt. 1, 363–369.

Weber, K., Rathke, P. C., Osborn, M., and Franke, W. W. (1976). *Exp. Cell Res.* **102**, 285–297.

Weber, K., Rathke, P. C., and Osborn, M. (1978). *Proc. Natl. Acad. Sci. U.S.A.* **75**, 1820–1824.

Webster, R. E., Osborn, M., and Weber, K. (1978). *Exp. Cell Res.* **117**, 47–61.

Wehland, J., Osborn, M., and Weber, K. (1977). *Proc. Natl. Acad. Sci. U.S.A.* **74**, 5613–5617.

Welsh, M. J., Dedman, J. R., Brinkley, B. R., and Means, A. R. (1978). *Proc. Natl. Acad. Sci. U.S.A.* **75**, 1867–1871.

Whalen, R. G., Butler-Browne, G. S., and Gros, F. (1976). *Proc. Natl. Acad. Sci. U.S.A.* **73**, 2018–2022.

White, J. G. (1968). *Blood* **31**, 604–622.

Whitehouse, J. M. A., and Holborow, E. J. (1971). *Br. Med. J.* **4**, 511–513.

Wickus, G., Gruenstein, E., Robbins, P. W., and Rich, A. (1975). *Proc. Natl. Acad. Sci. U.S.A.* **72**, 746–749.

Willingham, M. C., Ostlund, R. E., and Pastan, I. (1974). *Proc. Natl. Acad. Sci. U.S.A.* **71**, 4144–4148.

Willingham, M. C., Yamada, K. M., Yamada, S. S., Pouysségur, J., and Pastan, I. (1977). *Cell* **10**, 375–380.

Wilson, F. J., and Finck, H. (1971). *J. Biochem. (Tokyo)* **70**, 143–148.

Yang, Y., and Perdue, J. F. (1972). *J. Biol. Chem.* **247**, 4503–4509.

Yerna, M. J., Aksoy, M. O., Hartshorne, D. J., and Goldman, R. D. (1978). *J. Cell Sci.* **31**, 411–429.

Yerna, M. J., Dabrowska, R., Hartshorne, D. J., and Goldman, R. D. (1979). *Proc. Natl. Acad. Sci. U.S.A.* **76**, 184–188.

Zechel, K., and Weber, K. (1978). *Eur. J. Biochem.* **89**, 105–112.

Zucker-Franklin, D., Nachman, R. L., and Marcus, A. J. (1967). *Science* **157**, 945–946.

INTERNATIONAL REVIEW OF CYTOLOGY, VOL. 65

The Ultrastructural Visualization of Nucleolar and Extranucleolar RNA Synthesis and Distribution[1]

S. FAKAN

Swiss Institute for Experimental Cancer Research, Lausanne, Switzerland

E. PUVION

Institut de Recherches Scientifiques sur le Cancer, Villejuif, France

I. Introduction[2]

The aim of this article is to summarize briefly recent data concerning the ultrastructural localization of transcription sites and the subsequent distribution of newly synthesized RNA within the cell nucleus. One of our main goals will be to discuss the possible roles of RNP-containing nuclear structures with respect to the synthesis and processing of nucleolar and extranucleolar RNA.

[1]This article is dedicated to the memory of Dr. Wilhelm Bernhard.

[2]Abbreviations: ARG, autoradiography; CHO, Chinese hamster ovary; DNA, deoxyribonucleic acid; DNase, deoxyribonuclease; DNP, deoxyribonucleoprotein; EDTA, ethylenediaminetetraacetic acid; EM, electron microscope; GMA, glycol methacrylate; HnRNA, heterogeneous nuclear ribonucleic acid; IG, interchromatin granules; LM, light microscope; PCA, perchloric acid; PF, perichromatin fibrils; PG, perichromatin granules; PTA, phosphotungstic acid; RNA, ribonucleic acid; RNase, ribonuclease; RNP, ribonucleoprotein; snRNA, small-molecular-weight nuclear ribonucleic acid; UdR, uridine.

This article does not provide an exhaustive list of papers published on these subjects. Several reviews have recently been devoted to the fine morphology and cytochemistry of the cell nucleus (Bernhard and Granboulan, 1968; Bouteille *et al.*, 1974; Busch and Smetana, 1970; Franke, 1974; Franke and Scheer, 1974a; Lafontaine, 1974; Monneron and Bernhard, 1969; Smetana, 1974; Smetana and Busch, 1974), as well as to the EM ARG investigation of chromatin functions (Fakan, 1978). The action of physical and chemical agents on nuclear morphology at the ultrastructural level has also been repeatedly discussed and surveyed (Bernhard, 1971; Simard, 1970; Simard *et al.*, 1974). Several reviews have also discussed the methodological aspects of the application of ultrastructural cytochemistry or ARG to the study of the cell nucleus (Bouteille *et al.*, 1975; Gautier, 1976), some of them with particular attention to the EM ARG visualization of nucleic acids (Angelier *et al.*, 1976; Fakan, 1976; Geuskens, 1977). Consequently, we will refer only to the papers having a direct relationship with the topics discussed here.

The ultrastructural localization of RNA transcription within the eukaryotic cell nucleus has been rather extensively studied during the last 10–15 years. The main evidence comes from EM ARG after [^{3}H]UdR labeling. The fact that the transcribed molecules migrate throughout the nucleus after completion must be kept in mind, and consequently the labeling time should be as short as possible in order to ensure that the majority of labeled RNA is still in the course of transcription (for detailed discussion, see Fakan, 1978). The introduction of new cytochemical techniques, especially of the differential EDTA staining technique for nuclear nucleoproteins (Bernhard, 1969a) (Fig. 1), and their combination with EM ARG has permitted more precise localization of transcription sites with respect to nuclear components. While this question has been resolved, the subsequent distribution of the newly synthesized RNA within the nucleus after detachment from the DNA matrix, as well as the possible pathways or storage places of different RNA species in the nucleus, are still often a matter for speculation. We will try to confront these questions, whenever possible, with data provided by relevant molecular biological and biochemical studies concerning RNA processing and metabolism.

The comparison between biochemical and ultrastructural data is sometimes difficult, mainly due to completely different technical approaches and methods of evaluation of the results. Cytochemical evidence is often difficult to express in a quantitative way. When nuclear structures described *in situ* are compared with those observed in isolated nuclear fractions, the original form of a structure is often subjected to long and harsh chemical and physical treatments that finally preclude a direct comparison. All possible cytochemical as well as biochemical properties of the isolated particles in question have to be taken into consideration when an attempt to identify isolated structures with the structures *in situ* is to be made.

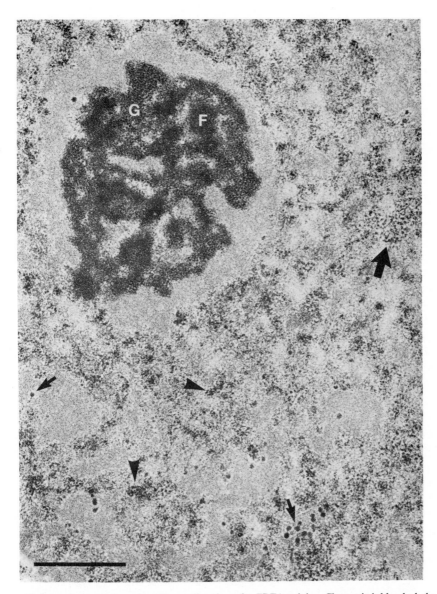

FIG. 1.   A general view of a rat liver cell nucleus after EDTA staining. Chromatin is bleached, the nucleolus, PF (arrowheads), PG (small arrows), and IG (large arrow) are contrasted. F, Fibrillar; G, granular nucleolar components. ×25,000. Bar, 1 μm. (Courtesy of Dr. W. Bernhard.)

II. Synthesis of Nucleolar RNA

The morphological aspects of the early stages of pre-rRNA formation in the nucleolus have been repeatedly studied and are now fairly well understood. The majority of relevant reports deal with EM ARG localization of [³H]UdR-labeled RNA and have been recently reviewed in detail (Bouteille et al., 1974; Fakan, 1978). Therefore we shall describe here only basic phenomena and discuss them in view of new findings related to nucleolar structure and functions.

A. SITES OF TRANSCRIPTION

1. Nucleolar Chromatin Regions

Direct observation of the topological relation between nucleolar chromatin areas and sites of RNA synthesis has been possible using various biological systems. In early embryos of Arbacia punctulata incubated with [³H]UdR, label was first revealed on the periphery of the fibrillar nucleoli where chromatin was associated with them (Karasaki, 1968). In nuclei of salivary gland explants from Smittia, radioactivity was detected within the inner portion of the nucleolus, where the intranucleolar chromatin is dispersed (Jacob, 1967). Such a distinction between nucleolar chromatin and RNP components, especially inside the nucleolar body, is relatively difficult in the majority of somatic or tissue culture cells after simple uranyl-lead staining of sections. When the EDTA staining technique (Bernhard, 1969a), which preferentially bleaches chromatin regions, is used, this problem can be overcome to a great extent. This technique was first applied to autoradiographs of sections of monkey kidney cells labeled for 2 minutes and permitted localization of radioactivity predominantly on the border of the nucleolar chromatin areas, within the intermediary zones between the chromatin and the fibrillar components (Fakan and Bernhard, 1971). This localization was later confirmed using isolated rat liver cells (Fakan et al., 1976) (Fig. 2) and synchronized Chinese hamster ovary (CHO) cells (Fakan and Nobis, 1978).

These experiments demonstrate that the transcription sites are associated with nucleolar chromatin areas. When chromatin is present in the form of clumps, the sites of transcription are observed on the periphery of these condensed areas in the transition regions between them and the fibrillar nucleolar component.

2. Fibrillar Centers

Recher et al. (1969) gave the name "fibrillar centers" to the clear, finely fibrillar intranucleolar areas surrounded by the RNP fibrillar component. These areas have previously been described in various types of cells as clear zones, nucleolar vacuoles, etc. (see Busch and Smetana, 1970; Goessens, 1976a).

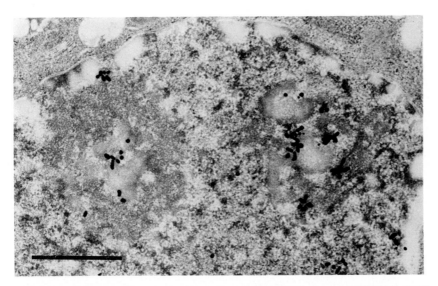

FIG. 2.   EM autoradiograph of an isolated rat liver cell labeled for 2 minutes with [³H]UdR. EDTA staining. Illustration of nucleolar transcription sites. Silver grains are localized predominantly over the periphery of the intranucleolar chromatin areas and the fibrillar component. ×24,000. Bar, 1 μm.

Goessens and Lepoint (1974) suggested that the fibrillar centers represent the nucleolar organizer regions, because, in Ehrlich ascites tumor cells, the prominent fibrillar centers persist in association with certain chromosomes during mitosis, and the nucleolar fibrillar component disappears in prophase and reappears in telophase. The structure of the fibrillar centers seen in metaphase and anaphase cells was similar to that of the nucleolar organizers described by others (Brinkley and Stubblefield, 1970; Lafontaine and Lord, 1974). When Ehrlich ascites tumor cells were labeled for 5 minutes with [³H]UdR, the silver grains detected by EM ARG appeared predominantly over the fibrillar components surrounding the fibrillar centers and rarely over the centers themselves (Goessens, 1974, 1976b; Lepoint and Goessens, 1978). Quail and mouse oocytes incubated with [³H]UdR during meiotic prophase I showed label over the peripheral RNP fibrillar component of the fibrillar center, whereas the center itself remained unlabeled (Mirre and Stahl, 1978a,b,c; Stahl et at., 1978). Similar findings were also reported after [³H]UdR pulse-chase labeling of cultured BSC₁ monkey cells (Fakan, 1971). The results of experiments where quail oocytes were prelabeled with [³H]UdR and then postincubated in the presence of inhibitors of transcription (cordycepin and ethidium bromide), or incubated together with the isotope and cordycepin, confirm the labeling of dense peripheral fibrils and the absence of radioactivity within the clear fibrillar centers (Mirre and Stahl, 1978b). These authors conclude that even if the ribosomal genes are present in both the dense peripheral fibrillar region and inside the center itself, transcription takes place only within the former region (Mirre and Stahl, 1978a,b,c).

3. Fibrillar Component

Granboulan and Granboulan (1965) originally demonstrated in primary cultures of monkey kidney cells that, after a 5 minute incubation with [^{3}H]UdR, the nucleolar label is situated preferentially over the fibrillar component. That this component is the first RNP nucleolar component to become labeled after a short pulse has been repeatedly confirmed in various cell systems (for review, see Bouteille et al., 1974; Fakan, 1978) (Fig. 3). It is only after longer labeling periods, or after a pulse–chase experiment, that the granular components also exhibit radioactivity (Fig. 4). (See Section II,B,1.) Biochemical studies have revealed the presence of 45 S pre-rRNA molecules in fractions containing nucleolar fibrillar component (Matsuura et al., 1974; Royal and Simard, 1975). The labeling kinetics followed in isolated fractions of the two nucleolar RNP components also showed that the fibrils became labeled before the granules (Daskal et al., 1974).

4. Correlations between Cellular and Molecular Levels of Visualization

The development by Miller and Beatty (1969a) of an elegant spreading technique permitting direct EM visualization of the transcription process within chromatin after nuclear lysis, has opened new possibilities for morphological analysis of transcriptional events. When this technique was first combined with EM ARG (Miller and Beatty, 1969b), it allowed confirmation of the RNA nature of the lateral fibrils appearing along the axes of the nucleolar transcription complexes in Xenopus oocytes. This technical combination has recently been applied to exponentially growing mouse P815 cells after a short labeling (3 or 5 minutes) with [^{3}H]UdR (Fakan et al., 1978; Villard and Fakan, 1978). The radioactive label appeared over densely aggregated fibrillar clusters in the areas where chromatin was not well dispersed. Within more dispersed chromatin, labeled gradients of transcribed RNA ("Christmas tree"-like forms) were often revealed on the periphery of the denser fibrillar areas. These pictures led us to propose a parallel between these dense fibrillar aggregates and the nucleolar fibrillar component observed in situ in sectioned material (Fakan, 1978; Villard and Fakan, 1978). Transcription complexes (Fig. 6) in spread preparations are visualized within the previously mentioned fibrillar aggregates (Fig. 5); this confirms the existence of an intimate contact between nucleolar chromatin and the nucleolar fibrillar component.

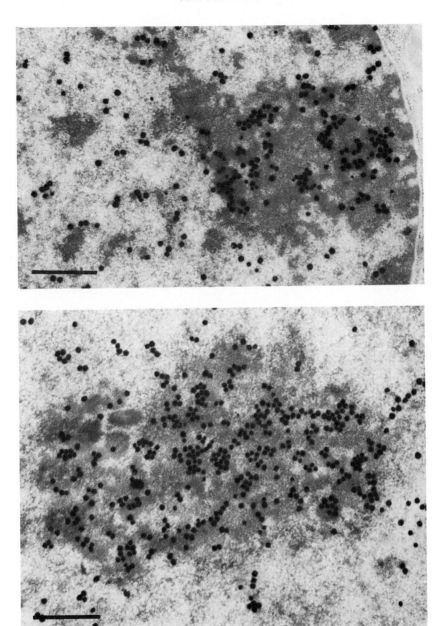

Figs. 3 and 4. Demonstration of a sequential labeling of the nucleolar components. Staining with uranyl acetate and lead citrate. Fig. 3 (top): After 5 minutes incubation with [³H]UdR, silver grains are situated exclusively over the fibrillar component. Fig. 4 (bottom): Following 30 minutes labeling, silver grains are distributed throughout the whole nucleolus. ×18,000. Bar, 1 μm.

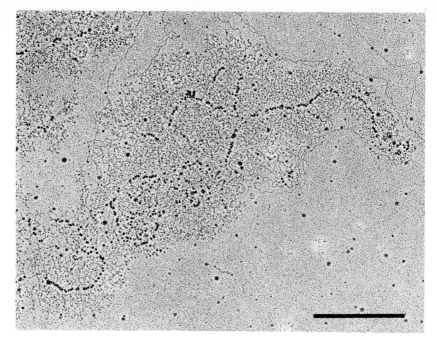

FIG. 5.

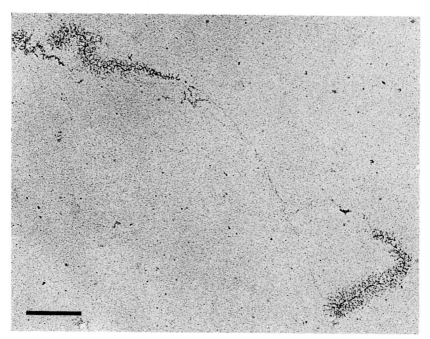

FIG. 6.

B. Further Distribution of Newly Synthesized RNA within the Nucleolus

1. Normal Processing

The sequence of labeling of the two nucleolar RNP components (fibrillar and granular), is now well established. The first report dealing with this question was published by Grandoulan and Granboulan (1965). They labeled cultured monkey kidney cells with [³H]UdR and found that whereas after 5 minutes of incubation with the isotope the fibrillar component was the only labeled nucleolar area, the granular component became gradually radioactive when the pulse was extended to 10 or 30 minutes. These results led the authors to conclude that the fibrils were the morphological precursors of the granules. They formulated the hypothesis that the transformation of fibrils into granules corresponded to the processing of 45 S pre-rRNA into 28 S and 18 S subunits. This sequential labeling of nucleolar components was later confirmed by many other investigators (for review, see Fakan, 1978). Supplementary evidence comes from experiments with actinomycin D-treated, cultured monkey cells (Geuskens and Bernhard, 1966). When cells were labeled for 5 minutes with [³H]UdR and then the label was chased for 30 or 120 minutes in the presence of actinomycin D in order to block further RNA synthesis, all radioactivity originally appearing in the fibrillar components was found in the granular region of the nucleoli, which progressively segregated into the four main constituents (see Bernhard, 1971). In addition, the labeling experiments were in agreement with an early report demonstrating morphological transition between nucleolar fibrils and granules (Marinozzi, 1964).

Biochemical analysis has revealed the presence of 36 S and 32 S RNA (Royal and Simard, 1975), as well as 28 S RNA (Koshiba et al., 1971), in isolated fractions of nucleolar granular components in cultured CHO cells and rat liver, respectively, and confirmed that the morphological transformation of the nucleolar RNP components is accompanied by processing of pre-rRNA. Similarly, in Urechis eggs, the pre-rRNA molecules were synthesized within the fibrillar nucleolar core and the products of processing accumulated in the granular cortex (Das et al., 1970).

2. Abnormal Processing

Several physical and chemical agents induce abnormalities in the conversion of labeled nucleolar RNA from the fibrillar to the granular components. Al-

FIGS. 5 AND 6. Spread ribosomal transcription complexes from isolated CHO cell nucleoli. Fig. 5 : Cluster of densely packed fibrillar material with recognizable DNP axial fibers. PTA staining, Pt shadowing. ×25,000. Fig. 6 : Well-dispersed area showing a sequence of two transcription complexes with a long spacer region in between. PTA staining. ×15,000. Bar, 1 μm. (Courtesy of Dr. F. Puvion-Dutilleul, Villejuif, France.)

though in all cases, altered processing of nucleolar RNA is involved, the morphological lesions of the nucleoli that are linked to this functional disturbance appear very different.

Simard and Bernhard (1967) have described striking nucleolar lesions consisting mainly in the disappearance of the nucleolar RNP granules and intranucleolar chromatin after incubation of cultured cells at temperatures varying from 38 to 45°C. Nucleolar RNA synthesis was almost completely inhibited after 1 hour treatment of cells at 43°C, whereas extranucleolar RNA synthesis was only slightly reduced. As rapidly as 15 minutes after the thermic shock, the remaining nucleoli consist of closely packed fibrils, with fading out of the normal nucleolonema. When returned to 37°C, the heated cells recovered, and a large amount of granular RNP accumulated in contact with the fibrils in areas where intranucleolar chromatin also reappeared. When cells were labeled at 37°C for 30 minutes with [^{3}H]UdR and chased either at 42–43 or 37°C for 1 hour, no appreciable difference in the total nucleolar labeling could be observed between the two lots of cells. While fibrils and granules were labeled in the controls, in the treated cells radioactivity was found in the fibrillar nucleolar remnants. The authors claimed that heat shock induces a reconversion of granules into fibrils and that the process of transformation of fibrils into granules is temperature dependent and reversible. Since no quantitative EM ARG evidence of accumulation of the total nucleolar label within the fibrillar remnants after the heat shock has been provided, this interpretation is purely speculative. Later on, morphological and autoradiographic experiments combined with biochemical studies (Amalric et al., 1969; Simard et al., 1969) on the effects of heat shock on Zajdela ascites hepatoma cells showed that supranormal temperature blocked the processing of nucleolar RNA synthesized before heating of the cells. The authors concluded that this inhibition was directly linked to the disappearance of the granules and their apparent transformation into fibrils. Waroquier and Scherrer (1969) confirmed the preferential sensitivity of processing of 45 S pre-rRNA in HeLa cells to supranormal temperature. In contradiction to the interpretation of Simard et al. (1969), the loss of the granules was accounted for by the metabolic breakdown of precursor rRNA molecules. A loss of the granular component occurring simultaneously with the appearance of large intranucleolar particles was observed in HEp-2 cells (Cervera, 1978) and in Zea mays kernels (Fransolet et al., 1979) after heat shock. Disappearance of the granular component was described in cultures of isolated hepatic cells submitted to low temperature (Puvion et al., 1979). After transfer to 0–4°C, the nucleoli became denser and exclusively fibrillar; at the same time, many RNP granules similar to perichromatin granules accumulated on the outer and inner side of perinucleolar chromatin (see Section III,B,1,e).

Whereas supranormal temperature affects first nucleolar RNA synthesis and nucleolar morphology, nucleolar lesions after hypothermal shock are preceded by extranucleolar changes that occur as early as 15 minutes (see Section III,B,1,d). As shown by Stevens and Amos (1971) in HeLa cells, suboptimal temperature induces abnormal processing of 45 S pre-rRNA prior to a complete arrest of nucleolar RNA synthesis. Thus, at both supraoptimal and suboptimal temperatures, the loss of the nucleolar granular component and the accumulation of labeling on the residual fibrils is primarily due to impaired nucleolar RNA processing.

Some drugs, in particular adenosine analogues, strongly reduce migration of labeled nucleolar RNA from the fibrillar to the granular component. In cultured mammalian cells, the antibiotic toyocamycin, when used at low doses (1 μg/ml or less), selectively inhibits processing of 45 S pre-rRNA synthesized in presence of the drug; at higher doses, it reduces overall RNA synthesis (Monneron et al., 1970). These authors showed that in cells labeled for 30 minutes with [^{3}H]UdR and then chased in the presence of toyocamycin (5 μg/ml) for up to several hours, radioactivity remained located over and at the border of the fibrillar component, and did not migrate toward the granular regions. Reduced nucleolar RNA synthesis was observed in addition to inhibition of processing of the 45 S pre-rRNA synthesized prior to the treatment with the drug. From a morphological point of view, toyocamycin induced a fragmentation of the nucleoli, but no detectable accumulation of the fibrillar component nor loss of granules could be detected. Another adenosine analogue, cordycepin, induced, in isolated rat hepatocytes, a similar retention of the incorporated label in the fibrillar component, and an atypical nucleolar segregation (microsegregation, representing segregation of nucleolar components occurring simultaneously in different areas of the same nucleolus; Svoboda and Higginson, 1968) was observed (Puvion et al., 1976). As in the case of toyocamycin, there was apparently no change in the amount of fibrils and granules. In the same cell system, the plant alkaloid camptothecin induced morphological and functional changes similar to those provoked by cordycepin. The mechanism of action of these different drugs on pre-rRNA synthesis and processing is far from being fully understood (for detailed discussion, see Gajkowska et al., 1977). It is important to stress that the blockage of the normal processing of pre-rRNA can be associated with different morphological nucleolar lesions such as loss of the granular component, nucleolar fragmentation and microsegregation, or appearance of abnormal PG-type granules. Therefore, no unique characteristic morphological change in nucleolar structure can be related to the abnormal processing of nucleolar RNA.

These papers also point out the fact that the presence of the fibrillar and granular components in altered nucleoli is not direct evidence of the presence of RNA within these components. This question was discussed by Monneron et al., (1970) with regard to the persistence of the granular component in spite of complete depletion of rRNA after toyocamycin treatment. This was also in concordance with findings on actinomycin D- (Unuma et al., 1972) and galactosamine- (Herzog and Farber, 1975) treated rat liver, where no RNA could be detected within the remaining fibrillar component.

C. Conclusion

Intranucleolar transcription takes place in the transition regions between intranucleolar chromatin areas and the adjacent fibrillar component. The labeled RNA then appears within 5 minutes, in the fibrillar regions that consequently represent the structures containing 45 S pre-rRNA. The close contact between nucleolar chromatin and the fibrillar component is clearly seen in spread preparations, where transcription complexes are revealed within apparently fibrillar aggregates (Figs. 5,6). After longer labeling periods or chase, label appears over granular components. It is now well established that RNA in the fibrillar component is the precursor of that detected in the granular regions, and biochemical evidence confirms the presence of intermediary products of rRNA processing, as well as of 28 S RNA, within the latter component. These conclusions are supported by numerous studies making use of physical or chemical agents affecting nucleolar RNA synthesis or processing.

As far as the function of fibrillar centers is concerned, they very probably represent the nucleolar organizer regions. RNA transcription occurs in the peripheral fibrillar area surrounding the fibrillar centers rather than within the centers themselves. The relation between the fibrillar centers and nucleolar activity remains to be elucidated.

III. Extranucleolar RNA Synthesis

A. Visualization of Transcription Sites

1. Transcription Sites and Chromatin

The inactivity of condensed chromatin with respect to RNA transcription was already stressed in several early reports. Littau *et al.* (1964) detected radioactivity only within dispersed chromatin areas after 30 minutes incubation of isolated calf thymus nuclei with [³H]UdR. Condensed chromatin was also found transcriptionally inactive after short incubations (5 to 30 minutes) of monkey kidney cells with this radioactive precursor (Granboulan and Granboulan, 1965). Whereas these authors did not mention a relation between the localization of silver grains and the periphery of condensed chromatin, Karasaki (1965) pointed out the role of this region in the transcription process. After incubation of early embryonic tissues explanted from *Triturus,* he first detected radioactivity on the periphery of condensed chromatin. Labeled RNA started to appear within the interchromatin areas after a period of cold chase following the labeling period (Karasaki, 1965). Another paper reported preferential localization of [^{3}H]UdR-labeled RNA in the perichromatin regions in phytohemagglutinin-transformed human lymphocytes or in mononuclear cells (Milner and Hayhoe, 1968). These authors proposed that RNA transcription occurs either in the perichromatin regions on the newly decondensed DNA, or in both perichromatin and euchromatin areas, and that the newly synthesized RNA then migrates towards euchromatin. Using Novikoff hepatoma ascites cells, Unuma et al., (1968) found the majority of silver grains over dispersed chromatin after 5 minutes to 6 hours incubations with [^{3}H]UdR, whereas about 20–25% were found over both the condensed chromatin and the junction between the two forms of chromatin. The difference in grain densities for different nuclear areas between this study and several other papers discussed later might be due, at least partially, to different definition of the respective nuclear areas with regard to the limits of EM ARG resolution. The perichromatin localization of newly synthesized RNA was also stressed by Kierszenbaum and Tres (1974a) in mouse spermatocytes during first meiotic prophase, and by Vazquez-Nin and Bernhard (1971) in Balbiani rings of Chironomus salivary glands labeled either in vivo or in vitro.

The combination of EM ARG with the regressive EDTA staining technique preferential for nuclear RNP structures (Bernhard, 1969a) has permitted more clear localization of transcription sites with respect to chromatin areas. It was first applied after [^{3}H]UdR pulse labeling (2 or 5 minutes) of cultured BSC$_1$ cells (Fakan and Bernhard, 1971). These authors found the majority of silver grains over the periphery of EDTA-bleached condensed chromatin regions. After a cold chase of 15 minutes to 3 hours, the interchromatin space became progressively labeled, demonstrating a shift of labeled RNA from perichromatin areas toward the interchromatin space. A similar perichromatin localization of RNA transcription sites was revealed in synchronized CHO cells (Fakan and Nobis, 1978) and in isolated rat liver cells (Fakan et al., 1976). An elegant confirmation of the preferential labeling of the condensed chromatin border after 15 minutes incubation of isolated rat hepatocytes with [^{3}H]UdR was recently made by Moyne (1977). He applied the "hypothetical grain" method (Blackett and Parry, 1973) to the quantitative EM ARG evaluation of transcriptional events using uranyl-lead stained preparations. The same quantitative method was used on rat liver cells in culture, first maintained for 24 hours without serum and then treated for 1 hour with cortisol in order to rapidly stimulate RNA synthesis. The cells were then labeled for 5 minutes with [^{3}H]UdR in the presence of the hormone. After staining with the EDTA method, about 70% of the total nuclear silver grains were found concentrated over the border of condensed chromatin. When the radioactive pulse was followed by 2 or 4 hours of chase in medium without serum or hormone but containing cold UdR, the radioactivity within the interchromatin space augmented 10–15-fold, whereas the label of the perichromatin region dropped to a very low level (Puvion and Moyne, 1978). These results confirmed, using precise statistical analysis, the migration of newly synthesized nucleoplasmic RNA toward the interchromatin space suggested by earlier observations (Fakan and Bernhard, 1971; Fakan *et al.*, 1976).

2. *Perichromatin Fibrils*

a. *Morphology and Cytochemical Nature of PF.* The use of the differential EDTA staining technique for nuclear structures (Bernhard, 1969a) allowed Monneron and Bernhard (1969) to describe a new class of nuclear RNP components (PF) in the nucleus of various mammalian cells. These could be identified mainly within the perichromatin regions as a rim of contrasted fibrils with a diameter varying between 30 to 50 Å, but which can measure up to 200 Å. They were sometimes observed to fold up into granules; in some cases their apparently granular form may be due to cross sectioning of fibrillar elements. Perichromatin fibrils (PF) can also be observed dispersed throughout the interchromatin space. Structural continuity between PF and IG, as well as connections between PF and PG, were occasionally seen (Monneron and Bernhard, 1969).

Perichromatin fibrils exhibit the same resistance against PCA extraction as PG or IG (Monneron and Bernhard, 1969), but they are more RNase sensitive when GMA sections are digested with the enzyme (Monneron and Bernhard, 1969; Petrov and Bernhard, 1971). When ultrathin frozen sections are examined after EDTA staining, it is interesting to notice that the contrast of PF is much lower than in Epon sections and that they are more difficult to identify (Fakan *et al.*, 1976; Puvion and Bernhard, 1975). Whether this lack of contrast is due to masking the RNA in PF by some proteins extracted during dehydration and plastic embedding, or to some other factors related to the absence of organic solvents in the cryoultramicrotomy procedure, is not clear. Structures similar to PF have also been revealed by a special cytochemical technique making use of pyronin treatment on blocks followed by negative staining with nonaqueous PTA on sections. These structures were RNase sensitive and their occurrence was prevented by actinomycin D (Miyawaki, 1974).

b. *Role of PF.* Monneron and Bernhard (1969) originally proposed that ''PF may represent the morphological expression of transcription along the active derepressed sites of nucleohistones.'' This hypothesis is compatible with all other findings concerning a possible function of PF in the nucleus. The preferential localization of transcription sites in the perichromatin area (see Section III,A,1), where PF are predominantly located, is an important but still indirect indication about the role of PF. However, as PF are often difficult to visualize clearly in the nuclei of tissue culture cells, probably because chromatin and PF are too dispersed, Fakan and Bernhard (1971) failed to identify them in their autoradiographs. The first direct demonstration of a relation between RNA synthesis and occurrence of PF was provided by Petrov and Bernhard (1971), who observed a significant decrease in the number of PF in hepatic cell nuclei of young rats after prolonged starvation or adrenalectomy. As early as 1 hour after refeeding or 15 minutes after cortisone administration to adrenalectomized animals, a high concentration of PF was detected around the condensed chromatin area. The same phenomenon was also described when isolated rat hepatocytes were cultured for 24 hours in the absence of serum and then cortisol was added to the medium for 1 hour (Nash *et al.*, 1975). Miyawaki (1974) employed the starvation–refeeding experimental schedule with mice and was able to visualize pyroninophilic structure corresponding to PF in the refed animals, whereas they were not detected in the 48 hour-fasted group. Petrov and Sekeris (1971) found that the reappearance of PF in starved–refed rats was strongly inhibited 1 hour after administration of α-amanitin, a specific inhibitor of RNA polymerase II. In normal animals, α-amanitin caused a significant decrease in the number of PF during the same period.

An interesting piece of information comes from experiments with rat liver regenerating after partial hepatectomy. Derenzini *et al.* (1977) attempted to correlate the localization of the dispersed chromatin visualized on the border of condensed chromatin areas with that of PF. Twenty-four hours after partial hepatectomy they observed, following classical uranyl-lead staining, a marked increase of loosened material, generally considered as decondensed chromatin in this kind of preparation. The application of the EDTA differential staining technique (Bernhard, 1969a) revealed that this material was composed mainly of PF. Perichromatin fibrils were generally detected only within areas containing unraveled chromatin fibers. The osmium ammine method specific for DNA (Cogliati and Gautier, 1973) demonstrated the presence of thin unraveled DNA fibers contributing to the formation of the loosened material. Administration of α-amanitin induced condensation of these loosened fibers and strongly reduced the number of PF. In a later paper, using the same experimental system, Derenzini *et al.* (1978) established more precisely the timing of the action of α-amanitin. They found the maximum of chromatin condensation and of inhibition of PF occurrence 2 hours after the administration of the toxin, whereas 30 minutes after injection of the drug, the morphological aspect of the two nuclear components was similar to that detected in control animals. The rate of RNA synthesis was about 50% of the control level for the two periods of time. The behavior of both the loosened chromatin and of PF was completely unaffected by the administration of cycloheximide to the hepatectomized rats. Orkisz and Bartel (1978) also reported increase of PF density around condensed chromatin areas after partial hepatectomy in rats.

c. *Labeling Features of PF.* The use of isolated rat hepatocytes in culture for studies of the localization of RNA transcription has two important advantages. The morphology of the nucleus, at least during the first 24 hours of culture, is perfectly comparable with that usually observed in hepatic tissue *in*

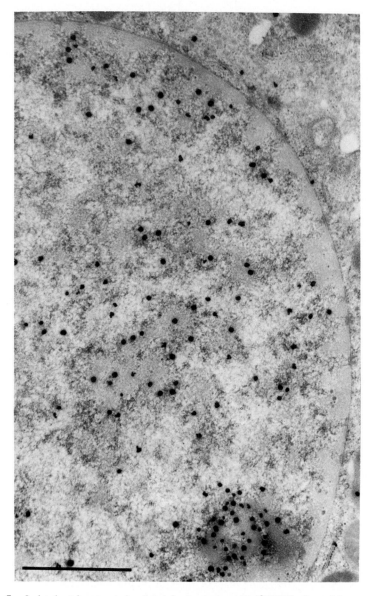

FIG. 7.   Isolated rat hepatocyte incubated for 5 minutes with [³H]UdR after a 1 hour cortisol treatment following a 24 hour culture without serum. EDTA staining. Most extranucleolar label appears over PF on the border of condensed chromatin areas. ×29,750. Bar, 1 μm. [From Nash *et al.* (1975) by permission of Academic Press.]

situ (Puvion *et al.*, 1974). At the same time, the culture conditions allow short labeling experiments or pulse–chase labeling as for established cell lines growing in monolayers. This cell system was used in parallel in two studies. Fakan *et al.* (1976) employed rat liver cells cultured under normal conditions in serum-supplemented medium. When the cells were incubated for 2 or 5 minutes with [³H]UdR, label was visualized in association with PF in the close vicinity of condensed chromatin regions (Fig. 8). Nash *et al.* (1975) maintained the cells for 24 hours in medium without serum; the number of PF considerably decreased and labeling after a 5 minute pulse of [³H]UdR was practically negligible. However, when cortisol was then added to the medium and, after 1 hour, the cells were labeled for 5 minutes with the isotope, a prominent rim of labeled PF appeared around condensed chromatin areas (Fig. 7).

  d. *PF and Migration of Newly Transcribed RNA.* Fakan *et al.* (1976) have observed, in isolated rat hepatocytes labeled *in vitro* for 5 minutes with [³H]UdR and then postincubated in cold medium for 2–4 hours, that interchromatin areas became more or less homogeneously labeled. This was interpreted as a shift of labeled RNA from perichromatin regions toward the interchromatin space (see Section III,A,1). Since the majority of label still seemed to be associated with

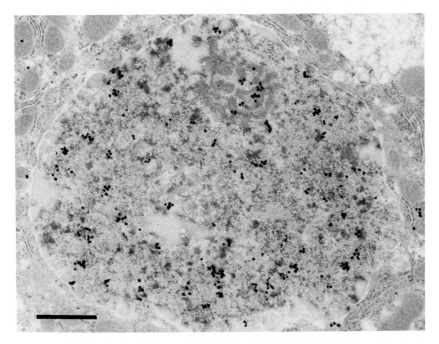

Fig. 8.   Isolated rat liver cell labeled for 2 minutes with [³H]UdR. Many silver grains are associated with PF in proximity of condensed chromatin areas. × 17,000. Bar, 1 μm.

PF, a migration of at least a part of the labeled PF is suggested (Fig. 9). This conclusion has recently been confirmed by Puvion and Moyne (1978) who conducted a detailed statistical analysis using the cell system employed previously by Nash *et al.* (1975, see preceding paragraph). They have been able to show that the shift of labeled RNA toward the interchromatin areas is accompanied by a migration of PF from the periphery of condensed chromatin regions in the direction of the interchromatin space. In terms of relative grain densities, about 70% of the total nuclear radioactivity was in the chromatin border area after a 5 minute pulse, whereas, after 4 hours of chase, only about 0.1% was found in this area and about 57% found within the interchromatin space.

e. *Morphological Correlations between Cellular and Molecular Levels of Visualization.* When extranucleolar RNA transcription is followed using preparations of cultured mouse cells incubated with [³H]UdR and then spread according to Miller and Bakken (1972), labeled material, dispersed to different degrees, can be identified (Fakan *et al.*, 1978; Villard and Fakan, 1978). Within less dispersed areas, it appears as contrasted RNP fibrils surrounded by chromatin

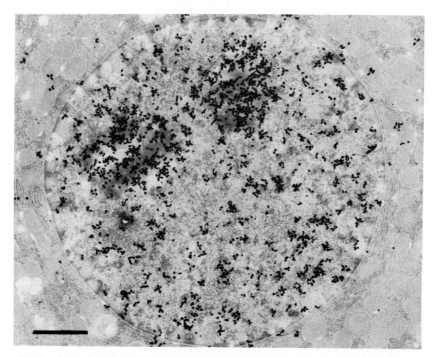

FIG. 9.   Isolated rat liver cell labeled for 5 minutes with [³H]UdR followed by 2 hours of cold UdR chase. Extranucleolar radioactivity, often associated with PF, is distributed throughout the nucleoplasm. ×14,000. Bar, 1 μm.

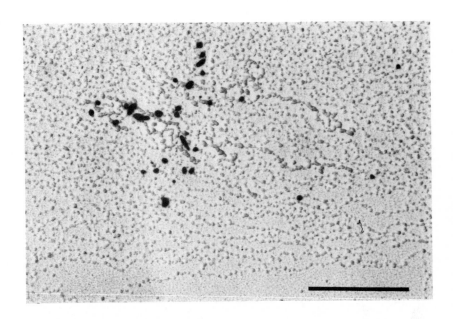

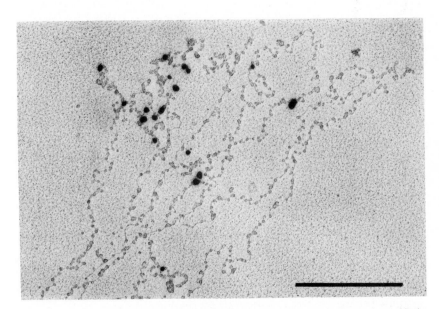

FIGS. 10 AND 11. Autoradiographs of mouse P815 cell chromatin spread after 5 minutes incubation of cells with [³H]UdR. Fig. 10 (top): Labeled RNP fibrils are surrounded by poorly dispersed chromatin. ×53,800. (From Villard and Fakan (1978) by permission of Gauthier-Villars.) Fig. 11 (bottom): Growing chains of extranucleolar RNP are labeled. ×57,800. Bar, 0.5 μm.

(Fig. 10). In well dispersed regions, RNA transcription complexes are seen mainly as individual units (Fig. 11) but sometimes also as gradients of RNP fibrils disposed along the DNA axis. Similarities in labeling properties, as well as in morphology, between PF, visualized *in situ* on EDTA stained autoradiographs of sectioned cells, and the dense, contrasted RNP fibrils in poorly dispersed chromatin regions after spreading have been noticed; a parallel between these two structures has been proposed (Fakan, 1978; Villard and Fakan, 1978).

    f. *Correlations between Morphological and Biochemical Data.* In a parallel EM ARG and biochemical study on isolated rat liver cells labeled with [³H]UdR, Fakan *et al.*, (1976) made a comparison between the localization of labeled extranucleolar RNA and the nature of these radioactive RNA molecules characterized by electrophoresis on polyacrylamide gels under denaturing conditions. After 5 minutes of radioactive pulse, the great majority of the label was represented by growing chains of pre-mRNA molecules. Following a cold chase of 10 minutes, there was predominantly nascent pre-mRNA of higher molecular weight, whereas after 4 hours of chase, the major part of the radioactive RNA consisted of intermediate size pre-mRNA. Bachellerie *et al.* (1975) isolated extranucleolar chromatin–RNP complexes containing rapidly labeled (3–15 minutes) RNA and identified structural elements corresponding to chromatin, PF, and PG, arranged similarly to the situation *in situ*. Labeled RNA was localized in the PF-like fibrillar material. The purpose of another study was to biochemically isolate PF from rat brain and to compare them with PF *in situ* (Devilliers *et al.*, 1977). Their morphology markedly resembled that of PF found *in situ* or in isolated nuclei. The isolated fibrils were not stained after application of Feulgen-type staining methods using thallium ethylate (Moyne, 1973) or osmium ammines (Cogliati and Gautier, 1973), but remained contrasted after staining with the EDTA regressive technique. They were rather resistant to RNase hydrolysis in GMA sections, but were digested when a proteolytic enzyme was used together with RNase, whereas particles in suspension were highly sensitive to RNase; this suggested the presence of proteins protecting the RNA moiety of the particles against digestion, especially after fixation: this is a conclusion compatible with the high protein–RNA ratio determined for these structures.

    Since the presence of protein particles of variable size has been visualized along growing RNP chains in spread transcription complexes (e.g., Miller and Bakken, 1972; Puvion-Dutilleul *et al.*, 1977, 1978; Villard and Fakan, 1978), it may be concluded that similarities exist between PF and the pre-mRNA informofer complexes originally described by Samarina *et al.* (1968). Biochemical and morphological properties of the informofer particles have been reviewed in detail by Georgiev (1974).

3. Conclusion

    Several conclusions can be drawn from the results discussed in the preceding paragraphs concerning the localization of extranucleolar RNA synthesis. It seems clear that chromatin in its condensed form is not transcriptionally active. A great number of related papers show that the peripheral region of condensed chromatin areas is the region where transcription sites are preferentially located. It is also within this chromatin border area where PF can be visualized after application of the regressive EDTA staining technique (Bernhard, 1969a). Perichromatin fibrils have been shown to be of RNP nature, and a direct quantitative relationship between occurrence of PF and the rate of extranucleolar RNA synthesis has been demonstrated. In addition, there is also an apparent relation between the density of PF and the degree of chromatin decondensation within the peripheral regions around condensed chromatin. Therefore, PF can be considered as the first morphological elements representing newly synthesized extranucleolar RNA identified *in situ* in sectioned material. Their striking resemblance to rapidly labeled growing RNP fibrils observed in spread transcribing chromatin, as well as with biochemically isolated fibrillar elements exhibiting similar labeling kinetics and cytochemical properties, are in favor of this conclusion. The observations indicating a migration of PF from transcription sites toward the inner nuclear regions, and the parallel diminution of the mean size of RNA molecules occurring during this process, suggest that at least a part of processing or turnover of extranucleolar RNA takes place when this RNA is in the form of PF.

B. FURTHER DISTRIBUTION OF EXTRANUCLEOLAR RNA

This part will deal with extranucleolar structures that are likely to be involved in the nuclear postsynthetic pathways of RNA. The two main candidates for this function are PG and IG.

1. *Perichromatin Granules*

a. *General Morphology.* Perichromatin granules (Swift, 1962; Watson, 1962) have been described in the nuclei of the majority of cell types so far studied. In dehydrated and plastic embedded cells, they generally appear as mainly individual granules, about 350–500 Å in diameter, surrounded by a clear halo about 200–500 Å thick. They are usually localized on the periphery of condensed chromatin areas, on the border of the nucleolus associated chromatin, and occasionally within the interchromatin space.

After differential EDTA staining (Bernhard, 1969a), their contrast is enhanced; the number of PG observed was about twice as high as in uranyl-lead stained sections (Monneron and Bernhard, 1969). Their substructure appeared to consist of twisted fibrils about 30 Å thick, sometimes emerging from their periphery; in some cases, small groups of granules seemed to be interconnected by thin filaments (Monneron and Bernhard, 1969). According to Vazquez-Nin and Bernhard (1971), PG are composed of even thinner fibrils with a mean diameter of 12–15 ± 5 Å, embedded in a diffuse matrix. In addition, no fine structural difference was found between PG and the Balbiani ring granules of Dipterans, except that the latter do not usually exhibit the clear halo.

Perichromatin granules (Rupec, 1974) or the Balbiani ring granules (Stevens and Swift, 1966) are only exceptionally found in the cytoplasm.

When ultrathin frozen sections of aldehyde-fixed cells are examined after EDTA staining or its modification using uranyl citrate solution (Puvion and Bernhard, 1975), several striking differences are observed when compared with plastic embedded material. The number of PG revealed after cryoultramicrotomy (Fig. 12) is much higher; they often appear in groups and can sometimes be

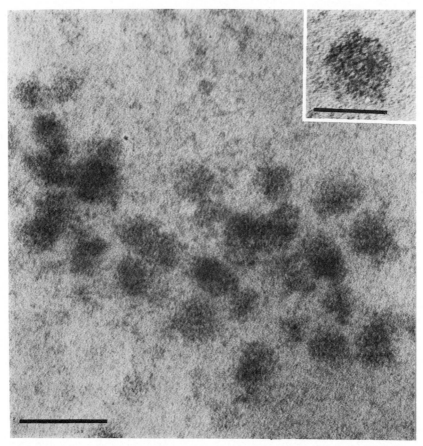

FIG. 12. Group of PG in the nucleoplasm of a rat hepatocyte sectioned by cryoultramicrotomy. Chromatin was bleached by staining with uranyl citrate solution. ×240,000. Bar, 0.1 μm. Inset : detail of a PG showing its composition of irregularly coiled filaments. ×400,000. Bar, 0.05 μm. (Courtesy of Dr. W. Bernhard.)

found within the condensed chromatin areas. Furthermore, a great size variation occurs and their shape tends to be more irregular. They occasionally form grape-like clusters where the grains are not usually interconnected (Fakan *et al.*, 1976; Puvion and Bernhard, 1975).

b. *Nature of PG.* Perichromatin granules are generally considered as structures containing RNP, although this fact has not yet been unequivocally proved. The numerous cytochemical studies attempting to elucidate their nature have made use of enzymatic or chemical extractions on one hand, and of specific or preferential staining techniques on the other.

Watson and Aldridge (1961, 1964) described a slightly decreased contrast of PG after PCA extraction and trivalent indium staining on blocks and concluded that PG contained RNA as well as DNA (Watson and Aldridge, 1964). Swift (1962) found PG to be sensitive to hot PCA. In another study (Monneron and Bernhard, 1969), PG were not destroyed by cold 10% PCA when applied to blocks, but often disappeared after treatment of aldehyde-fixed Epon sections with 35% PCA. They were scarcely visible after the combination of PCA extraction and blocking reduction and acetylation procedure of Watson and Aldridge (1964). When extraction with ethanolic sodium hydroxide was used on sections in order to remove preferentially RNP-containing structures, the contrast of PG appeared to be less distinct, and their morphology and density was variable. In addition, the presence within PG of dense filaments about 23 Å thick was reported. These were similar to elementary chromatin filaments, sensitive to combined pepsin–DNase digestion, and connections between PG and adjacent chromatin by these filaments were observed (Smetana, 1977). The NaOH extraction method was, however, found unsuccessful for PG in another report (Vazquez-Nin and Bernhard, 1971).

Extensive studies performed on GMA sections of various animal tissues (Monneron, 1966; Monneron and Bernhard, 1969; Vazquez-Nin and Bernhard, 1971) treated with Pronase or pepsin, alone or combined with RNase, indicate the presence of RNA within PG. Their contrast generally decreased considerably, but not completely, only after protease pretreatment prior to RNase, although nucleolar RNP and cytoplasmic ribosomes were totally extracted under the same conditions. Perichromatin granules were not attacked by DNase (Monneron and Bernhard, 1969) nor by a sequential Pronase–DNase treatment (Vazquez-Nin and Bernhard, 1971).

Other arguments supporting a probable RNP nature of PG are provided by application of specific or preferential staining methods permitting differential visualization of nucleic acid containing nuclear structures. The EDTA technique (Bernhard, 1969a) reveals PG as highly contrasted structures in all kinds of cells where they have been observed under routine double-staining conditions. In concordance with these data, various specific procedures for DNA detection such as ruthenium red, quinacrine, or Nile blue-sulfate of Gautier and Schreyer (1970)

(Vazquez-Nin and Bernhard, 1971), thallium ethylate (Moyne, 1973) or osmium ammine complex of Cogliati and Gautier (1973) (Derenzini *et al.*, 1977); Moyne *et al.*, 1977) did not stain PG. Similarly they were not revealed by the PTA negative staining combined with actinomycin D to demonstrate DNA (Miyawaki, 1977).

c. *PG during the Cell Cycle and Cell Differentiation.* Perichromatin granules have been repeatedly observed in association with the periphery of chromosomes in a variety of dividing cells. The morphology and EDTA staining properties of perichromosomal PG do not differ from those found in interphase nuclei (Moyne *et al.*, 1974; Moyne and Garrido, 1976; Fakan and Nobis, 1978). The number of PG within mouse cerebellum granular cells has been quantitatively analyzed throughout mitosis (Yamamoto *et al.*, 1969). The PG density was found to be always higher in interphase cells than during different stages of division. Moreover, their frequency during interphase in the outer layer granular cells capable of mitosis was lower than that in the inner granular nondividing cells. Erlandson and De Harven (1971) showed that PG became increasingly prominent at the periphery of nucleoli and condensed chromatin from late S phase up to early prophase in synchronized HeLa cells.

Interesting data are provided by studies of nuclear changes during the meiotic prophase of oocytes or spermatocytes. Palombi and Viron (1977) have recently examined PG during the meiotic prophase I using ultrathin frozen sections of mouse oocytes. They observed a correlation between the increasing degree of chromatin despiralization and the augmentation of PG concentration; both were in accordance with an increasing level of RNA transcription on lampbrush chromosomes. A high amount of EDTA-positive granules of two distinct size classes were described in late prophase tipulid spermatocytes (Fuge, 1976). Particles associated with condensed chromosomes were about 630 Å in diameter, and those found scattered within the chromatin-free karyoplasm were about 540 Å. They seemed to be aggregates of 200 Å subunits.

Increases in concentration of PG during differentiation of human sebaceous cells were recently described (Karasek *et al.*, 1973). This report is in accordance with another paper showing augmentation of PG number by a factor of about two within nuclei of human epidermal spinous cells during differentiation from the basal layer cells (Nagy *et al.*, 1977). In contrast, these authors described a decrease of the number of PG in the course of differentiation of rat bone marrow erythroblasts, which was independent of the age of animals. In addition, an age-dependent decrease of PG was observed in brain cortex, cerebellum, and liver cells, whereas no change was noticed in heart muscle cells. A similar age-dependent decrease in number of PG was also noticed in rat adrenocorticotrophin-producing cells. The number of PG returned almost completely to the level of young animals, when adult and even old rats were adrenalectomized (Del Moro and Nagy, 1977).

d. *Induced Changes of PG Frequency.* Variations of incubation temperature change the number of PG in cultured mammalian cells. Exposure of HeLa-S_3 cells to 42–43°C partially inhibited synthesis of 45 S pre-rRNA, as well as processing of molecules formed at 37°C, and caused a partial loss of the nucleolar granular component with retention of the fibrils. This treatment gave rise to a marked increase in the number of PG and to appearance of threadlike structures of the approximate diameter and density of PG (Heine *et al.*, 1971). A 15 minute to 2 hour exposure of isolated rat hepatocytes to 0–4°C, which is known to strongly decrease global RNA and protein synthesis, provoked an increase in the number of PG. This increase took place as early as 15 minutes after the beginning of the treatment, whereas nucleolar lesions occurred only after 45 minutes (Puvion *et al.*, 1977). After 2 hours cold treatment of cells previously cultured for 24 hours in serum-free medium and then stimulated with cortisol prior to cold shock, PG were three times more abundant than in untreated cells. Simultaneously, the migration of PF toward the interchromatin space seemed to be retarded and the nucleolar granular component disappeared. The cortisol effect itself is expressed only by a slight increase in number of PG (Moyne *et al.*, 1977) in contrast to the strong stimulation of PF number (Nash *et al.*, 1975).

Various drugs can influence the frequency of PG within the nucleus (for review, see Simard *et al.*, 1974). A significant increase in number of PG has been detected in rat liver nuclei after administration of cycloheximide (an inhibitor of protein and rRNA synthesis and processing) (Daskal *et al.*, 1975). A similar effect was obtained when isolated rat hepatocytes were incubated *in vitro* with the same drug (Moyne et al., 1977). Other antimetabolites (preferentially blocking nucleolar RNA synthesis), the hepatocarcinogens aflatoxin and lasiocarpin, were used for experiments *in vivo* with rats. They induced an increase of free PG and development of spherical structures consisting of PG-type granules embedded in fibrillar proteinaceous matrix (Monneron *et al.*, 1968). Injection of α-amanitin into rats and mice also increased the number of PG in the nuclei of liver cells at the same time as the nucleolus became segregated (Marinozzi and Fiume, 1971; Petrov and Sekeris, 1971). On the contrary, prolonged fasting decreases PG number in rat hepatic nuclei in comparison with normally fed animals (Petrov and Bernhard, 1971). Administration of thioacetamide, which induces nucleolar enlargement and a striking increase of nucleolar RNA synthesis in liver cells, did not increase the number of PG when compared with control mice (Yamamoto *et al.*, 1969).

A marked increase in PG number has also been observed in various cells after viral infection. It has been suggested that these granules, when they appear in human glial cells in culture after infection with herpes HSV 2 virus, contain viral pre-mRNA (Dupuy-Coin *et al.*, 1978; see also for more detailed discussion about this question).

e. *PG and the Nucleolus.* Perichromatin granules are not a usual component of the nucleolus. However, several papers have recently reported the presence of granules having the same morphology as PG within the nucleolar area under various experimental conditions. Appearance of coarse granular elements within nucleoli of tissue culture cells was described simultaneously by Recher (1970) and Monneron *et al.* (1970) after incubation of cells with adenosine. Similar results were observed after treatment of BSC cells with toyocamycin, which, at low doses, inhibits processing of 45 S RNA, whereas at higher concentrations, it strongly reduces overall RNA synthesis (Monneron *et al.*, 1970). These granules, about 300 Å in diameter, were situated around spherical, protease-sensitive, and EDTA-bleached structures called "nodules" (Recher, 1970) or "spherules" (Monneron *et al.*, 1970). In addition, when toyocamycin-treated cells were labeled with [³H]UdR, nucleolar radioactivity was localized preferentially within the fibrillar component or on the periphery of the "spherules" where the 300 Å large granules were observed. Ethionine, which induces ATP deficiency, also provoked the appearance of granules of similar size within fragmented nucleoli of liver cells *in vivo* (Shinozuka *et al.*, 1968).

An interesting piece of information has been provided by experiments performed on isolated rat liver cells treated with cordycepin (Puvion *et al.*, 1976), which is known to preferentially affect pre-rRNA synthesis and processing (Siev *et al.*, 1969) as well as polyadenylation of pre-mRNA (Darnell *et al.*, 1971; Brawerman and Diez, 1975), but has also recently been reported as an inhibitor of mRNA synthesis (Beach and Ross, 1978). This antimetabolite induced striking nucleolar lesions. The nucleoli were segregated, and EDTA-positive "spherical bodies", surrounded by clusters of PG-type granules, appeared within the nucleolar region. Neither of these two structures was revealed after DNA-specific staining with osmium ammine (Cogliati and Gautier, 1973). A morphometrical study of the size of nucleoplasmic PG and of nucleolar PG-type granules showed diameters of 377 Å and 344 Å, respectively, with a statistically significant difference between the two values, whereas their morphology was identical. A parallel EM ARG study demonstrated inhibition of the shift of [³H]UdR-labeled RNA from the fibrillar toward the granular component after 3 hours of cold chase in the presence of the drug (Puvion *et al.*, 1976). The same cell system was employed in experiments using the plant alkaloid camptothecin (Gajkowska *et al.*, 1977). This antitumor substance is known to affect synthesis of 45 S pre-rRNA and its processing, as well as transport of 28 S RNA towards the cytoplasm (Kumar and Wu, 1973). It provoked, after only 15 minutes of treatment, a reversible appearance of PG-type granules in the nucleolar area. These granules were in connection with the fibrillar component of gradually disorganized nucleolonema (Fig. 13). They became particularly numerous within microsegregated nucleoli after a prolonged treatment, whereas the number of PG in the nucleoplasm remained unchanged. When [³H]UdR pulse-chase labeling was performed in presence of the drug, the fibrillar component

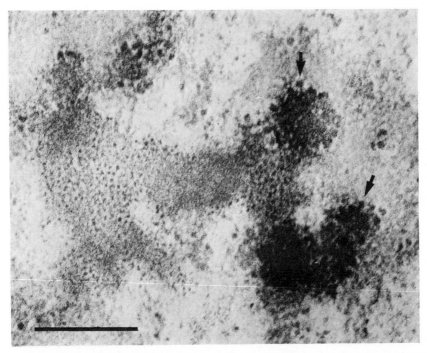

FIG. 13.   Nucleolus of a rat liver cell in culture after 30 minutes treatment with camptothecin (20 μg/ml). EDTA staining. PG-type granules (arrows) are observed in the immediate vicinity of the fibrillar component. ×56,000. Bar, 0.5 μm.

was again, as in the case of cardycepin, the only labeled nucleolar area. However in cells prelabeled under control conditions and then chased in camptothecin-containing medium, radioactivity was observed solely over the perinucleolar chromatin. This labeling, as well as nucleolar appearance of PG-type granules, could be prevented by actinomycin D pretreatment. These data are in partial disagreement with results of Recher *et al.* (1972) obtained on cultured human ME-180 cells. These authors observed the same type of lesions, but they reported a shift of the label from the fibrillar toward the granular nucleolar components under similar experimental conditions. The presence of PG-type granules in the nucleolar region has also been detected in CHO cells (Kedinger and Simard, 1974) and isolated rat hepatocytes (Derenzini and Moyne, 1978) after incubation with α-amanitin. The former authors have reported a strong fragmentation of the nucleoli, which however were still able to normally synthesize pre-rRNA. Moreover the processing of this RNA seemed only slightly affected by the treatment. The latter paper demonstrated, using serial sections, a structural continuity between the PG-type granules and the nucleolus. They were more irregular in shape than PG and their appearance could be prevented by preincubation of cells with low doses of actinomycin D prior to α-amanitin treatment. Reversible occurrence of PG-type granules within the nucleolar body of rat liver cells *in vitro* was also detected after hypothermal shock (Puvion *et al.*, 1977). A parallel disappearance of the nucleolar granular component was noticed. Similar changes were described in HEp-2 cells as a result of the effects of supranormal temperature (Cervera, 1978).

It is interesting to mention some results obtained with plant cells. When nuclei of *Zea mays* embryos are studied during the first several hours of germination, while the nucleolar synthetic activity and pre-rRNA processing are very low (Van de Walle *et al.*, 1976), PG-type granules can be observed in intimate contact with the nucleolar organizer regions (Deltour *et al.*, 1979). In maize embryos treated with cordycepin, these granules were shown even during later developmental periods to be one of the consequences of the impairment of pre-rRNA synthesis (Greimers, 1978). The presence of "macrogranules" was reported in interphase nucleoli of cells incubated with the same drug (Gimenez-Martin *et al.*, 1973), as well as on the periphery of telophase prenucleolar bodies in *Allium cepa* root cells (Risueno *et al.*, 1976). Exposure of *Zea mays* embryos to heat shock, inducing an inhibition of pre-rRNA synthesis, provoked the appearance of large granules within the nucleolar body. This change occurred simultaneously with a loss of the nucleolar granular component (Fransolet *et al.*, 1979).

f. *Possible Role of PG.* Stevens and Swift (1966) originally suggested that the Balbiani ring granules represent a product of pre-mRNA synthesis on the Balbiani ring and that they transport mRNA into the cytoplasm directly through the nuclear pores. This idea is supported by similar results of Vazquez-Nin and Bernhard (1971) who, using EM ARG after [³H]UdR labeling, found strong radioactivity within the puffing zones in the areas where many Balbiani granules can be observed.

A similar role was proposed for PG (Monneron and Bernhard, 1969), but the evidence here is much more limited. In fact, PG have never been seen to pass through a nuclear pore. They were visualized close to the pores and seemed to uncoil toward the pore, possibly by transformation of a PG into a bundle of filaments, which then penetrate through the nuclear pore into the cytoplasm (Monneron and Bernhard, 1969). There are, however, no other more direct data supporting this hypothesis. In addition, the small size of PG, as well as their individual distribution, unfortunately preclude use of EM ARG because of the limits of resolution.

Results obtained on ultrathin frozen sections of mouse oocytes (Palombi and Viron, 1977) and early mouse embryos (Fakan and Odartchenko, 1980), in which PG are particularly abundant and where storage of mRNA as well as rRNA takes place, are in favor of a possible role of PG in the storage of nuclear RNA.

These two ideas are also supported by recent experiments on isolated rat endometrial cells treated *in vitro* with estradiol (Vazquez-Nin *et al.*, 1979). In addition to the stimulation of nucleolar and extranucleolar RNA synthesis, a transitory increase in the transport of RNA from the nucleus to the cytoplasm is observed. This very early effect of the hormone is accompanied by a simultaneous temporary decrease of PG density in the nucleus.

These two hypotheses about the functions attributable to PG seem to be the most probable ones. How do they correlate with the data discussed in the preceding paragraphs? As far as the changes of PG frequency are concerned, it seems that several correlations can be made with changes of activity of RNA transcription or protein synthesis. The density of PG decreases in many tissues as a function of the age of experimental animals (Nagy *et al.*, 1977). Similarly it drops in late erythroblast maturation stages, where RNA and protein synthetic activity is markedly slowing down. However, it increases in skin cells such as differentiating sebaceous cells (Karasek *et al.*, 1973), as well as in spinous granular cells during differentiation from basal layer stem cells (Nagy *et al.*, 1977). In the two latter cases, the more differentiated cell stages exhibit a higher rate of protein synthesis, and consequently, PG might represent a pool of mRNA precursors as suggested by Nagy *et al.* (1977).

In experiments where physical or chemical agents induce augmentation of PG density within the nucleoplasm, several facts must be pointed out. The majority of these treatments affect, to different extents, extranucleolar RNA synthesis; in most instances, nucleolar RNA synthesis and processing are preferentially impaired and nucleolar lesions occur. Several hypotheses to account for this increase of PG have been proposed. In the case of cycloheximide, either increased nuclear RNA synthesis as a result of specific inhibition of translation was claimed or accumulation of PG resulting from the failure of their further processing (Daskal *et al.*, 1975).

In this context, it seems interesting to mention results concerning nucleocytoplasmic translocation of RNA in *Tetrahymena* (Eckert *et al.*, 1975). After treatment with actinomycin D or cycloheximide at concentrations inhibitory to RNA synthesis, both antibiotics effectively inhibited the nucleocytoplasmic transport of preexisting nuclear RNA and resulted in relative accumulation of stable RNA and especially of rRNA precursors in the macronucleus. These results also stressed the stability of apparent rRNA precursor molecules, indicating that neither considerable processing nor degradation occurred during 90 minutes of chase with the drugs.

The hypothesis that cycloheximide-induced PG represent accumulation of incompletely processed extranucleolar RNP was also suggested by Moyne *et al.* (1977). Monneron *et al.* (1968) and Monneron (1971) speculated, in the case of aflatoxin and lasiocarpin, which preferentially inhibit nucleolar RNA synthesis, that PG increase might be due to persistent synthesis of a certain kind of RNP and to a block of its transport towards the cytoplasm.

Similarly, the results of Heine *et al.* (1971) on heat shock-treated cells favor the idea of a possible arrest of mRNA transport from the nucleus to the cytoplasm. A rise in PG number after treatment with α-amanitin (Marinozzi and Fiume, 1971; Petrov and Sekeris, 1971), which inhibits mRNA as well as rRNA synthesis under *in vivo* conditions may also represent an accumulation of RNP formed prior to treatment, which undergoes storage and/or degradation.

It is, however, not excluded that in some of these experimental situations, or even under normal conditions, nucleolar RNA would participate in the formation of PG. A possibility that some PG might contain abortively processed rRNA or a degradation form of this RNA as a consequence of the action of inhibitors cannot be ruled out. The fact that PG-type granules are found within or in close relation with the nucleolus in certain cell systems or under some experimental conditions adds an important new element to the discussion about the role of these structures in the nucleus. In practically all reports, this phenomenon occurred while pre-rRNA synthesis and processing were arrested or reduced to a very low level. Appearance of experimentally induced PG-type nucleolar granules took place, indeed, in parallel with the disappearance of the nucleolar granular component or inhibition of migration of [³H]UdR-labeled RNA from the fibrillar component into the granular one. It is very likely that the PG-type granules consequently represent abnormal storage forms of rRNA, the processing and transport of which was blocked by the experimental treatment (Gajkowska *et al.*, 1977; Puvion *et al.*, 1977). In the cell systems where these granules appear spontaneously, they might be a temporary storage form of nucleolar RNA, which disappears when complete nucleolonema is developed (Deltour *et al.*, 1979).

g. *Conclusion*   The majority of reports on staining and cytochemical properties of PG are in favor of the hypothesis that these structures are carriers of RNA and that they do not contain DNA. It has been shown that there are two types of PG, at least with respect to their topological origin, one related to the nucleolus, the other to the nucleoplasm, whereas their fine morphology is identical. Their accumulation, which often occurs when inhibition or decrease of RNA formation takes place, is interpreted as an expression of either normal and/or abortive RNA processing. The general role of PG would consequently imply storage and transport of RNA transcribed in the respective nuclear regions where they appear. However, a presence of some rRNA within the nucleoplasmic PG cannot be completely excluded.

2. Interchromatin Granules

a. *Morphology and Cytochemical Nature.*   Interchromatin granules (Swift, 1959; Granboulan and Bernhard, 1961) are structures present in practically all interphase nuclei (Fig. 14). They are strongly contrasted after a standard double staining, as well as after the EDTA staining preferential for RNP (Bernhard, 1969a), and appear in clusters distributed apparently randomly within the nuc-

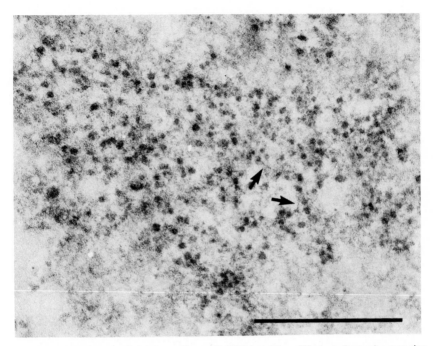

Fɪɢ. 14.   Aggregate of IG from a rat liver cell. EDTA staining. Filaments (arrows) connecting granules are occasionally seen. ×80,000. Bar, 0.5 μm. (Courtesy of Dr. M. Wassef, Villejuif, France.)

leoplasm. Their mean diameter is about 200–250 Å and they are interconnected by fibrils apparently forming a network (Monneron and Bernhard, 1969). Smetana *et al.* (1963) proposed, on the basis of LM and EM observations of isolated nuclei or cells treated with different concentrations of NaCl, the existence of an RNP network in the nucleus, continuous between nucleoli and the nuclear membrane and with IG as a part of it. That IG are located within a more or less contiguous network was also found in relatively thick ultrathin frozen sections stained with the EDTA technique (Puvion and Bernhard, 1975). The latter authors also reported the apparently finely fibrillar substructure of IG. Some of them formed a chain-like network as already observed before (Monneron and Bernhard, 1969). Such a chain-like arrangement of IG was also observed both in sections and in whole mount preparations of mouse pachytene spermatocytes (Kierszenbaum and Tres, 1974b).

The occurrence of IG-like EDTA-positive structures was recently reported in the cytoplasm of mitotic CHO cells (Fakan and Nobis, 1978), in agreement with earlier data of Swift (1963).

The prominent contrast of IG after the application of the EDTA staining procedure suggests the presence of RNP within these structures. The results of many cytochemical investigations making use of enzymatic or chemical extractions, as well as of different staining methods are, however, rather contradictory. Monneron and Bernhard (1969) performed a complex cytochemical study on interphase nuclear structures. When GMA sections of various mammalian cells were treated with RNase alone or preceded by Pronase, the size of IG was decreased, but their contrast remained sharp. This was in agreement with previous results obtained on liver and pancreatic cells (Monneron, 1966). DNase digestion had no effect on IG. Interchromatin granules did not disappear after prolonged treatment of blocks with 10% PCA, whereas treatment of Epon sections with 35% PCA made them disappear completely. The resistance of IG to nuclease and PCA extraction confirmed earlier data from frozen sections of rat liver (Swift, 1962). When the blocking reactions according to Watson and Aldridge (1961), consisting of reduction and acetylation, were applied, and the sections then stained with the EDTA method without indium poststaining, IG themselves were strongly contrasted, whereas the reticular network associated with them was no longer revealed. Monneron and Bernhard (1969) reached the conclusion that IG consist of RNA strongly protected by proteins against enzymatic hydrolysis.

The second group of related papers concerned direct enzymatic treatments of material before embedding in plastic media. Swift (1963) found the IG of Ehrlich ascites cells resistant to RNase digestion following formol fixation. Shankar Narayan *et al.* (1967) used unfixed isolated nuclei from Walker tumor and rat liver. After RNase treatment, the contrast of IG decreased and they appeared to be only partially destroyed. Their density and size were also reduced by pepsin digestion. Most IG were, however, destroyed when nuclei were incubated in pepsin after RNase predigestion. In addition, IG were found to be resistant to DNase, as well as to incubation with $2 M$ NaCl. They were consequently considered to be composed largely of RNP. A similar conclusion was drawn in another paper where formalin fixed human granulocytes were investigated (Smetana *et al.*, 1971). Interchromatin granules of 100–300 Å diameter were observed, intermingled as small clusters with fine filaments about 15–30 Å thick. Treatment of cells with RNase alone did not produce marked changes in either of the above components. Pepsin digestion alone removed only the finely filamentous network, whereas, when followed by RNase, a complete loss of both components occurred.

How can the results of these studies be interpreted? There is a disagreement between a certain number of reports showing resistance of IG to combined protease–RNase treatment performed on GMA sections, whereas when cells or isolated nuclei are digested before embedding, IG are sensitive to such a treatment. The embedding procedure consequently decreases the sensitivity or accessibility of RNA in IG to RNase. Most authors have suggested the existence of a rather resistant protein moiety in IG that protects their RNA against RNase hydrolysis.

When specific DNA staining methods were applied to reveal IG, all attempts were negative (e.g., Cogliati and Gautier, 1973; Moyne, 1973; Puvion and Bernhard, 1975).

It is interesting to mention some data concerning localization of enzymatic activities in IG. ATPase and GTPase, using ultrathin frozen sections (Vorbrodt and Bernhard, 1968), and ATPase, using isolated nuclei (Raikhlin *et al.*, 1976; Buchwalow and Unger, 1977), were localized in IG areas. β-Glycerophosphatase (Raikhlin *et al.*, 1976; Buchwalow and Unger, 1977) and NAD-pyrophosphatase (Buchwalow and Unger, 1977) were also identified within clusters of IG.

Other results using potassium pyroantimonate fixation and indicating the presence of antimonate precipitable inorganic cations in IG are also worth mentioning. Stockert and Schuchner (1972) showed reaction deposits in IG and nucleoli of mouse liver and ovarian cells, and speculated about a possible relation between this phenomenon and the presence of orthophosphates within the respective nuclear regions. Tres *et al.* (1972) reported similar results for mouse seminiferous epithelium during meiotic prophase and Kierszenbaum *et al.* (1972) during mouse spermiogenesis. A recent cytochemical study of the nature of IG, making use of various reactions to reveal different types of proteins, indicated the presence of phosphoproteins within these structures (Wassef, 1979).

b. *Changes in the Amount of IG.*   Clumping of IG, as well as increases in their number, have been observed in many pathological situations as a result of treatments with various drugs or physical agents acting at different metabolic levels (for review, see Simard, 1970; Simard *et al.*, 1974). These changes were interpreted as representing part of a nonspecific reaction to cytotoxicity (Simard, 1970). They were also reported for some cancer cells (see Bernhard and Granboulan, 1963), but they did not show constant specific differences when compared to normal cells (Bernhard, 1969b). Accumulation of IG was also described in cells after infection with different viruses. Singer (1975) showed large clumps of IG in cultured human NB cells infected with H-1 parvovirus. He found a close association of IG with fragmenting nucleoli and observed apparent transformation of the nucleolar granular component into IG. Recently, Dupuy-Coin *et al.* (1978) detected a marked increase in IG number within nuclei of cultured glial cells following infection with herpes virus HSV 2. A marked IG accumulation was also found in pigeon pachytene spermatocytes (Nebel and Coulon, 1962a,b), in mouse spermatocytes (Tres *et al.*, 1972; Kierszenbaum and Tres, 1974b), as well as in mouse Sertoli cells (Kierszenbaum, 1974).

c. *Labeling Features of IG.*   When cultured BSC_1 cells were studied by EM ARG following incubation with [^{3}H]UdR for 5 minutes to 1 hour and a chase of up to 3 hours, it was observed that IG were very weakly labeled or unlabeled, whereas strong radioactivity was detected within neighboring interchromatin areas. When silver grains were found over IG, they were identified especially on the periphery of the clusters; clearly labeled IG aggregates were rare (Fakan and Bernhard, 1971). This question was later reexamined using the same experimental system and labeling for 1 to 6 hours followed by postincubation without isotope for up to 96 hours. A similar labeling pattern of IG was observed. The only incubation period showing a slight increase in labeled IG clusters was 6 hours followed by 24 hours of chase (Fakan and Bernhard, 1973). A low level of radioactivity within IG was also reported after 2 to 4 hours of chase in 5 minutes-labeled cultured rat hepatocytes, using either Epon or ultrathin frozen sections (Fakan *et al.*, 1976). Labeled IG clusters were also found rather rarely in the nuclei of synchronized CHO cells throughout interphase after 1 or 3 hours of [^{3}H]UdR incubation, sometimes followed by a cold chase of 13 hours (Fakan and Nobis, 1978). These data are in agreement with results of Kierszenbaum (1974) on *in vivo* [^{3}H]UdR-labeled mouse Sertoli cells. Three hours after precursor administration, certain IG clusters became partially labeled, whereas after 24 hours, labeling was dropping, and after 5 days it disappeared. Recher *et al.* (1976) identified labeled aggregates of IG in cultured ME 180 cells incubated for 45 minutes with [^{3}H]UdR and subsequently treated with actinomycin D for 4 hours. This finding is in disagreement with previously reported data, where after 1 hour of labeling and 12 hours of actinomycin D chase, the majority of clusters of IG were weakly labeled or unlabeled, whereas the granular component of segregated nucleoli remained radioactive (Fakan and Bernhard, 1973). Monneron and Kerros (1970) were able to visualize radioactivity in IG areas 3 hours after [^{3}H]UdR injection in rats previously treated with lasiocarpin, which predominantly inhibits nucleolar RNA synthesis (Monneron and Kerros, 1970).

In this context, it is interesting to mention some results obtained on cultured amphibian cells treated with actinomycin D (Simard and Duprat, 1969). With progressive development of nucleolar segregation, numerous intranuclear inclusions started to appear. These EDTA-positive inclusions resembled IG clusters, but consisted of fibrils 150–200 Å thick and 400–600 Å long. They disappeared during mitosis and reappeared at the end of telophase. No significant labeling was detected in these structures whether [^{3}H]UdR was added to the cultures before or after actinomycin treatment.

d. *Possible Role of IG.* In spite of the fact that IG are one of the most common structural elements of the cell nucleus, their function is practically unknown. Several hypotheses have been formulated as to their role with respect to RNA metabolism, but none of them has really been confirmed. Smetana *et al.* (1963) proposed, on the basis of the apparent existence of a continuous RNP network connecting nucleoli, IG clusters, and the nuclear periphery, that they might represent nucleolar RNA on the pathway of migration towards the cyto-plasm. Singer (1975), after infection of cultured cells with H-1 parvovirus, observed transformation of nucleolar RNP granules into IG, which accumulated in great numbers in areas adjacent to the remnants of fragmented nucleoli, and considered IG as byproducts of nucleolar disintegration. Similarly Recher *et al.* (1976) suggested, following a study of the effect of actinomycin on tissue culture cells, that IG derive from the nucleolar granular components and that they are likely to contain 28 S rRNA. On the contrary, Monneron and Kerros (1970) claimed to detect [^{3}H]UdR label within areas corresponding to IG while nucleolar RNA synthesis was strongly reduced by lasiocarpin. Consequently, these authors proposed rejection of the hypothesis that IG are future ribosomal subunits (Smetana *et al.*, 1963), and favored the idea of independence of nucleoli and IG.

The use of various chemical or physical agents that has been very useful in elucidating functions of other nuclear structures is not really helpful in the case of IG. The structural response to these treatments of IG is, in fact, too heterogeneous and apparently nonspecific, so that no clear conclusions can be drawn from these experiments. The number of IG generally increases when RNA formation is impaired, but the number of IG also increases in certain cells in which RNA is synthesized at a very high rate (Kierszenbaum, 1974; Kierszenbaum and Tres, 1974b).

The results of labeling experiments indicate that IG contain a small amount of RNA. On the basis of the kinetic features of IG labeling, it was proposed that IG contain rather slowly labeled RNA, or RNA species with a slow turnover (Fakan and Bernhard, 1971, 1973), or nuclear pools of RNP species synthesized elsewhere in the nucleus and accumulated in IG (Kierszenbaum, 1974; Kierszenbaum and Tres, 1974b). The location of pyroantimonate precipitates coinciding with clusters of IG and indicating the presence of inorganic cations within IG led Tres *et al.* (1972) to suggest that RNA polymerases are concentrated in IG areas and might be responsible for synthesis of slowly labeled RNA species. This suggestion is, however, purely speculative and needs more direct experimental evidence.

e. *Morphological and Biochemical Correlations.* Monneron and Moulé (1968) examined morphological and cytochemical properties of biochemically isolated 40 S nuclear RNP particles from rat liver, and found striking similarities between these particles and IG observed *in situ*. The 40 S particles did not exhibit filamentous interconnections, and they consisted of 11% RNA and 89% proteins (Moulé and Chauveau, 1968). The presence of rapidly labeled RNA within these particles is, however, not compatible with the labeling pattern of IG *in situ* (see Section III,B,2,c). The presence of poly(A)-rich HnRNA in IG was also recently suggested (Tata and Baker, 1975). As no precise morphological definition was given to characterize the so-called IG fraction, it is not clear whether it actually corresponded to IG.

Clusters of IG seem to be present within biochemically isolated nuclear skeletons (Berezney and Coffey, 1977; Miller *et al.*, 1978a). As most species of small-molecular-weight nuclear RNA (snRNA) have been shown to be associated with the skeletons (Zieve and Penman, 1976; Berezney and Coffey, 1977; Miller *et al.*, 1978b), and several of them were characterized as metabolically highly stable (Weinberg and Penman, 1969; Ro-Choi and Busch, 1974; Goldstein *et al.*, 1977; Howard, 1978), one is tempted to speculate about the presence of some snRNA in IG. Benecke and Penman (1977) were, however, unable to visualize an association of an snRNA synthesized by the type I RNA polymerase with IG, using EM ARG of HeLa cell nuclei, where label was distributed throughout the nucleoplasm.

In addition, snRNA was also demonstrated in nuclear RNP particles carrying HnRNA (informofers) and was considered as a possible structural component of these particles (Sekeris and Niessing, 1975; Deimel *et al.*, 1977). It has also been shown that proteins in nuclear HnRNA-containing particles and those associated with cytoplasmic mRNA are different (see Samarina *et al.*, 1973; Egly and Stevenin, 1977). Another tempting hypothesis for the role of IG is that they represent a pool of snRNA–protein particulate complexes that become associated with nascent HnRNA and dissociate again once their HnRNA moiety has broken down or the mRNA has left the nucleus.

f. *Conclusion.* IG represent morphologically well-defined components of the nucleus of the majority of eukaryotic cells so far studied. It is rather rare that they are confused with other nuclear structures or considered as general interchromatin substance (e.g., Brown and Locke, 1978). As follows from sometimes controversial cytochemical studies and from [^{3}H]UdR labeling experiments, it seems probable that they contain a rather limited amount of RNA. In spite of several hypotheses formulated to account for their possible role in the nucleus (see Sections III,B,2,d and e), the kind of RNA that IG contain, as well as their function in metabolic events taking place in the nucleus, still remain questions for further investigation.

IV. RNA Transcription and Mitosis

A. DISTRIBUTION OF PERICHROMOSOMAL RNA

Papsidero and Braselton (1973) studied *Cyperus* mitotic chromosomes using the EDTA differential staining (Bernhard, 1969a). They demonstrated a layer of EDTA-positive and RNase-sensitive perichromosomal fibrils surrounding metaphase and anaphase chromosomes. The presence of these RNP fibrils was subsequently confirmed in other plant cells (Moreno Diaz de la Espina *et al.*, 1976), as well as animal cells (Moyne *et al.*, 1974; Moyne and Garrido, 1976; Fakan and Nobis, 1978), in mitosis. The latter papers also reported the appearance of PG-type granules on the periphery of chromosomes.

Moyne and Garrido (1976) investigated the localization of radioactive label during the following mitosis after 30 minutes incubation of synchronized hamster cells with [³H]UdR. They found preferential localization of silver grains over the periphery of chromosomes in cells that were labeled during and particularly at the end of G2 phase, whereas G1 or S-labeled cells displayed radioactivity dispersed throughout the whole cell. These authors proposed, especially on the basis of the morphological analogy between the perichromosomal fibrils and PF, that RNA synthesized in the late G2 phase or at the beginning of prophase and then observed on the periphery of mitotic chromosomes represents HnRNA. The recent study of Lepoint and Goessens (1978), however, favors the idea that perichromosomal RNP, at least in part, consists of nucleolar RNA. This conclusion is based on observations of Ehrlich ascites tumor cells labeled for 5 minutes with [³H]UdR and then postincubated in nonradioactive medium for 1 hour. Metaphase cells were then harvested by shaking and either fixed or allowed to grow in medium for 45 minutes more to obtain anaphase–telophase cells. In metaphase cells, silver grains were localized near the periphery of chromosomes as well as in the cytoplasm, whereas in the posttelophase daughter cells, the majority of the label was concentrated over the new nucleoli. A possible reutilization of breakdown products of labeled RNA, or of labeled precursor molecules from the intracellular pool during the chase period must, however, be taken into consideration when these results are interpreted. The presence of nucleolar material on the surface of chromosomes, detected by means of cytochemical methods, has also been demonstrated (e.g., Moreno Diaz de la Espina et al., 1976). The fact that actinomycin D inhibition of nucleolar RNA synthesis before mitosis prevents formation of nucleoli immediately after the mitotic division is worth mentioning in this context (Phillips and Phillips, 1973). These morphological data are supported by biochemical evidence. Experiments with metaphase-arrested HeLa and Chinese hamster cells showed that pre-rRNA synthesis and processing ceased when cells entered metaphase and resumed as cells left mitosis. The 45 S and 32 S rRNA intermediates were stable and persisted in metaphase arrested cells and were found associated with the chromosome fraction. They seemed to be relatively tightly bound to the metaphase chromosomes (Fan and Penman, 1971). Association of previously synthesized snRNA with partially condensed chromosomes in Amoeba proteus during prophase and anaphase has recently been shown. This RNA was, however, not associated with fully condensed metaphase chromosomes (Goldstein et al., 1977).

B. LOCALIZATION OF TRANSCRIPTION SITES

Simmons et al. (1973) investigated the restitution of RNA synthesis at the end of mitosis in HeLa cells synchronized after a double thymidine block followed by 40 minutes of [³H]UdR incubation of harvested metaphase cells. Using EM ARG, label was first detected on the periphery of chromosomes in association with the reforming nuclear membrane. The authors claimed that initiation of RNA synthesis in early telophase took place at sites on the nuclear membrane. This conclusion is not in agreement with results obtained from CHO cells synchronized by mitotic selection without use of inhibitors (Fakan and Nobis, 1978). After 2–20 minutes of [³H]UdR labeling of cells, the first rare labeling sites were already detected at the chromosomal periphery of apparently metaphase–anaphase or anaphase cells. In telophase cells, nuclear membrane was reforming around the chromosomal masses, but radioactivity appeared around them independently of the nuclear membrane. The occurrence of some RNA synthesis as early as at metaphase–anaphase is in concordance with the finding of synthesis, at a relatively high rate, of two small RNA species in metaphase arrested HeLa cells (Zylber and Penman, 1971).

C. Conclusion

The first rare transcription sites found in the course of mitosis have been detected in apparently metaphase–anaphase cells independently of the nuclear membrane, but the nature of the newly transcribed RNA is not known. As far as the nature of perichromosomal RNP fibrillar and granular structures in metaphase and anaphase is concerned, there are some indications that rRNA and snRNA participate in their formation; whether HnRNA is also present within these structures is not clear. This question will therefore demand further investigation in order to be completely elucidated.

V. General Comments

The studies discussed in the previous chapters provide much important information about the nature and functions of various nuclear structural elements with regard to RNA synthesis and metabolism. Ribonucleic acid transcription in the nucleolar, as well as extranucleolar, areas takes place predominantly within the regions situated on the border of condensed chromatin. Perichromatin fibrils, considered as the first *in situ* detected structures containing newly transcribed extranucleolar RNA, originates in these perichromatin regions, and the nucleolar fibrillar components are in intimate contact with intranucleolar chromatin areas. DNA replication sites also appear predominantly within this nuclear region (Fakan and Hancock, 1974; for review, see Fakan, 1978). This particular nuclear area consequently seems to play a key role in the expression of chromatin functions and these findings lead to the conclusion that the layer of chromatin decondensing on the periphery of dense chromatin areas represents the most active chromatin fraction in the nucleus. The fact that replication and transcription processes occur within the same general nuclear area does not, of course, mean that these two processes should be topologically related to each other, even though evidence of transcription on DNA within replication loops has recently been provided (McKnight et al., 1978). There are several complementary data stressing the importance of the perichromatin area for nuclear functions. It has been repeatedly shown that newly synthesized proteins appear in the nucleus and are localized predominantly within these nuclear areas. This was the case in antibody forming cells where [^{3}H]leucine-labeled proteins migrated very rapidly toward the nucleus and accumulated preferentially in the nucleolus and in the perichromatin regions (Bouteille, 1972a,b). The presence of ^{55}Fe-labeled hemoglobin within the junctional region between the condensed and dispersed chromatin in erythroid cell nuclei has also been demonstrated (Parry and Blackett, 1973), and the occurrence of PF in this same region was simultaneously revealed (Fakan and Odartchenko, 1975). As the above proteins are accumulated within areas where synthesis of DNA and RNA, as well as assembly of these molecules with proteins, take place, some of these proteins are probably related to the formation of nucleoproteins and may play regulatory or other roles in nucleic acid synthesis.

The existence of a nuclear RNP network bound to the nuclear membrane and containing HnRNA has recently been described, and the role of the nuclear membrane in the control of genetic expression in eukaryotes was suggested (Faiferman and Pogo, 1975). Such a role, if any, must, however, be expressed in an indirect way, as it has been shown that transcription sites appear independently of the nuclear membrane (Goldstein, 1970; Fakan and Bernhard, 1971; Fakan et al., 1976; Fakan and Nobis, 1978).

While the topological aspects of the distribution of transcription sites are now fairly well known, the pathways of different RNA species after their synthesis are not yet fully understood. As far as the nucleolar area is concerned, it has been shown that the fibrillar component contains pre-rRNA, and the granular one, the intermediates of its processing and 28 S RNA. In some experimental situations where processing and transport of rRNA are impaired, the granular component becomes reduced and PG-type granules appear within the nucleolar area. When the inhibiting effect ceases, e.g., in the case of camptothecin, previously accumulated 28 S RNA again becomes available and migrates toward the cytoplasm. This coincides with a rapid dispersion in the nucleoplasm of PG-type nucleolar granules (Gajkowska et al., 1977). The PG-type granules may, therefore, in some particular situations, fulfill a role of storage of rRNA. In other cases, they also might be the sites of abortive processing or degradation of this RNA (see Section III,B,1,e).

According to a number of indications, PG found in normal conditions are probably involved in storage and/or transport of HnRNA. This possible function of PG would be supported by their relative resistance to enzyme extractions suggesting a protective role by their protein moiety of the RNA contained in the granules. Whether PG also contain some rRNA is not clear. The pathways of rRNA, with regard to nuclear structures, during its migration from the nucleolus towards the cytoplasm, are, at present, unknown.

It has been demonstrated that, after their formation, PF migrate toward the interchromatin space (Puvion and Moyne, 1978). As newly synthesized pre-mRNA is associated with these fibrils and its size simultaneously decreases during the migration period (Fakan et al., 1976), it seems obvious that at least a part of pre-mRNA undergoes processing or degradation while associated with PF. PF are more sensitive to enzymatic extractions than PG or IG (Monneron and Bernhard, 1969). It seems probable that they represent a transitory form in the pathways of HnRNA rather than one of the storage structural elements. As far as the nucleocytoplasmic transport of RNP is concerned, an extensive review has recently been devoted to this topic (Franke and Scheer, 1974b).

In spite of the fact that IG represent very common nuclear structures, it is for the moment difficult to attribute a specific function to these structures. The findings that they are rather resistant to different extraction treatments and that only a low percentage of the clusters of IG become radioactive after [³H]UdR labeling suggest several hypotheses as to their possible role in nuclear RNP pathways. In view of the above properties, it is tempting to speculate about IG as storage places of some stable snRNA, possibly associated with proteins, which is involved in assembly of HnRNA–protein complexes (informofers) and their migration throughout the nucleus. This idea would be in agreement with recent evidence of the presence of certain snRNA within HnRNP complexes (see Section III,B,2,e). If snRNA are present in IG, the question arises whether they are synthesized in association with these structures or accumulated within IG after they have been synthesized elsewhere in the nucleus.

Biochemical isolation of nuclear protein skeletons has revealed a new nuclear structural element that may be involved in different nuclear functions. Association of HnRNA with nuclear skeletons has been repeatedly demonstrated (Herman et al., 1976; Miller et al., 1978a). The former authors speculated about a class of large, poly(A)-terminated HnRNA molecules with a relatively long half-life, which would attach to both the skeleton and the chromatin. Miller et al. (1978a) revealed an association of rapidly labeled HnRNA with the skeleton from rat liver nuclei and claimed that, after removing chromatin, this RNA was present in the form of a fibrogranular RNP network. A clear distinction of nuclear structural elements within these preparations was, however, difficult. Further work will be necessary to define a more precise topological relationship between different types of RNP containing nuclear structures and the nuclear skeleton network.

From the preceding discussion, it becomes obvious that a simultaneous application of both the biochemical and the electron microscopic approaches to the study of nuclear functions represents the only way that will allow elucidation of these complex processes. The former permits isolation and characterization of various nuclear components, but fails to localize them precisely within the nucleus. Ultrastructural cytochemistry and EM ARG provide information about the nature and precise topological aspects of the nuclear structures and complements in a fundamental way the data obtained by biochemical and molecular biological investigations.

ACKNOWLEDGMENTS

The authors are much indebted to Drs. R. Hancock, D. Hughes, and Mrs. M. Hughes for constructive discussions and corrections during the preparation of this review. They thank Dr. G. Moyne for critical reading of the manuscript and Mrs. L. Morand and Miss D. Chiapparelli for its preparation. This work was supported by the Swiss National Science Foundation.

REFERENCES

Amalric, F., Simard, R., and Zalta, J. P. (1969). *Exp. Cell Res.* **55**, 370–377.
Angelier, N., Bouteille, M., Charret, R., Curgy, J. J., Delain, E., Fakan, S., Geuskens, M., Guelin, M., Lacroix, J. C., Laval, M., Steinert, G., and Van Assel, S. (1976). *J. Microsc. Biol. Cell.* **27**, 215–230.
Bachellerie, J. P., Puvion, E., and Zalta, J. P. (1975). *Eur. J. Biochem.* **58**, 327–337.
Beach, L. R., and Ross, J. (1978). *J. Biol. Chem.* **253**, 2628–2632.
Benecke, B.-J., and Penman, S. (1977). *Cell* **12**, 939–946.
Berezney, R., and Coffey, D. S. (1977). *J. Cell Biol.* **73**, 616–637.
Bernhard, W. (1969a). *J. Ultrastruct. Res.* **27**, 250–265.
Bernhard, W. (1969b). *In* "Handbook of Molecular Cytology" (A. Lima-de-Faria, ed.), pp. 687–715. North-Holland Publ., Amsterdam.
Bernhard, W. (1971). *Adv. Cytopharmacol.* **1**, 49–67.
Bernhard, W., and Granboulan, N. (1963). *Exp. Cell Res., Suppl.* **9**, 19–53.
Bernhard, W., and Granboulan, N. (1968). *In* "The Nucleus" (A. J. Dalton and F. Haguenau, eds.), pp. 81–149. Academic Press, New York.
Blackett, N. M., and Parry, D. M. (1973). *J. Cell Biol.* **57**, 9–15.
Bouteille, M. (1972a). *Exp. Cell Res.* **74**, 343–354.
Bouteille, M. (1972b). *Acta Endocrinol. (Copenhagen), Suppl.* **168**, 11–34.
Bouteille, M., Laval, M., and Dupuy-Coin, A. M. (1974). *In* "The Cell Nucleus" (H. Busch, ed.), Vol. 1, pp. 3–71. Academic Press, New York.
Bouteille, M., Dupuy-Coin, A. M., and Moyne, G. (1975). *In* "Methods in Enzymology" (B. W. O'Malley and J. G. Hardman, eds.), Vol. 40, pp. 3–41. Academic Press, New York.
Brawerman, G., and Diez, J. (1975). *Cell* **5**, 271–280.
Brinkley, B. R., and Stubblefield, E. (1970). *Adv. Cell Biol.* **1**, 119–185.
Brown, G. L., and Locke, M. (1978). *Tissue & Cell* **10**, 365–388.
Buchwalow, I. B., and Unger, E. (1977). *Exp. Cell Res.* **106**, 139–150.
Busch, H., and Smetana, K. (1970). "The Nucleolus." Academic Press, New York.
Cervera, J. (1978). *J. Ultrastruct. Res.* **63**, 51–63.
Cogliati, R., and Gautier, A. (1973). *C. R. Hebd. Seances Acad. Sci., Ser. D* **276**, 3041–3044.

296   S. FAKAN AND E. PUVION

Darnell, J. E., Philipson, L., Wall, R., and Adesnik, M. (1971). *Science* **174**, 507–510.
Das, N. K., Micou-Eastwood, J., Ramamurthy, G., and Alfert, M. (1970). *Proc. Natl. Acad. Sci. U.S.A.* **67**, 968–975.
Daskal, Y., Prestayko, A. W., and Busch, H. (1974). *Exp. Cell Res.* **88**, 1–14.
Daskal, Y., Merski, J. A., Hughes, J. B., and Busch, H. (1975). *Exp. Cell Res.* **93**, 395–401.
Deimel, B., Louis, C., and Sekeris, C. E. (1977). *FEBS Lett.* **73**, 80–84.
Del Moro, M., and Nagy, Z. I. (1977). *J. Submicrosc. Cytol.* **9**, 403–408.
Deltour, R., Gautier, A., and Fakan, J. (1979). *J. Cell Sci.* **40**, 43–62.
Derenzini, M., and Moyne, G. (1978). *J. Ultrastruct. Res.* **62**, 213–219.
Derenzini, M., Lorenzoni, E., Marinozzi, V., and Barsotti, P. (1977). *J. Ultrastruct. Res.* **59**, 250–262.
Derenzini, M., Novello, F., and Pession-Brizzi, A. (1978). *Exp. Cell Res.* **112**, 443–454.
Devilliers, G., Stevenin, J., and Jacob, M. (1977). *Biol. Cell.* **28**, 215–220.
Dupuy-Coin, A. M., Arnoult, J., and Bouteille, M. (1978). *J. Ultrastruct. Res.* **65**, 60–72.
Eckert, W. A., Franke, W. W., and Scheer, U. (1975). *Exp. Cell Res.* **94**, 31–46.
Egly, J.-M., and Stevenin, J. (1977). *Pathol. Biol.* **25**, 741–754.
Erlandson, R. A., and De Harven, E. (1971). *J. Cell Sci.* **8**, 353–397.
Faiferman, I., and Pogo, A. O. (1975). *Biochemistry* **14**, 3808–3816.
Fakan, S. (1971). *J. Ultrastruct. Res.* **34**, 586–596.
Fakan, S. (1976). *J. Microsc. (Oxford)* **106**, 159–171.
Fakan, S. (1978). *In* "The Cell Nucleus" (H. Busch, ed.), Vol. 5, pp. 3–53. Academic Press, New York.
Fakan, S., and Bernhard, W. (1971). *Exp. Cell Res.* **67**, 129–141.
Fakan, S., and Bernhard, W. (1973). *Exp. Cell Res.* **79**, 431–444.
Fakan, S., and Hancock, R. (1974). *Exp. Cell Res.* **83**, 95–102.
Fakan, S., and Nobis, P. (1978). *Exp. Cell Res.* **113**, 327–337.
Fakan, S., and Odartchenko, N. (1975). *J. Microsc. Biol. Cell.* **23**, 203–206.
Fakan, S., and Odartchenko, N. (1980). *Biol. Cell.*, in press.
Fakan, S., Puvion, E., and Spohr, G. (1976). *Exp. Cell Res.* **99**, 155–164.
Fakan, S., Villard, D., and Hughes, M. E. (1978). *Electron Microsc., Proc. Int. Congr., 9th, 1978* Vol. II, pp. 218–219.
Fan, H., and Penman, S. (1971). *J. Mol. Biol.* **59**, 27–42.
Franke, W. W. (1974). *Int. Rev. Cytol., Suppl.* **4**, 71–236.
Franke, W. W., and Scheer, U. (1974a). *In* "The Cell Nucleus" (H. Busch, ed.), Vol. 1, pp. 219–347. Academic Press, New York.
Franke, W. W., and Scheer, U. (1974b). *Symp. Soc. Exp. Biol.* **28**, 249–282.
Fransolet, S., Deltour, R., Bronchart, R., and van de Walle, C. (1979). *Planta* **146**, 7–18.
Fuge, H. (1976). *Chromosoma* **56**, 363–379.
Gajkowska, B., Puvion, E., and Bernhard, W. (1977). *J. Ultrastruct. Res.* **60**, 335–347.
Gautier, A. (1976). *Int. Rev. Cytol.* **44**, 113–191.
Gautier, A., and Schreyer, M. (1970). *Electron Microsc., Proc. Int. Congr., 7th, 1970* Vol. 1, pp. 559–560.
Georgiev, G. P. (1974). *In* "The Cell Nucleus" (H. Busch, ed.), Vol. 3, pp. 67–108. Academic Press, New York.
Geuskens, M. (1977). *In* "Principles and Techniques of Electron Microscopy" (M. A. Hayat, ed.), Vol. 7, pp. 163–201. Van Nostrand-Reinhold, Princeton, New Jersey.
Geuskens, M., and Bernhard, W. (1966). *Exp. Cell Res.* **44**, 579–598.
Gimenez-Martin, G., Risueno, M. C., Fernandez-Gomez, M. E., and Ahmadian, P. (1973). *Cytobiologie* **7**, 181–192.
Goessens, G. (1974). *C.R. Hebd. Seances Acad. Sci., Ser. D* **279**, 991–993.

Goessens, G. (1976a). *Cell Tissue Res.* **173.** 315-324.

Goessens, G. (1976b). *Exp. Cell Res.* **100,** 88-94.

Goessens, G., and Lepoint, A. (1974). *Exp. Cell Res.* **87,** 63-72.

Goldstein, L. (1970). *Exp. Cell Res.* **61,** 218-222.

Goldstein, L., Wise, G. E., and Ko, C. (1977). *J. Cell Biol.* **73,** 322-331.

Granboulan, N., and Bernhard, W. (1961). *C.R. Seances Soc. Biol. Ses. Fil.* **155,** 1767.

Granboulan, N., and Granboulan, P. (1965). *Exp. Cell Res.* **38,** 604-619.

Greimers, R. (1978). Mémoire de licence, Université de Liège, Belgium.

Heine, U., Sverak, L., Kondratick, J., and Bonar, R. A. (1971). *J. Ultrastruct. Res.* **34,** 375-396.

Herman, R., Zieve, G., Williams, J., Lenk, R., and Penman, S. (1976). *Prog. Nucleic Acid Res. Mol. Biol.* **19,** 379-401.

Herzog, J., and Farber, J. L. (1975). *Exp. Cell Res.* **93,** 502-505.

Howard, E. F. (1978). *Biochemistry* **17,** 3228-3236.

Jacob, J. (1967). *Exp. Cell Res.* **48,** 276-282.

Karasaki, S. (1965). *J. Cell Biol.* **26,** 937-958.

Karasaki, S. (1968). *Exp. Cell Res.* **52,** 13-26.

Karasek, J., Hrdlicka, A., and Smetana, K. (1973). *J. Ultrastruct. Res.* **42,** 234-243.

Kedinger, C., and Simard, R. (1974). *J. Cell Biol.* **63,** 831-842.

Kierszenbaum, A. L. (1974). *Biol. Reprod.* **11,** 365-376.

Kierszenbaum, A. L., and Tres, L. L. (1974a). *J. Cell Biol.* **60,** 39-53.

Kierszenbaum, A. L., and Tres, L. L. (1974b). *J. Cell Biol.* **63,** 923-935.

Kierszenbaum, A. L., Tres, L. L., and Tandler, C. J. (1972). *J. Cell Biol.* **53,** 239-243.

Koshiba, K., Thirumalachary, C., Daskal, Y., and Busch, H. (1971). *Exp. Cell Res.* **68,** 235-246.

Kumar, A., and Wu, R. S. (1973). *J. Mol. Biol.* **80,** 265-276.

Lafontaine, J.-G. (1974). *In* "The Cell Nucleus" (H. Busch, ed.), Vol. 1, pp. 149-185. Academic Press, New York.

Lafontaine, J.-G., and Lord, A. (1974). *J. Cell Sci.* **16,** 63-93.

Lepoint, A., and Goessens, G. (1978). *Exp. Cell Res.* **117,** 89-94.

Littau, V. C., Allfrey, V. G., Frenster, J. H., and Mirsky, A. E. (1964). *Proc. Natl. Acad. Sci. U.S.A.* **52,** 93-100.

McKnight, S. L., Bustin, M., and Miller, O. L., Jr. (1978). *Cold Spring Harbor Symp. Quant. Biol.* **42,** 741-754.

Marinozzi, V. (1964). *J. Ultrastruct. Res.* **10,** 433-456.

Marinozzi, V., and Fiume, L. (1971). *Exp. Cell Res.* **67,** 311-322.

Matsuura, S., Morimoto, T., Tashiro, Y., Higashinakagawa, T., and Muramatsu, M. (1974). *J. Cell Biol.* **63,** 629-640.

Miller, O. L., and Bakken, A. H. (1972). *Acta Endocrinol. (Copenhagen), Suppl.* **168,** 155-173.

Miller, O. L., and Beatty, B. R. (1969a). *Science* **164,** 955-957.

Miller, O. L., and Beatty, B. R. (1969b). *Genetics* **61,** Suppl., 133-143.

Miller, T. E., Huang, C. Y., and Pogo, A. O. (1978a). *J. Cell Biol.* **76,** 675-691.

Miller, T. E., Huang, C. Y., and Pogo, A. O. (1978b). *J. Cell Biol.* **76,** 692-704.

Milner, G. R., and Hayhoe, G. J. (1968). *Nature (London)* **218,** 785-787.

Mirre, C., and Stahl, A. (1978a). *J. Cell Sci.* **31,** 79-100.

Mirre, C., and Stahl, A. (1978b). *Biol. Cell.* **32,** 9a.

Mirre, C., and Stahl, A. (1978c). *J. Ultrastruct. Res.* **64,** 377-387.

Miyawaki, H. (1974). *J. Ultrastruct. Res.* **47,** 255-271.

Miyawaki, H. (1977). *Biol. Cell* **29,** 7-16.

Monneron, A. (1966). *J. Micros. (Paris)* **5,** 583-596.

Monneron, A. (1971). *Adv. Cytopharmacol.* **1,** 131-144.

Monneron, A., and Bernhard, W. (1969). *J. Ultrastruct. Res.* **27,** 266-288.

Monneron, A., and Kerros, N. (1970). *Int. J. Cancer* **5**, 55–63.
Monneron, A., and Moulé, Y. (1968). *Exp. Cell Res.* **51**, 531–554.
Monneron, A., Lafarge, C., and Frayssinet, C. (1968). *C.R. Hebd. Seances Acad. Sci., Ser. D* **267**, 2053–2056.
Monneron, A. Burglen, J., and Bernhard, W. (1970). *J. Ultrastruct. Res.* **32**, 370–389.
Moreno Diaz de la Espina, S., Risueno, M. C., Fernandez-Gomez, M. E., and Tandler, C. J. (1976). *J. Microsc. Biol. Cell.* **25**, 265–278.
Moulé, Y., and Chauveau, J. (1968). *J. Mol. Biol.* **33**, 465–481.
Moyne, G. (1973). *J. Ultrastruct. Res.* **45**, 102–123.
Moyne, G. (1977). *Cytobiologie* **15**, 126–134.
Moyne, G., and Garrido, J. (1976). *Exp. Cell Res.* **98**, 237–247.
Moyne, G., Garrido, J., and Bernhard, W. (1974). *C.R. Hebd. Seances Acad. Sci., Ser. D* **278**, 1385–1388.
Moyne, G., Nash, R. E., and Puvion, E. (1977). *Biol. Cell.* **30**, 5–16.
Nagy, V. Z., Bertoni-Freddari, C., Nagy, I. Z., Pieri, C., and Giuli, C. (1977). *Gerontology* **23**, 267–276.
Nash, R. E., Puvion, E., and Bernhard, W. (1975). *J. Ultrastruct. Res.* **53**, 395–405.
Nebel, B. R., and Coulon, E. M. (1962a). *Chromosoma* **13**, 272–291.
Nebel, B. R., and Coulon, E. M. (1962b). *Chromosoma* **13**, 292–299.
Orkisz, S., and Bartel, H. (1978). *Histochemistry* **57**, 87–92.
Palombi, F., and Viron, A. (1977). *J. Ultrastruct. Res* **61**, 10–20.
Papsidero, L. D., and Braselton, J. P. (1973). *Cytobiologie* **8**, 118–129.
Parry, D. M., and Blackett, N. M. (1973). *J. Cell Biol.* **57**, 16–26.
Petrov, P., and Bernhard, W. (1971). *J. Ultrastruct. Res.* **35**, 386–402.
Petrov, P., and Sekeris, C. E. (1971). *Exp. Cell Res.* **69**, 393–401.
Phillips, D. M., and Phillips, S. G. (1973). *J. Cell Biol.* **58**, 54–63.
Puvion, E., and Bernhard, W. (1975). *J. Cell Biol.* **67**, 200–214.
Puvion, E., and Moyne, G. (1978). *Exp. Cell Res.* **115**, 79–88.
Puvion, E., Garrido, J., and Viron, A. (1974). *C.R. Hebd. Seances Acad. Sci., Ser. D* **279**, 509–512.
Puvion, E., Moyne, G., and Bernhard, W. (1976). *J. Microsc. Biol. Cell.* **25**, 17–32.
Puvion, E., Viron, A., and Bernhard, W. (1977). *Biol. Cell.* **29**, 81–88.
Puvion-Dutilleul, F., Bachellerie, J. P., Bernadac, A., and Zalta, J. P. (1977). *C.R. Hebd. Seances Acad. Sci., Ser. D* **284**, 663–666.
Puvion-Dutilleul, F., Puvion, E., and Bernhard, W. (1978). *J. Ultrastruct. Res.* **63**, 118–131.
Raikhlin, N. T., Buchvalov, I. B., and Unger, E. (1976). *Fal. Histochem. Cytochem.* **14**, 217–222.
Recher, L. (1970). *J. Ultrastruct. Res.* **32**, 212–225.
Recher, L., Whitescarver, J., and Briggs, L. (1969). *J. Ultrastruct. Res.* **29**, 1–14.
Recher, L., Chan, H., Briggs, L., and Parry, N. (1972). *Cancer Res.* **32**, 2495–2501.
Recher, L., Sykes, J. A., and Chan, H. (1976). *J. Ultrastruct. Res.* **56**, 152–163.
Risueno, M. C., Moreno Diaz de la Espina, S., Fernandez-Gomez, M. E., and Gimenez-Martin, G. (1976). *J. Microsc. Biol. Cell.* **26**, 5–18.
Ro-Choi, T. S., and Busch, H. (1974). *In* "The Cell Nucleus" (H. Busch, ed.), Vol. 3, pp. 151–208. Academic Press, New York.
Royal, A., and Simard, R. (1975). *J. Cell Biol.* **66**, 577–585.
Rupec, M. (1974). *Arch. Dermatol. Forsch.* **249**, 21–27.
Samarina, O. P., Lukanidin, E. M., Molnar, J., and Georgiev, G. P. 1968). *J. Mol. Biol.* **33**, 251–263.
Samarina, O. P., Lukanidin, E. M., and Georgiev, G. P. (1973). *Acta Endocrinol.* suppl. **180**, 130–160.

Sekeris, C. E., and Niessing, J. (1975). *Biochem. Biophys. Res. Commun.* **62**, 642–650.

Shankar Narayan, K., Steele, W. J., Smetana, K., and Busch, H. (1967). *Exp. Cell Res.* **46**, 65–77.

Shinozuka, H. P., Goldblatt, P. J., and Farber, E. (1968). *J. Cell Biol.* **36**, 313–328.

Siev, M., Weinberg, R., and Penman, S. (1969). *J. Cell Biol.* **41**, 510–520.

Simard, R. (1970). *Int. Rev. Cytol.* **28**, 169–211.

Simard, R., and Bernhard, W. (1967). *J. Cell Biol.* **34**, 61–76.

Simard, R., and Duprat, A. M. (1969). *J. Ultrastruct. Res.* **29**, 60–75.

Simard, R., Amalric, F., and Zalta, J. P. (1969). *Exp. Cell Res.* **55**, 359–369.

Simard, R., Langelier, Y., Mandeville, R., Maestracci, N., and Royal, A. (1974). *In* "The Cell Nucleus" (H. Busch, ed.), pp. 447–487. Academic Press, New York.

Simmons, T., Heywood, P., and Hodge, L. (1973). *J. Cell Biol.* **59**, 150–164.

Singer, I. L. (1975). *Exp. Cell Res.* **95**, 205–217.

Smetana, K. (1974). *Acta Fac. Med. Univ. Brun.* **49**, 155–197.

Smetana, K. (1977). *Biol. Cell.* **30**, 207–210.

Smetana, K., and Busch, H. (1974). *In* "The Cell Nucleus" (H. Busch, ed.), Vol. 1, pp. 73–147. Academic Press, New York.

Smetana, K., Steele, W. J., and Busch, H. (1963). *Exp. Cell Res.* **31**, 198–201.

Smetana, K., Lejnar, J., Vlastiborova, A., and Busch, H. (1971). *Exp. Cell Res.* **64**, 105–112.

Stahl, A., Mirre, C., Hartung, M., Knibiehler, B., and Navarro, A. (1978). *Ann. Biol. Anim., Biochim., Biophys.* **18**, 399–408.

Stevens, B. J., and Swift, H. (1966). *J. Cell Biol.* **31**, 55–77.

Stevens, R. H., and Amos, H. (1971). *J. Cell Biol.* **50**, 818–829.

Stockert, J. C., and Schuchner, E. B. (1972). *Exp. Cell Res.* **70**, 250–253.

Svoboda, D., and Higginson, J. (1968). *Cancer Res.* **28**, 1703–1733.

Swift, H. (1959). *Brookhaven Symp. Biol.* **12**, BNL (C22), 134.

Swift, H. (1962). *Symp. Int. Soc. Cell Biol.* **2**, 21.

Swift, H. (1963). *Exp. Cell Res., Suppl.* **9**, 54–67.

Tata, J. R. and Baker, B. (1975). *Exp. Cell Res.* **93**, 191–201.

Tres, L. L., Kierszenbaum, A. L., and Tandler, C. J. (1972). *J. Cell Biol.* **53**, 483–493.

Unuma, T., Arendell, J. P., and Busch, H. (1968). *Exp. Cell Res.* **52**, 429–438.

Unuma, T., Senda, R., and Muramatsu, M. (1972). *J. Electron Microsc.* **21**, 60–70.

Van de Walle, C., Bernier, G., Deltour, R., and Bronchart, R. (1976). *Plant Physiol.* **157**, 632–639.

Vazquez-Nin, G., and Bernhard, W. (1971). *J. Ultrastruct. Res.* **36**, 842–860.

Vazquez-Nin, G. H., Echeverria, O. M., and Pedron, J. (1979). *Biol. Cell.* **35**, 221–228.

Villard, D., and Fakan, S. (1978). *C.R. Hebd. Seances Acad. Sci., Ser. D* **286**, 777–780.

Vorbrodt, A., and Bernhard, W. (1968). *J. Microsc. (Paris)* **7**, 195–204.

Warocquier, R., and Scherrer, K. (1969). *Eur. J. Biochem.* **10**, 362–370.

Wassef, M. (1979). *J. Ultrastruct. Res.* **69**, 121–133.

Watson, M. L. (1962). *J. Cell Biol.* **13**, 162–167.

Watson, M. L., and Aldridge, W. G. (1961). *J. Biophys. Biochem. Cytol.* **11**, 257–272.

Watson, M. L., and Aldridge, W. G. (1964). *J. Histochem. Cytochem.* **12**, 96–103.

Weinberg, R., and Penman, S. (1969). *Biochim. Biophys. Acta* **190**, 10–29.

Yamamoto, H., Shiraiwa, S., Ashida, T., and Ito, Y. (1969). *J. Electron Microsc.* **18**, 57–62.

Zieve, G., and Penman, S. (1976). *Cell* **8**, 19–31.

Zylber, E. A., and Penman, S. (1971). *Science* **172**, 947–949.

INTERNATIONAL REVIEW OF CYTOLOGY, VOL. 65

Cytological Mechanisms of Calcium Carbonate Excavation by Boring Sponges

SHIRLEY A. POMPONI[1]

University of Miami, Rosenstiel School of Marine and Atmospheric Science, Miami, Florida

I. Introduction

A. BIOLOGICAL MECHANISMS OF CALCIUM CARBONATE REMOVAL

A variety of organisms excavate calcified substrates by chemomechanical mechanisms. In marine environments, bacteria, fungi, algae, and invertebrates penetrate calcium carbonate substrates by secreting acids, chelators, or enzymes (Carriker *et al.,* 1969; Bromley, 1970; Warme, 1975). These may function either singly or in combination (Carriker, 1978). Boring bacteria, fungi, algae, and sponges penetrate calcium carbonate by chemical dissolution at the cellular level. These cellular mechanisms, particularly those utilized by sponges, will be discussed in detail.

Although several hypotheses have been proposed for the cytological mode of penetration of calcium carbonate by boring bacteria, fungi, and algae, little is known of the chemical substance involved in substrate dissolution. Bacteria

---

[1]Present address: Horn Point Environmental Laboratories, University of Maryland, Center for Environmental and Estuarine Studies, P.O. Box 775, Cambridge, Maryland 21613.

isolated from coral skeletons are capable of digesting chitin *in vitro* (DiSalvo, 1969), suggesting that the mode of carbonate breakdown is via the organic matrix of skeletal carbonates (Warme, 1975).

Marine fungi penetrate calcium carbonate by first roughening and pitting the surface, and then extending hyphae throughout the substrate (Korringa, 1951; Kohlmeyer, 1969; Golubic *et al.*, 1975). Soil fungi penetrate dead bone (calcium phosphate) by simultaneously dissolving both calcium phosphate and organic matrix (Marchiafava *et al.*, 1974). Resorption occurs at the site of contact of the fungal membrane with bone. Marchiafava and others (1974) found increased decalcification when degenerative changes were visible in hyphae, and concluded that these metabolic changes either produced substances that were decalcifying or that reduced the capacity to control decalcification.

Marine fungi do not bore into inorganic substrates. They only invade such substrates after substrate colonization by algae (Golubic *et al.*, 1975). Although fungi are capable of penetrating inorganic substrates, they do not, because they are heterotrophic and depend on organic material in the substrate for food.

Penetration of calcium carbonate by chlorophytes and cyanophytes is a chemical process (Golubic, 1969; Golubic *et al.*, 1975; Fogg, 1973; Alexandersson, 1975; Risk and MacGeachy, 1979), and may involve an acid or chelator (LeCampion-Alsumard, 1975). The site of dissolution by filamentous blue-green algae is at the tips of the filaments (Golubic *et al.*, 1975), and may be effected by terminal cells (Golubic, 1969) or thread-like external projections that etch small grooves (Alexandersson, 1975). Morphology of algal borings is controlled by biological properties of the algae, whereas orientation of borings is partially influenced by substrate mineralogy (Golubic, 1969; Golubic *et al.*, 1975; Risk and MacGeachy, 1979). Initial penetration of Iceland spar (a mineralogically well-defined calcium carbonate) is strongly controlled by substrate mineralogy and occurs along crystal cleavage planes where solubility is highest (Golubic, 1969; Golubic *et al.*, 1975; Kobluk, 1976; Risk and MacGeachy, 1979). Borings become more randomly oriented as their densities within the substrate increase (Risk and MacGeachy, 1979). In skeletal carbonates, substrate control of boring patterns is difficult to interpret (Golubic *et al.*, 1975) and may not influence boring patterns at all (Alexandersson, 1975).

Boring sponges excavate calcium carbonate substrates by chemical and mechanical methods. The chemical phase of excavation is effected at the cellular level by localized secretion at the cell–substrate interface (Nassonov, 1883, 1924; Cotte, 1902; Warburton, 1958a; Cobb, 1969, 1971, 1975; Rützler and Rieger, 1973; Pomponi, 1977a,b, 1979a,b,c) (Figs. 1 and 2). Chemical etching detaches a chip of calcium carbonate (about 40–60 μm diameter), which is then mechanically removed from the substrate through the sponge tissue and out through the excurrent canal system of the sponge. Only 2–3% of the substrate is dissolved (Rützler and Rieger, 1973); the rest is removed as characteristically shaped chips.

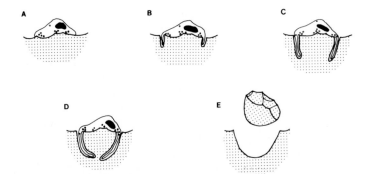

FIG. 1. Schematic drawing of the etching process. (A) Etching cell in contact with substrate (stippled). (B) Initial penetration of cell processes by chemical etching of substrate. (C,D) Continued dissolution of substrate. (E) Etched chip removed from pitted substrate. (After Hatch, 1975.)

Studies of the mechanism of sponge boring define spatial relationships that sponge cells establish with the substrate (Cobb, 1971), describe the architectural nature of substrate destruction (Cobb, 1969, 1971, 1975; Rützler and Rieger, 1973; Pomponi, 1976, 1977b; Ward and Risk, 1977), identify the cell type responsible for chemical etching (Rützler and Rieger, 1973), and localize chemi-

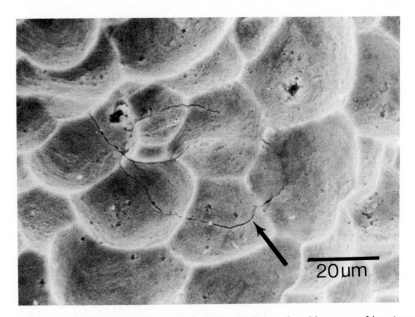

FIG. 2. Scanning electron micrograph of pitted coral skeleton bored by sponge. Lines (arrow) within pits represent initial penetration of substrate by etching cells. (From Pomponi, 1976.)

cal substances that may function in substrate dissolution (Hatch, 1975; Pomponi, 1977b, 1979b,c). These investigations will be discussed later.

Several morphological and functional similarities exist between etching cells of boring sponges and vertebrate osteoclasts (Pomponi, 1977a,b, 1979a,b,c). Osteoclasts are found on bone that is undergoing normal or pathological resorption (Hancox, 1972). The distinguishing characteristic of the osteoclast is the ruffled border, which consists of numerous cytoplasmic processes (Hancox, 1972). The ruffled border is the site of bone resorption; the area beneath it is characterized by resorption pits. At least two mechanisms are operative in osteoclastic bone resorption: (1) secretion of acids and enzymes resulting in dissolution of both inorganic and organic elements (Vaes, 1968; Lucht, 1971; Holtrop and King, 1977); and (2) uptake and digestion within osteoclasts of organic and inorganic components of bone being resorbed (Lucht, 1972a,b,c; Göthlin and Ericsson, 1976; Holtrop and King, 1977).

Similarities between osteoclasts and sponge etching cells suggest valuable experimental potential (see Section III, C).

B. Significance of Calcium Carbonate Excavation by Boring Sponges

Boring sponges occur in tropical to temperate coastal waters (Pang, 1973; Rützler, 1974). In temperate waters, they bore into oyster and clam shells. Although the sponges do not attack living tissue, the damage they inflict on the mollusc shells can often kill the bivalves. They are, thus, an important economic problem in these fisheries.

In the tropics, boring sponges can account for at least 90% of the total boring in reef corals (MacGeachy, 1977). This results in weakening of coral skeletons, making them more susceptible to physical and chemical destruction (Goreau and Hartman, 1963; Yonge, 1963; Neumann, 1966).

Since most of the calcium carbonate removed by boring sponges is particulate (Rützler and Rieger, 1973), sponge bioerosion produces a significant amount of sediment in the form of fine-grained chips (Goreau and Hartman, 1963; Fütterer, 1974). Estimates of the amount of sediment produced by boring sponges in reef environments range from 0.2 to 22 kg/m²/year (Neumann, 1966; Rützler, 1975; Moore and Shedd, 1977; Stearn and Scoffin, 1977). Boring sponges are, therefore, important ecologically and geologically in reworking and stabilizing coral reef environments.

II. Mechanisms of Sponge Boring

Boring sponges are structurally and functionally similar to other sponges of the class Demospongiae. However, most of the tissue is endolithic, living within the calcium carbonate substrate. Contact with the environment is maintained through incurrent and excurrent papillae, or, less commonly, through an epilithic encrustation of tissue. Some species (e.g., *Cliona celata*) can reach a massive epilithic (gamma) stage when the substrate is completely excavated (Vosmaer, 1933–1935).

Excavations (galleries, chambers, and tunnels) are usually small, about 1–5 mm in diameter (Pang, 1973; Rützler, 1974; S. A. Pomponi, unpublished). Extensive boring results in a honeycombed appearance of the substrate. An exception to this gross pattern of boring is found in the genus *Siphonodictyon*. Species of this genus excavate large spherical cavities from 5 to 10 cm in diameter in massive corals (Rützler, 1971).

There is no evidence to suggest that any nutritional requirements of boring sponges are satisfied by ingestion or absorption of the organic matrix or organic materials deposited in excavated skeletal carbonates (Goreau and Hartman, 1963; Hatch, 1975).

Reproduction and development of boring sponges is poorly understood. It is generally agreed that clionids are oviparous and that the larva is a solid, ciliated parenchymella (Topsent, 1900, 1928; Dendy, 1921; de Laubenfels, 1936; Lévi, 1956; Warburton, 1958b, 1966). Larvae settle on calcium carbonate substrates that are not heavily encrusted. Frequently, new boring sponge growth occurs in the dead centers of living coral colonies that have been killed by an algal infection (Pomponi, 1977b). Larvae do not normally settle on the surface of encrusting organisms or on living coral (MacGeachy, 1977). Exceptions to this include *Siphonodictyon coralliphagum* and *S. cachacrouense,* whose larvae may settle among living coral polyps (Rützler, 1971). These species produce large quantities of mucus, and it is suggested that the mucus may protect the larva during settlement and promote coral polyp death (Rützler, 1971).

A. EARLY THEORIES

Mechanisms of sponge boring have been investigated for over 100 years. Three hypotheses emerge: (1) chemical dissolution of the substrate (Bate, 1849; Revelle and Fairbridge, 1957; Cloud, 1959); (2) mechanical removal of substrate (Hancock, 1849, 1867; Fischer, 1868; Beale, 1871; Topsent, 1887; Letellier, 1894); and (3) a combination of chemical and mechanical processes by which the substrate is partially dissolved and then mechanically removed (Priest, 1881; Nassonov, 1883; Cotte, 1902; Vosmaer, 1933–1935; Warburton, 1958a). Recent studies support the third hypothesis (Cobb, 1969, 1971, 1975; Rützler and Rieger, 1973; Hatch, 1975; Pomponi, 1976, 1977a,b, 1979a,b,c).

B. CURRENT HYPOTHESES

1. *Spatial Relationships between Boring Sponge Cells and Calcified Substrate*

Cobb (1971) defined the spatial relationships sponge cells established with the substrate and explained the chemomechanical process at the level of optical microscopy. Penetration by cultures of *Cliona celata* into Iceland spar begins within three to four weeks with the clustering of cells into a rosette formation on the crystal surface. Boring is triggered by a single cell, which produces a circular etching. Rosettes increase in size as additional rings of cells join the original formation. Cells in the center of the rosette, which begin etching, move deeper into the substrate than the cells around the periphery. Thus, an orderly temporal and spatial sequence of boring is established. Eventually, cavities and tunnels are formed; these increase in diameter and depth with the excavation of more chips.

Cobb (1969, 1971) observed that etchings form along the cell edge, beginning at one or more points where the edge comes in contact with the substrate and spreading progressively around the periphery. The cell edge moves down perpendicular to the initial etching, dissolving a crevice and constricting until the crevice becomes pit-shaped and a chip is etched free (Figs. 1 and 2). The nucleus remains in the center of the initial etching and only the cell edge moves into the substrate. The overall dimensions of the cell increase to accommodate the depth of penetration by the cell edge.

Cobb (1971) described three cell types associated with etching. Etching cells (32 × 40 μm, estimated from etching outlines) have a spherical to ovoid, nucleolate nucleus, surrounded by numerous basophilic granules. There are no granules in the perimeter of the cell body or in the cell edges penetrating the substrate. Cobb (1971) suggests that the granules could be lysosomes containing acid hydrolases capable of digesting organic components of the substrate. However, he did not present evidence for this and could find no correlation between depth of etching and number of granules. (Cytochemical evidence for an acid hydrolase is presented later.)

Other cells (40 × 50 μm) in the vicinity of the etching cells are similar, but have, in addition, short filamentous extensions. These cells occur singly or in groups of two or three. They were not observed to participate in the etching process, but due to their morphologic similarity to etching cells, Cobb (1971) designated them as intermediate-type cells.

The third cell type is found attached to the substrate along cavity walls where there is no boring activity. These cells are polygonal or irregular, with broad and filamentous cytoplasmic extensions. There is a spherical nucleus with a nucleolus half the size of that in etching cells. Basophilic granules are rare. The cells are loosely organized into a reticular sheet and there are large intercellular spaces. This cell was never observed etching and was designated as a nonetching-type cell (Cobb, 1971). This cell type is probably the same as that observed by Warburton (1958a) to be attached to a cover glass. Warburton described these as etching cells, but they were later (Rützler and Rieger, 1973) defined as collencytes and assigned a connective function.

Cobb (1971) suggests that the increase in size of the nucleolus and intensity of cytoplasmic basophilia indicates a transition to a more active state of protein synthesis and that the three cell types are different functional states in the development of etching cells. Experimental evidence to support this hypothesis is lacking.

Rützler and Rieger (1973) used scanning and transmission electron microscopy to study *Cliona lampa* and its excavations. Initial etchings in Iceland spar crystals measure $39 \times 27 \mu m$ and penetrate to a depth of $12-17 \mu m$ (Fig. 2). The width of the crevices is $0.15-0.25 \mu m$, except at the mouth of the crevice, where the width is $0.3-0.8 \mu m$. Excavation of chips of calcium carbonate results in the formation of a series of chambers within the substrate. The first chamber forms after the sponge penetrates $0.2-0.5$ mm. Surface erosion does not proceed further once the sponge has established chambers and galleries in the new substrate.

2. *Ultrastructural Examinations of Etching Cells*

Etching cells occur in the peripheral parts of bored tunnels and galleries. Rützler and Rieger (1973) describe etching cells as elongate, twisted, and situated perpendicular to the substrate, widest at the distal end, and tapering toward the base (dimensions $30 \mu m$ long, $5 \mu m$ at base, $15 \mu m$ at distal end). The nucleus is large, circular or oval, with a distinct nucleolus. A Golgi apparatus and large, ovoid mitochondria are present; and a prominent rough endoplasmic reticulum extends throughout the distal portion of the cell. Small vacuoles, multivesicular bodies, and a few phagosomes are present, medium dense droplets without membranes are scattered, and glycogen granules occur distally. Numerous filopodial processes extend from the distal end of the cell into the etched crevice, forming a meshwork or "filopodial basket" that eventually envelopes the chip it etches from the substrate. Filopodia from several adjacent etching cells occur in the same crevice, and often several adjacent chips are removed simultaneously (Rützler and Rieger, 1973) (Fig. 3).

Cytoplasmolysis of the etching cell is observed to begin soon after the first filaments penetrate the substrate (Rützler and Rieger, 1973). The cell contents loosen, and mitochondria, rough endoplasmic reticulum, and vesicles associated with the Golgi complex become very prominent. A flocculent secretory product, perhaps the etching agent (Rützler and Rieger, 1973), extends into the filopodia. The cell body inflates and organelles begin to degenerate: first the nucleolus, then phagosomes, mitochondria, and the nuclear membrane (Rützler and Rieger, 1973).

Increased activity of the Golgi complex as well as the presence of a prominent nucleolus and rough endoplasmic reticulum suggest the synthesis of a chemical etching agent (Rützler and Rieger, 1973). Etching is localized at the cell processes. No further etching occurs after the chip is dislodged from the substrate. Removal of chips is suggested, on the basis of SEM observations, to be effected by displacement of old etching cells by new etching cells and other cells moving into the newly bored space (Rützler and Rieger, 1973).

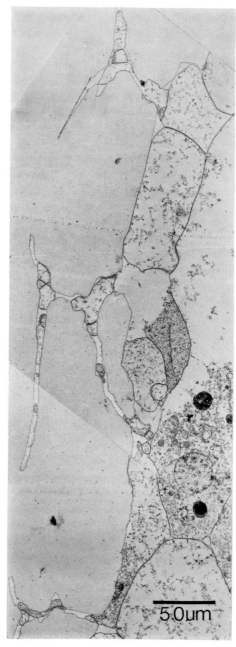

Fɪɢ. 3.   Two adjacent chips being excavated by processes of etching cells. (From Rützler and Rieger, 1973.)

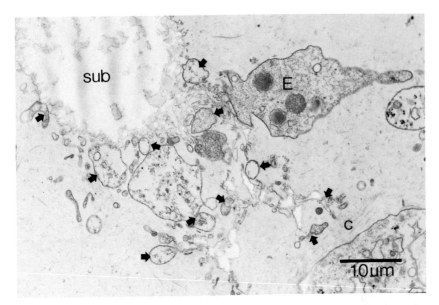

FIG. 4. Etching cell bodies (E) and processes (arrows) adjacent to decalcified substrate (sub). Note large intercellular spaces with collagen fibrils (c). (From Pomponi, 1979a.)

Current studies of the ultrastructure and cytochemistry of the etching area of boring sponges define the functional etching cell as the etching cell body and numerous associated cell processes (Pomponi, 1977a,b, 1979a,b,c). Cell processes occur in an area adjacent to the calcium carbonate substrate that is characterized by large intercellular spaces (Fig. 4). This description is in contrast with Cobb's (1969, 1971) interpretation of boring as the penetration of the entire cell edge. The abundance and size (0.1–1.5 μm diameter) of the cell processes (Pomponi, 1977b, 1979a) suggest that Cobb's (1969, 1971) interpretation of etching cell structure and function may have been hampered by the limits of resolution of optical microscopy. Cytochemical assays (Pomponi, 1977b, 1979b), discussed later, support the hypothesis that dissolution of substrate is localized at the membranes of etching cell processes (Nassonov, 1883; Rützler and Rieger, 1973; Cobb, 1969, 1971).

3. Ultrastructure of Associated Cells

Amoebocytes, gray cells, spherulous cells, and cells with osmiophilic inclusions occasionally occur in the etching area. Amoebocytes regulate the equilibrium among the cell population and can differentiate into other cell types. Etching cells are probably derived from amoebocytes (Rützler and Rieger, 1973).

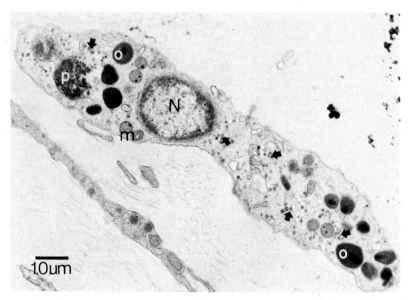

Fig. 5.   Gray cell. Note abundant glycogen granules (arrows) and osmiophilic inclusions (o).
N,Nucleus; m,mitochondria; p,phagosome. (From Pomponi, 1979a.)

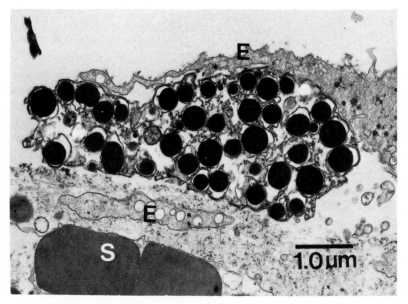

Fig. 6.   Cell with osmiophilic inclusions located near etching cells (E) and spherulous cell (S).
(From Pomponi, 1979a.)

Gray cells (Fig. 5) synthesize, accumulate, store, and transport glycogen (Boury-Esnault, 1977). Spherulous cells (Fig. 6) are filled with large, membrane-bound, granular inclusions and are probably secretory (Donadey and Vacelet, 1977). Cells with osmiophilic inclusions (Fig. 6) are larger than spherulous cells and the inclusions are more homogeneous than granular. These cells could be a different type or functional state of spherulous cells (J. Vacelet, personal communication). Both cell types may play an auxiliary role in calcium carbonate dissolution by etching cells. Indeed, what Cobb (1971) described as etching cells may actually be the associated cells with osmiophilic inclusions. The staining affinity, size, and abundance of the basophilic inclusions in Cobb's etching cells closely correspond with ultrastructural analyses of the osmiophilic inclusions in cells that occur in the etching area (Pomponi, 1977b, 1979a). Clearly, further analyses of these cells are desirable to determine what role, if any, they play in calcium carbonate dissolution.

C. THE CHEMICAL MECHANISM

Chemical dissolution of skeletal carbonates can occur via a chelator, an acid, an enzyme, or any combination of these (Carriker, 1978).

1. Chelators and Acids

Andersen (1978) extracted a brominated indole derivative, tetracetyl clionamide, from *Cliona celata*, and suggested that the triphenolic portion of the derivative might be a good calcium chelator. This has not been tested, and no other investigations of the role of chelators in calcium carbonate dissolution by boring sponges have been made.

It has been suggested that the primary mechanism for the excavation of calcium carbonate substrates by boring sponges is via the production of an acid (Nassonov, 1883, 1924; Cotte, 1902; Vosmaer, 1933–1935; Warburton, 1958a), although attempts at demonstrating an increase in hydrogen ion concentration have been unsuccessful (Cotte, 1902; Warburton, 1958a).

2. Enzymes

a. *Carbonic Anhydrase.* Excavation of calcium carbonate substrates may be achieved alternatively through the secretion of carbonic anhydrase and the resulting shift in carbonate solubility product. The involvement of carbonic anhydrase in the excavation of calcium carbonate substrates has been demonstrated in the accessory boring organs of gastropods (Chétail and Fournié, 1969; Smarsh et al., 1969; Bundy, 1977). Bundy (1977) speculates that the enzyme is likely to occur in boring sponges.

Carbonic anhydrase activity has been demonstrated in *Cliona celata* by electrometric assay (Hatch, 1975). Differential centrifugation localized the enzyme in the mitochondrial-sized fraction. Treatment of extracts with isobutanol and sonication released the enzyme activity into solution, indicating that it is membrane- or particle-bound. Furthermore, inhibition of boring by *C. celata* in the presence of acetazolamide, an inhibitor of carbonic anhydrase, has been demonstrated. There is a positive correlation between the excavating activity of the sponge and the content of carbonic anhydrase in the sponge tissues (Hatch, 1975).

Paradoxically, the highest activity of carbonic anhydrase is concentrated in the cortical tissue of the gamma growth form of *C. celata,* which has completely excavated the calcium carbonate substrate and is no longer boring. Enzyme activity is 1.5–2 times that of actively boring forms (Hatch, 1975). Hatch (1975) postulates that the high concentration of enzyme represents the potential for boring, and that the cortical tissues are most likely to come into contact with new calcium carbonate substrates. This explanation is unsatisfactory, since it is unlikely that the sponge would synthesize enzyme in anticipation of future requirements.

This assay has been repeated with *Cliona delitrix* and *C. caribbaea* and similar, ambiguous results were obtained (Pomponi, 1977b). The highest activity of enzyme occurs in the sponge tissues not involved with boring.

The significance of these results is difficult to interpret satisfactorily. One problem associated with the biochemical assay is that it cannot distinguish between different functions of the enzyme (Maren, 1977). Since carbonic anhydrase has a broad spectrum of cellular functions, a positive assay may not necessarily correlate with boring activity. Interpretation of function depends on the proper use of inhibitors (Maren, 1977). The observation that acetazolamide inhibits boring does confirm that one function of the enzyme in boring sponges is related to the excavation of calcium carbonate substrates (Hatch, 1975).

The enzyme could also have a physiological role when present in amounts below the limit of detection. This situation is further compounded because it is usually necessary to dilute the tissue when extracting for the assay (Maren, 1977). Furthermore, the assay attempts to measure activity of an enzyme believed to be functional in a very small population of cells, the etching cells. These cells constitute a monolayer less than 25 μm thick, and techniques had not been perfected that could isolate the etching cells. (However, see Section III,C.) Carbonic anhydrase activity has been found in ectosomal and choanosomal cells (Hatch, 1975; Pomponi, 1977b). It is likely that any carbonic anhydrase activity in etching cells has been diluted or masked by the large proportion of nonetching cells in the sponge extract. Clearly, direct correlation of boring activity with carbonic anhydrase activity on the basis of biochemical assay awaits isolation or concentration of the etching cells.

The basic hypothesis, however, is supported by contemporary studies of the enzyme and may still apply. The mechanism of calcium carbonate dissolution may involve a shift in the carbonate solubility product in the microenvironment of the etching cell, mediated through the activity of carbonic anhydrase.

Cytochemical assays for carbonic anhydrase in etching areas of boring sponges (Pomponi, 1977b, 1979c) localized the enzyme in cytoplasmic bodies within etching cells, on the membranes of etching cell bodies, cell processes, and cytoplasmic bodies, and in the extracellular spaces between etching cell processes (Fig. 7). The mode of action of carbonic anhydrase could be regulation of ion transport or of acid secretion at the membranes of the etching cell processes, resulting in the dissolution of calcium carbonate. The enzyme could also provide a pH optimum for the action of acid phosphatase, which has been localized in etching cells.

b. *Acid Phosphatase*. Ultrastructural analyses demonstrate that the etching cell contains a well-developed rough endoplasmic reticulum, Golgi complex, lysosomes, phagosomes, and abundant cytoplasmic bodies, clear vesicles, and tubular elements (Pomponi, 1977b, 1979a) (Fig. 8). Thus, the etching cell has the capacity for synthesis of proteins, secretion of substances into extracellular spaces, absorption of substances from extracellular spaces, and digestion of absorbed substances intracellularly.

Functional implications of etching cell morphology are confirmed by cytochemical assays for acid phosphatase (Pomponi, 1977b, 1979b).

Acid phosphatase, a lysosomal marker, is active on the rough endoplasmic reticulum, Golgi complex, lysosomes, and phagosomes, indicating that enzyme is being synthesized by the endoplasmic reticulum and Golgi complex, and that

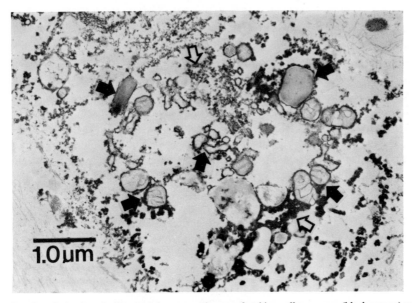

Fig. 7.   Carbonic anhydrase activity on membranes of etching cell processes (black arrows) and in extracellular spaces (open arrows). (From Pomponi, 1979c.)

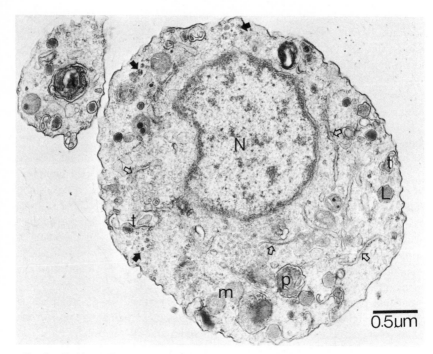

FIG. 8. Etching cell. Note well-developed rough endoplasmic reticulum (open arrows), lysosomes (L), phagosomes (p), tubular elements (t), and glycogen (black arrows). N,Nucleus; m,mitochondria. (From Pomponi, 1979a.)

lysosomes are being formed. Enzyme activity in phagosomes indicates that intracellular digestion of their contents is occurring, probably through fusion with lysosomes. Acid phosphatase is most intense on the outer surfaces of the membranes of etching cell processes, where the enzyme is released (Fig. 9). This provides evidence to support previous hypotheses (Cobb, 1971, 1975; Rützler and Rieger, 1973) that a chemical etching agent is localized at the cell membrane. Further evidence of secretion of the enzyme is provided by the localization of enzyme activity in the extracellular spaces between etching cell processes, as well as in the clear vesicles and tubules, which represent the deepest invaginations of the extracellular channels (Pomponi, 1977b, 1979b) (Fig. 9).

Pomponi (1977b, 1979b,c) suggests that etching cells have the morphologic and enzymatic capacity to effect chemical dissolution of calcium carbonate substrates by two mechanisms: (1) enzymatic digestion of organic components of skeletal carbonates, both intracellularly and extracellularly, via the lysosomal system and the membranes of etching cells processes; and (2) solubilization of mineral components via carbonic anhydrase regulation of hydrogen ion concentration at the etching cell membrane.

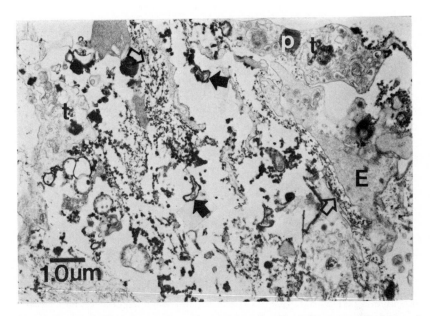

FIG. 9. Acid phosphatase activity on outer membranes of etching cell processes (black arrows), in extracellular channels between etching cell processes (open arrows), in tubular elements (t), and in phagosomes (p). E,Etching cell.

III. Similarities to Osteoclastic Bone Resorption

Etching cells of boring sponges are characterized by the same features found in active osteoclasts. These similarities include a region of numerous cell processes (the ruffled border) and a system of cytoplasmic bodies, vesicles, and vacuoles, which are structurally connected with a tubular system and with extracellular channels (Pomponi, 1979a; Hancox, 1972; Lucht, 1972c; Holtrop and King, 1977).

In both cell types, a lysosomal system is operative (Pomponi, 1977b, 1979b). This lysosomal system has been implicated as the primary mechanism in calcified tissue destruction by osteoclasts (Vaes, 1968; Göthlin and Ericsson, 1976; Holtrop and King, 1977).

A. Osteoclastic Bone Resorption—Modes of Action

The mechanisms proposed for osteoclastic bone resorption suggest the following modes of action:

1. The osteoclast synthesizes lysosomal enzymes, some of which are secreted to the extracellular resorption zone. There they hydrolyze the noncollagenous

organic matrix components of bone (Vaes, 1968; Göthlin and Ericsson, 1971; Lucht and Nørgaard, 1976). Other lysosomal enzymes are responsible for intracellular digestion of partially digested, noncollagenous, organic matrix, which is absorbed by endocytosis (Vaes, 1968; Daems et al., 1969; Lucht, 1972a). Collagenous components of the organic matrix are phagocytosed and digested not by osteoclasts, but by mononuclear cells (Heersche, 1978).

2. The mineral component may be dissolved by acid secreted into the resorption zone by the osteoclast (Vaes, 1968). Parathyroid hormone simultaneously stimulates osteoclastic bone resorption and aerobic glycolysis, resulting in the release of citrate and lactate from the osteoclast (Vaes, 1968). Acid secretion could solubilize the mineral component and also provide a pH optimum for the action of lysosomal acid hydrolases (Vaes, 1968).

3. Other cell types may also be involved. For example mononuclear cells from lymphoid organs may participate in bone resorption by either elaborating an osteoclast-activating factor, or by transforming into osteoclasts (Marks and Schneider, 1978; Schneider, 1978).

The suggested sequence of events in bone resorption is (1) dissolution of the mineral component, and then (2) hydrolysis of noncollagenous organic matrix by osteoclasts (Holtrop and King, 1977). After the osteoclasts become detached from the demineralized area, (3) mononuclear cells phagocytose collagenous components of the organic matrix (Heersche, 1978).

B. Calcium Carbonate Dissolution by Boring Sponges—Modes of Action

Boring sponges are not restricted to excavation of skeletal calcium carbonate, i.e., calcium carbonate with organic components. They can bore just as rapidly in a completely inorganic calcium carbonate substrate such as Iceland spar (Rützler, 1975; S. A. Pomponi, unpublished). This suggests that the mode of action does not necessarily depend on hydrolysis of an organic matrix, although the presence of a lysosomal system (Pomponi, 1977b, 1979b) indicates that etching cells do have the capacity to digest organic matrix when present.

Present evidence suggests, therefore, that the mode of action of calcium carbonate removal by etching cells is via dissolution of the inorganic component of the substrate. If any enzyme is involved, its mode of action is most likely through the mediation of bicarbonate and hydrogen ion concentrations. The presence of carbonic anhydrase in etching cells (Pomponi, 1977b, 1979c) provides evidence for the involvement of one such enzyme.

It is also possible that the cells associated with etching cells (gray cells, spherulous cells, and cells with osmiophilic inclusions) may secrete an acid or chelator. Gray cells or "glycocytes" (Boury-Esnault and Doumenc, 1979) could secrete products of aerobic glycolysis such as citrate and lactate, or the cells could transfer glycogen to the etching cells (Boury-Esnault, 1977), which, in turn, could secrete an acid. These cellular relationships are poorly understood and more detailed analyses of their role in calcium carbonate dissolution are necessary.

C. Boring Sponges as a Model System

Boring sponges offer a unique multicellular system that can be used as an experimental model of osteoclastic bone resorption, particularly resorption of the mineral component of bone. Structural and functional similarities have been demonstrated in etching cells and osteoclasts (Pomponi, 1977b, 1979a,b,c). These may be exploited in experimental studies employing the relatively simple multicellular system of the sponge to develop an experimental model of the more complex mammalian system of bone resorption.

Sponges lend themselves to *in vitro* analyses of the intact organism. Cultures of boring sponges in calcium carbonate (Iceland spar) can be maintained under controlled conditions (S. A. Pomponi, unpublished). Rates of excavation can be monitored and conditions affecting the physiology of boring can be examined (S. A. Pomponi, unpublished). The primitive multicellularity of sponges provides opportunities for the application of cell sorting techniques. With this methodology, sponge cells can be separated into distinct cell types using density gradient centrifugation (DeSutter and Van de Vyver, 1977; DeSutter and Buscema, 1977). More precise techniques for separating the relatively small percentage of etching cells from the other cell types now make it possible to characterize the etching cell biochemically and physiologically.

Acknowledgments

This work was supported by grants from the National Science Foundation (OCE 76-16886) and the U.S. Navy Office of Naval Research (N00014-79-C-0395). I thank Dr. Dennis L. Taylor for critically reading the manuscript and Susan M. Markley for preparation of the line drawing.

References

Alexandersson, E. T. (1975). *Nature (London)* **254,** 212, 237–238.
Andersen, R. J. (1978). *Tetrahedron Lett.* **29,** 2541–2544.
Bate, C. S. (1849). *Br. Assoc. Adv. Sci., Rep.* **19,** 73–75.
Beale, L. S. (1871). *J. Quekett Microsc. Club* **2,** 279–280.
Boury-Esnault, N. (1977). *Cell Tissue Res.* **175,** 523–539.
Boury-Esnault, N., and Doumenc, D. (1979). *Colloq. Int. C.N.R.S.* **291** (in press).
Bromley, R. G. (1970). *Geol. J., Spec. Issue* **3,** 49–90.

Bundy, H. F. (1977). *Comp. Biochem. Physiol. B* **57**, 1-7.

Carriker, M. R. (1978). *Mar. Biol.* **48**, 105-134.

Carriker, M. R., Smith, E. H., and Wilce, R. T., eds. (1969). *Am. Zool.* **9**, 629-1020.

Chétail, M., and Fournié, J. (1969). *Am. Zool.* **9**, 983-990.

Cloud, P. E. (1959). *U.S., Geol. Surv., Prof. Pap.* **280K**, 361-445.

Cobb, W. R. (1969). *Am. Zool.* **9**, 783-790.

Cobb, W. R. (1971). Ph.D. Dissertation, University of Rhode Island, Kingston.

Cobb, W. R. (1975). *Trans. Am. Microsc. Soc.* **94**, 197-202.

Cotte, J. (1902). *C.R. Seances Soc. Biol. Ses. Fil.* **54**, 636-637.

Daems, W., Wisse, E., and Brederoo, P. (1969). *In* "Lysosomes in Biology and Pathology" (J. T. Dingle and H. B. Fell, eds.), Vol. 1, pp. 64ff. Am. Elsevier, New York.

de Laubenfels, M. W. (1936). *Carnegie Inst. Washington Publ.* **467**, 1-225.

Dendy, A. (1921). *Trans. Linn. Soc. London* **18**, 16-164.

DeSutter, D., and Buscema, M. (1977). *Wilhelm Roux' Arch. Entwicklungsmech. Org.* **183**, 149-153.

DeSutter, D., and Van de Vyver, G. (1977). *Wilhelm Roux' Arch. Entwicklungsmech. Org.* **181**, 151-161.

DiSalvo, L. H. (1969). *Am. Zool.* **9**, 735-740.

Donadey, C., and Vacelet, J. (1977). *Arch. Zool. Exp. Gen.* **118**, 273-284.

Fischer, P. (1868). *Lyons Mus. Hist. Nat. Nouv. Arch.* **4**, 117-172.

Fogg, G. E. (1973). *In* "The Biology of Blue-Green Algae" (N. G. Carr and B. A. Whitton, eds.), pp. 368-378. Univ. of California Press, Berkeley.

Fütterer, D. K. (1974). *J. Sediment. Petrol.* **44**, 79-84.

Golubic, S. (1969). *Am. Zool.* **9**, 747-751.

Golubic, S., Perkins, R. D., and Lukas, K. J. (1975). *In* "The Study of Trace Fossils" (R. W. Frey, ed.), pp. 229-259. Springer-Verlag, Berlin and New York.

Goreau, T. F., and Hartman, W. D. (1963). *In* "Mechanisms of Hard Tissue Destruction" (R. F. Sognnaes, ed.), Publ. No. 75. pp. 25-54. Am. Assoc. Adv. Sci., Washington, D.C.

Göthlin, G., and Ericsson, J.L.E. (1971). *Histochemie* **28**, 337-344.

Göthlin, G., and Ericsson, J.L.E. (1976). *Clin. Orthop. Relat. Res.* **120**, 201-231.

Hancock, A. (1849). *Ann. Mag. nat. Hist.* [2] **3**, 321-348.

Hancock, A. (1867). *Ann. Mag. Nat. Hist.* [3] **19**, 229-242.

Hancox, N. M. (1972). *In* "The Biochemistry and Physiology of Bone" (G. H. Bourne, ed.), Vol. 1, pp. 45-67. Academic Press, New York.

Hatch, W. I., Jr. (1975). Ph.D. Dissertation, Boston University, Boston, Massachusetts.

Heersche, J.N.M. (1978). *Calcif. Tissue Res.* **26**, 81-84.

Holtrop, M. E., and King, G. J. (1977). *Clin. Orthop. Relat. Res.* **123**, 177-196.

Kobluk, D. R. (1976). Ph.D. Dissertation, McMaster University, Hamilton, Ontario.

Kohlmeyer, J. (1969). *Am. Zool.* **9**, 741-746.

Korringa, P. (1951). *Rapp. P.-V. Reun., Cons. Int. Explor. Mer* **128**, 50-54.

LeCampion-Alsumard, T. (1975). *Cah. Biol. Mar.* **16**, 177-185.

Letellier, M. (1894). *C.R. Hebd. Seances Acad. Sci.* **118**, 986-989.

Lévi, C. (1956). *Arch. Zool. Exp. Gen.* **93**, 1-181.

Lucht, U. (1971). *Histochemie* **28**, 103-117.

Lucht, U. (1972a). *Histochemie* **29**, 274-286.

Lucht, U. (1972b). *Z. Zellforsch. Mikrosk. Anat.* **135**, 211-228.

Lucht, U. (1972c). *Z. Zellforsch. Mikrosk. Anat.* **135**, 229-244.

Lucht, U., and Nørgaard, J. O. (1976). *Cell Tissue Res.* **168**, 89-99.

MacGeachy, J. K. (1977). *Proc. Int. Coral Reef Symp., 3rd, 1977* Vol. 2, pp. 477-483.

Marchiafava, V., Bonucci, E., and Ascenzi, A. (1974). *Calcif. Tissue Res.* **14**, 195-210.

Maren, T. H. (1977). *Am. J. Physiol.* **232**, F291–F297.

Marks, S. C., Jr., and Schneider, G. B. (1978). *Am. J. Anat.* **152**, 331–342.

Moore, C. H., Jr., and Shedd, W. W. (1977). *Proc. Int. Coral Reef Symp., 3rd, 1977* Vol. 2, pp. 499–505.

Nassonov, N. (1883). *Z. Wiss. Zool.* **39**, 295–308.

Nassonov, N. (1924). *Dokl. Biol. Sci. (Engl. Transl.)* pp. 113–115.

Neumann, A. C. (1966). *Limnol. Oceanogr.* **11**, 92–108.

Pang, R. K. (1973). *Postilla* **161**, 1–75.

Pomponi, S. A. (1976). *Scanning Electron Microsc.* **2**, 569–575.

Pomponi, S. A. (1977a). *Proc. Int. Coral Reef Symp., 3rd, 1977* Vol. 2, pp. 485–490.

Pomponi, S. A. (1977b). Ph.D. Dissertation, University of Miami, Miami, Florida.

Pomponi, S. A. (1979a). *J. Mar. Biol. Assoc. U.K.* **59**, 777–784.

Pomponi, S. A. (1979b). *J. Mar. Biol. Assoc. U.K.* **59**, 785–789.

Pomponi, S. A. (1979c). *Colloq. Int. C.N.R.S.* **291** (in press).

Priest, B. W. (1881). *J. Quekett Microsc. Club* **6**, 269–271.

Revelle, R., and Fairbridge, R. (1957). *Mem., Geol. Soc. Am.* **67**, 239–296.

Risk, M. J., and MacGeachy, J. K. (1979). *Rev. Biol. Trop.* (in press).

Rützler, K. (1971). *Smithson. Contrib. Zool.* **77**, 1–37.

Rützler, K. (1974). *Smithson. Contrib. Zool.* **165**, 1–32.

Rützler, K. (1975). *Oecologia* **19**, 203–216.

Rützler, K., and Rieger, G. (1973). *Mar. Biol.* **21**, 144–162.

Schneider, G. B. (1978). *Am. J. Anat.* **153**, 305–320.

Smarsh, A., Chauncey, H. H., Carriker, M. R., and Person, P. (1969). *Am. Zool.* **9**, 967–982.

Stearn, C. W., and Scoffin, T. P. (1977). *Proc. Int. Coral Reef Symp., 3rd, 1977* Vol. 2, pp. 471–476.

Topsent, E. (1887). *Arch. Zool. Exp. Gen.* **5**, 1–165.

Topsent, E. (1900). *Arch. Zool. Exp. Gen.* **8**, 1–331.

Topsent, E. (1928). *Result. Campagnes Sci. Monaco* **74**, 1–376.

Vaes, G. (1968). *J. Cell Biol.* **39**, 676–697.

Vosmaer, G.C.J. (1933–1935). "The Sponges of the Bay of Naples," Vols. I–III. Nijhoff, The Hague.

Warburton, F. E. (1958a). *Can. J. Zool.* **36**, 555–562.

Warburton, F. E. (1958b). *Nature (London)* **181**, 493–494.

Warburton, F. E. (1966). *Ecology* **47**, 672–674.

Ward, P., and Risk, M. J. (1977). *J. Paleontol.* **51**, 520–526.

Warme, J. E. (1975). *In* "The Study of Trace Fossils" (R. W. Frey, ed.), pp. 181–227. Springer-Verlag, Berlin and New York.

Yonge, C. M. (1963). *In* "Mechanisms of Hard Tissue Destruction" (R. F. Sognnaes, ed.), Publ. No. 75, pp. 1–24. Am. Assoc. Adv. Sci., Washington, D.C.

INTERNATIONAL REVIEW OF CYTOLOGY, VOL. 65

Neuromuscular Disorders with Abnormal Muscle Mitochondria

Z. Kamieniecka and H. Schmalbruch

Institute of Neurophysiology, University of Copenhagen, Copenhagen, Denmark.

I. Introduction

Mitochondria are the site of oxidative phosphorylation of ADP to ATP. The mechanical work of muscle fibers, i.e., shortening of myofibrils, depends on an adequate supply of ATP, and metabolic disorders of mitochondria can be assumed to impair the function of the muscle fiber.

Since 1962 (Luft *et al.*, 1962), several patients presenting neuromuscular disorders with structural abnormalities of muscle mitochondria have been reported. The clinical findings in these patients vary; attempts to identify a metabolic defect have given differing results. This indicates that different metabolic disorders are covered by the term "mitochondrial myopathy." The study of these disorders is complicated by the facts that (1) few patients have identical clinical and biochemical findings, (2) many reports lack relevant clinical data, and (3) no animal models exist.

The aim of this article is to describe the morphology of mitochondria in skeletal muscle fibers of mammals under normal and experimental conditions and in patients with "mitochondrial myopathies," and to classify these disorders according to clinical and biochemical parameters.

The biochemistry of mitochondria in normal muscle cells and in some "mitochondrial myopathies" has been reviewed (DeHaan *et al.*, 1973; Williamson, 1979; DiMauro *et al.*, 1974; DiMauro, 1979).

II. The Architecture of the Chondrioma of Mammalian Muscle Fibers and Its Modifications

The chondrioma of skeletal muscle fibers differs from that in other cells because of the spatial distribution of cell organelles. Mitochondria form networks that transverse the fiber and encircle the myofibrils at the level of the I-bands. Each sarcomere contains two mitochondrial "grids"; the many grids in a fiber are connected by longitudinal branches. Thus, the chondrioma forms a three-dimensional framework that stretches throughout the entire muscle fiber (Bubenzer, 1966) (Figs. 1 and 2).

Longitudinal sections of muscle fibers reveal cross sections through bars of these grids, which appear as circular profiles between myofibrils on both sides of the Z-line (Figs. 1a and 2b). In human muscle fibers, the diameter of these circular profiles is 0.2–0.5 μm. In addition to the framework of mitochondria, muscle fibers may contain clusters of spherical mitochondria that are localized beneath the sarcolemma, often close to capillaries. The diameter of subsarcolemmal mitochondria rarely exceeds 0.7 μm. Triglyceride droplets occur together with mitochondria, both in the interior of the fiber and within subsarcolemmal clusters of mitochondria. In man, their diameter is at most 1 μm, but usually less than 0.5 μm.

Skeletal muscle fibers of mammals are specialized with respect to speed of

Fig. 1. Longitudinal section (a) and cross section (b) of slow twitch muscle fibers from the medial gastrocnemius of cat to show the spatial arrangement of mitochondria. The circular profiles of mitochondria seen in longitudinal sections at the level of the I-bands are branches of a transverse network. Note electron-translucent lipid droplets among mitochondria. ×18,000.

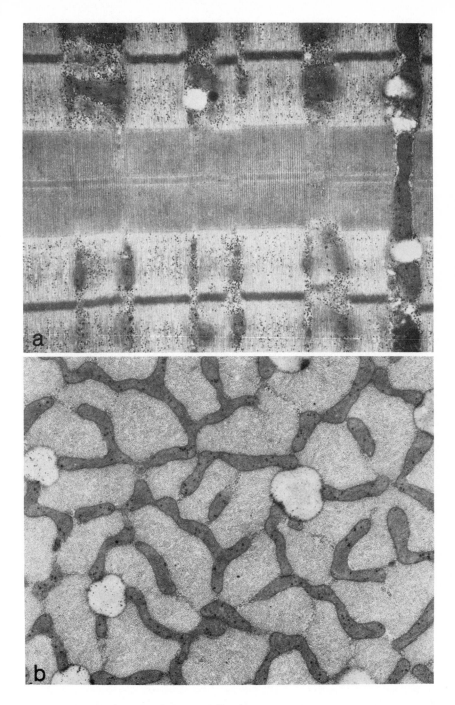

Fig. 1

Fig. 2. Longitudinal freeze–fracture (a) through a fast twitch fiber rich in mitochondria from the anterior tibial muscle of rat. A longitudinal section of a similar fiber is placed in register (b). In (a), parts of four transverse mitochondrial networks, two longitudinal connections across the A-band (middle), and one connection across a Z-line (top right) are shown. Also visible are t-tubules, terminal cisternae, longitudinal tubules, and the fenestrated collar of the sarcoplasmic reticulum. Glycogen granules that are prominent in the section are not visible in the freeze fracture. ×29,000.

contraction and endurance. The endurance of a fiber depends on its preferred pathways of metabolism. Fibers designed for chronic activity must rely on a constant supply of oxygen and substrates and need many mitochondria; fibers designed for intermittent activity can generate energy from anaerobic glycolysis of stored glycogen and may be poor in mitochondria. The concentration of mitochondrial cristae is correlated with the rate of oxidation. [For the metabolic differences in specialized muscle fibers demonstrated by histochemistry, see Beckett and Bourne (1973).]

Since fast contracting muscles are activated at higher frequencies than slowly contracting muscles, the blood supply to fast muscles is compromised during contraction more than that of the slow twitch muscles (Folkow and Halicka, 1968). This is probably the reason that, in skeletal muscles of cat and man, slow twitch fibers are rich in mitochondria ("red") and fast twitch fibers are poor in mitochondria ("white"). Nevertheless, fast fibers are not invariably poor in mitochondria, because in muscles of small mammals (rat: Close, 1967) and in a small specialized muscle (vocal cord: Hall-Craggs, 1968), fast fibers with high oxidative capacity also have been found. Fast twitch fibers vary in resistance to fatigue; the resistance to fatigue decreases within this group of fibers with decreasing oxidative capacity, i.e., with mitochondrial content (Henneman and Olson, 1965; Burke *et al.*, 1971, 1973). In fibers poor in mitochondria, the transverse networks are coarser and longitudinal branches and subsarcolemmal clusters are less frequent than in fibers rich in mitochondria. Lipid droplets are most numerous in fibers rich in mitochondria. In fast twitch fibers of the human quadriceps femoris muscle, 1.1% of the fiber volume is occupied by mitochondria; for slow twitch fibers, the value is 3.4% (B. R. Eisenberg, personal communication, 1978). Corresponding values for experimental animals are: rat extensor digitorum longus muscle, 5 and 25% (Schiaffino *et al.*, 1970); guinea pig vastus muscle, 1.9 and 8.2% (Eisenberg and Kuda, 1976); cat gastrocnemius muscle, 1.6 and 4.9% (Schmalbruch, 1979b).

The general pattern of mitochondria (clusters of spherical mitochondria in the periphery and a framework of mitochondria in the interior of the fiber) may be modified in different muscles and under different conditions. In heart muscle cells (Fawcett and McNutt, 1969) and in laryngeal muscles (Schmalbruch, 1971), a spatial arrangement and networks are absent, and all mitochondria are spheres or ovoids. This is also the case in young fibers in developing muscles (Kelly and Zacks, 1969) and in regenerating adult muscles (Schmalbruch, 1976). Some extraocular muscles lack mitochondria between myofibrils and contain only a rim of mitochondria beneath the sarcolemma (Mayr, 1971). Incompletely regenerated fibers show, in cross sections, strands of individual mitochondria passing through the interior of the fiber. These have probably been subsarcolemmal mitochondria that now mark the site of lateral fusion of adjacent developing fibers (Schmalbruch, 1979a).

When a muscle fiber is denervated, the internal mitochondrial framework decomposes; this allows, 24 hours after denervation, denervated and innervated fibers to be distinguished (rat diaphragm: Miledi and Slater, 1968).

An increase in the number and size of subsarcolemmal clusters of mitochondria was reported in cortisone-induced myopathy of rat (Walsh *et al.*, 1971): a finding not confirmed in a later study (Stern *et al.*, 1972).

III. The Membrane Configuration of Muscle Mitochondria and Its Experimental Modification

Mitochondria consist of two concentric membranes; the inner one forms cristae that are sheet- or finger-like. The compartment between the inner and outer membrane, the envelope space, is continuous with the intracristal space. The existence of an intracristal space *in vivo* has been disputed by Sjöstrand and Bernhard (1976), who conclude from frozen sections that both leaflets of the cristae are closely apposed. In thin sections of plastic-embedded specimens of normal muscle fibers, intracristal space and envelope space are 5–20 nm wide; the content appears less electron-dense than the matrix enclosed by the inner mitochondrial membrane. The matrix contains few matrix granules; these granules are 10–30 nm in diameter and very electron-dense. In subsarcolemmal mitochondria, cristae are usually more densely packed than in mitochondria forming transverse grids.

Isolated mitochondria from heart and liver cells show changes in the configuration of their cristae that are related to the metabolic state (Penniston *et al.*, 1968; Green *et al.*, 1968). This is not so in mitochondria isolated from skeletal muscle fibers (Kuner and Beyer, 1970). The difference may be due to the fact that isolation of mitochondria from skeletal muscle involves fragmentation of the mitochondrial networks. Though the fragments of muscle mitochondria are sealed off and appear as spheres in the pellets, they may be more damaged than mitochondria from other sources isolated more or less intact.

Hülsmann *et al.* (1968) have isolated two types of mitochondria from sketal muscles of rat; mild fragmentation releases mitochondria with a more loosely coupled state of oxidative phosphorylation than those released after a more thorough homogenization of the muscle. The mitochondria isolated by mild fragmentation are supposed to originate from the subsarcolemmal area. The authors speculate whether these mitochondria are more aged than those from the interior of the fiber, which are fragments of the mitochondrial framework. The type of homogenizer used should not affect this sequence of release. Even when the muscle fibers are chopped transversely during the initial stage into segments with a length of 5–10 times their width (Worsfold *et al.*, 1973), spherical mitochondria are more likely to be isolated by mild fragmentation than mitochondria forming a large framework.

When mitochondria of muscle fibers are studied by electron microscopy, one has to be aware of the fact that these organelles are most apt to change their structure during fixation or during the unavoidable delay between interruption of the blood supply and fixation of a fiber. Since this delay differs for fibers in the periphery and in the center of a specimen, the preservation of mitochondria varies in the same sample. Homogeneous fixation can only be achieved when the living muscle is fixed by vascular perfusion. Nonoptimal fixation causes loss of intra-mitochondrial granules, the electron contrast of the membranes decreases, mitochondria acquire a "mushy" appearance, the inner compartment swells, the matrix becomes electron-translucent, and cristae disappear. Since these are unspecific signs of damage, they may be present under certain conditions *in vivo* as well. Nevertheless, this is difficult to ascertain. Extreme swelling of muscle mitochondria was reported after exhaustion (Gollnick and King, 1969) and ischemia (Strock and Majno, 1969).

Hanzlíková and Schiaffino (1977) have demonstrated mitochondrial changes after 6–24 hours of ischemia. Subsarcolemmal mitochondria increase in size to diameters of 4–5 μm; the cristae in these giant mitochondria are densely packed. The intracristal space often contains an amorphous substance and plate-like inclusions with a crystalline pattern. These plates are localized midway between the two membrane leaflets. The authors suggest that giant mitochondria result from fusion of small mitochondria and that the crystalline structures are polymerized enzymes present in the intracristal space. The plate-like inclusions are identical with those observed by Reznik and Hansen (1969) and Karpati *et al.* (1974) in necrotizing muscle fibers.

Plate-like inclusions of the same type, together with an increase in the total mass of mitochondria, have been produced in rat soleus muscle by perfusion with 2,4-dinitrophenol for 3 hours (Melmed *et al.,* 1975). 2,4-Dinitrophenol uncouples the mitochondrial oxidative phosphorylation. Chloramphenicol, an inhibitor of mitochondrial protein synthesis, prevents the increase in the mass of mitochondria, but not the formation of plate-like inclusions. The histochemical changes after 2,4-dinitrophenol treatment resemble those in "mitochondrial myopathies" of man. Nevertheless, plate-like inclusions are different from intra-mitochondrial crystalloids in human muscle (Section IV,A). Cross sections through plate-like inclusions show two lines running parallel at a distance of 6nm, the subunits of intramitochondrial crystalloids show four lines with the same spacing. The two-line structures may represent "precursors" of crystalloids. The only observation of intramitochondrial crystalloids in animal muscle identical with those in human muscles was made by A. M. Kelly (personal communication, 1978) in the soleus muscle of a 3-year-old mouse.

IV. The Morphology of Mitochondria in "Mitochondrial Myopathy"

A. Intramitochondrial Crystalloids

Mitochondria containing crystalloids (Luft *et al.*, 1962; Gruner, 1963) are seen mainly within large subsarcolemmal clusters of mitochondria, but may also occur singly in the center of a fiber. The crystalloids are up to 3 μm long and usually rectangular in shape. They are enclosed by a membrane that is continuous with the inner mitochondrial membrane, i.e., crystalloids are localized in the outer mitochondrial compartment, either in the intracristal space or in the envelope space (Fig. 3a,b).

Each crystalloid consists of several, usually four, subunits. These subunits are 30 nm wide; the gap between adjacent subunits is 8 nm wide. Electron micrographs show four straight lines running parallel at a distance of 6 nm and isolated from each other; or, if the plane of sectioning is different, four lines connected by bridges spanning the gaps at a periodicity of 6 nm (Fig. 3a); or no subunits and crystalloids that appear more or less homogeneous (Fig. 3b,c).

It is difficult to derive the three-dimensional structure from these views, since the thickness of the section is 5–10 times the periodicity of the lines and connecting bridges. Chou (1969) has proposed a model according to which the crystalloids consist of lamellae formed by hollow granules in a double helical arrangement. For the two-layered plate-like inclusions produced in rat muscles by ischemia, Hanzlíková and Schiaffino (1977) suggest that they are one row of parallel tubules formed by helically arranged filaments. We present a simpler model to explain the appearance in different sections (Fig. 4). Each subunit consists of four lamellae connected by parallel ribbons. Lamellae alone, or lamellae with bridges, are visible only when the electron beam coincides with two of the main axes (Figs. 3a and 4); when the lamellae run parallel with the plane of sectioning, the crystalloid appears homogeneous (Fig. 3b). Oblique sections show narrower spacings and, if the crystalloid is tilted along two axes, a herringbone pattern may be produced (Fig. 3c).

Fig. 3. Intramitochondrial crystalloids in muscle fibers from patients with "mitochondrial myopathy." (For the three-dimensional model, see Fig. 4 and text). (a) Two crystalloids consisting of four subunits each are surrounded by membranes continuous with the inner mitochondrial membrane. Each subunit shows four parallel lines, mainly separate in the right crystalloid and often connected by periodic bridges in the left crystalloid. ×200,000. (b) Crystalloid within the envelope space of a mitochondrium. The interior appears homogeneous because the plane of sectioning is parallel to the lamellae shown in (a). Note one large matrix granule. ×160,000. (c) Obliquely sectioned crystalloid displaying some sort of herringbone pattern. ×160,000.

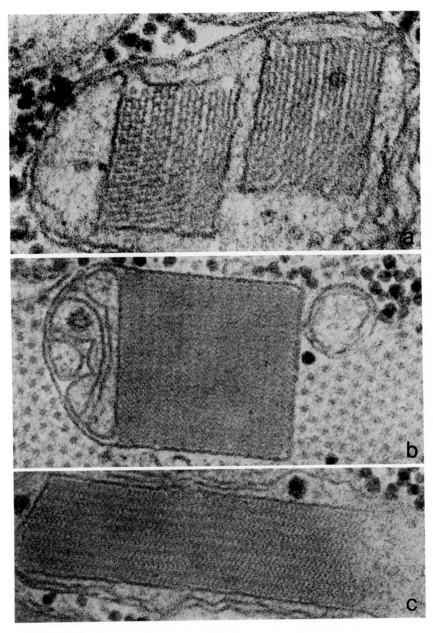

Fig. 3

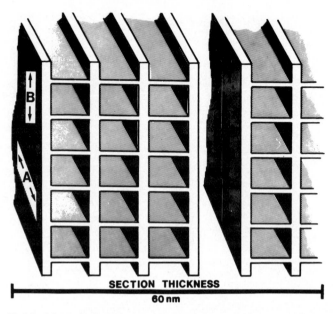

SECTION THICKNESS
60 nm

FIG. 4. Model of intramitochondrial crystalloid to explain the different patterns observed in sections. One and one-half subunits are shown. Each subunit consists of four lamellae connected by parallel ribbons. The length of the bar (60 nm) corresponds to the average thickness of a section for electron microscopy and shows that even in very thin sections several periodicities are superimposed. When the electron beam coincides with the B-axis, the section will show parallel lines only (Fig. 3a, right crystalloid); when it coincides with the A-axis, these lines will appear connected by parallel bridges (Fig. 3a, left crystalloid); when sectioned parallel to the plane of the lamellae, the crystalloid will appear homogeneous (Fig. 3b); when both A and B axes are tilted toward the plane of sectioning, a herringbone pattern (Fig. 3c) may arise because several slightly displaced periodicities become superimposed.

B. INCREASE OF MATRIX

In some mitochondria, the inner mitochondrial compartment is dilated and stuffed with moderately electron-dense matrix. Cristae are rare and sometimes lacking; then the mitochondrion consists of two concentric membranes only (Figs. 5a,b and 6).

FIG. 5. Mitochondria from muscle fibers of patients with "mitochondrial myopathy." (a) Mitochondria belonging to the transverse network contain an increased amount of matrix. Intramitochondrial crystalloids are seen on the left, sarcolemma and endothelial cells on the right. ×33,000. (b) Subsarcolemmal mitochondria stuffed with matrix. Each mitochondrium contains only one crista that runs parallel with the outer mitochondrial membrane. ×40,000. (c) Large mitochondrion with concentric cristae. Note dense matrix granules. ×34,000. (d) Large subsarcolemmal mitochondrion with parallel cristae. This is possibly a mitochondrion with concentric cristae that has been cut longitudinally. Note dense matrix granules. ×50,000.

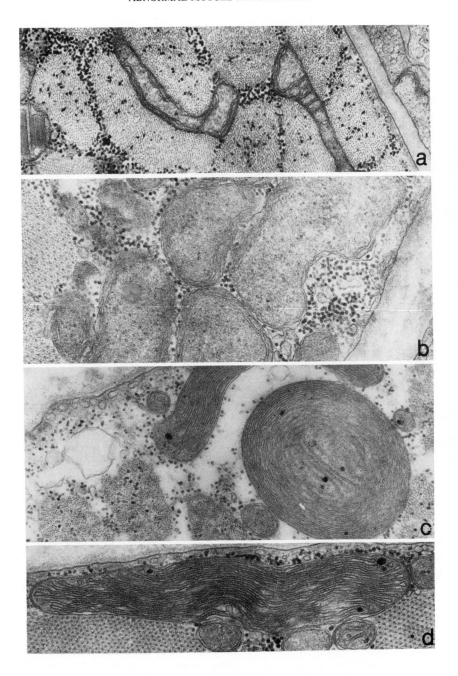

Fig. 5

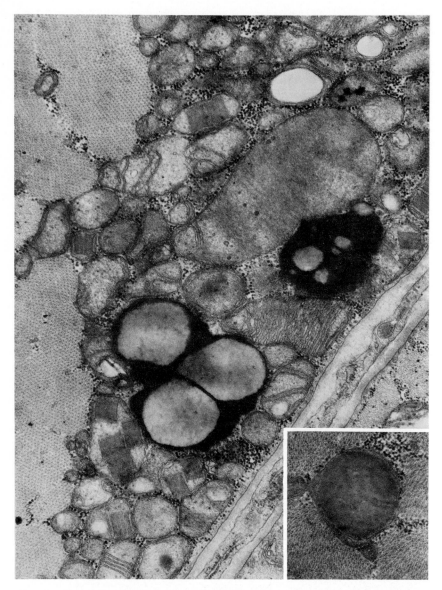

FIG. 6.   Cluster of subsarcolemmal mitochondria of a patient with "mitochondrial myopathy."
Various mitochondrial abnormalities are present. Several mitochondria contain crystalloids; some are
stuffed with matrix and cristae are rarified. A dense matrix body is present in two mitochondria
(bottom and inset). The giant mitochondrion in the middle appears homogeneous, probably because
stacked sheet-like cristae run parallel with the plane of sectioning. Two residual bodies (lipofuscin
granules) are present as well. ×25,000 (inset ×34,000).

C. CONCENTRIC CRISTAE AND GIANT MITOCHONDRIA

Mitochondria may be abnormally large or small. Small mitochondria do not show typical cristae, but consist of two to six concentric membranes. The density of the gaps alternates, and that between the two most peripheral membranes is always light; it corresponds to the envelope space. The light gaps localized more centrally correspond to intracristal spaces. Thus cristae have the form of concentric tubules. In large mitochondria, which may be up to 5 μm in diameter, concentric cristae are frequent (Fig. 5c). In some of the large mitochondria, cristae appear straight (Fig. 5d). These are either longitudinal sections of mitochondria with concentric cristae (Fig. 5c) or mitochondria containing stacks of sheet-like cristae (Fig. 6). When straight cristae run parallel to the section, the interior of the mitochondrion appears homogeneous (Fig. 6). This should not be confused with mitochondria stuffed with matrix. The abnormal membrane pattern of mitochondria was studied in freeze–fracture preparations by Ketelsen *et al.* (1978).

D. INCLUSION BODIES

Spherical bodies occur within the matrix of mitochondria with sparse cristae. These bodies are 0.05–0.1 μm in diameter and are either electron-dense or electron-translucent (Chou, 1969). An internal structure is absent, as is often a sharp delineation from the matrix (Fig. 6). The electron-translucent bodies are probably lipid droplets. The electron-dense inclusions that are not dissolved during the embedding procedure are moderately osmiophilic and probably consist of protein.

E. INTRAMITOCHONDRIAL GRANULES

All normal muscle mitochondria contain matrix granules 10–30 nm in diameter. They are probably binding sites for cations and disappear during anoxia before swelling of the mitochondria takes place. In "mitochondrial myopathies," many of the giant mitochondria contain prominent matrix granules up to 80 nm in diameter and are often edge-shaped like crystals (Figs. 4, 6, and 7).

F. STRUCTURES THAT MAY BE ASSOCIATED WITH ABNORMAL MITOCHONDRIA

Abnormal mitochondria are associated with intracellular *lipid droplets* larger and more numerous than in normal muscle fibers. After fixation by glutaraldehyde and osmium, these droplets, which probably contain triglycerides, appear electron-translucent, since most of the lipid is dissolved during the embedding procedure. Their shape is spherical (Fig. 1). After fixation by osmium alone, they are irregularly shaped and electron-dense.

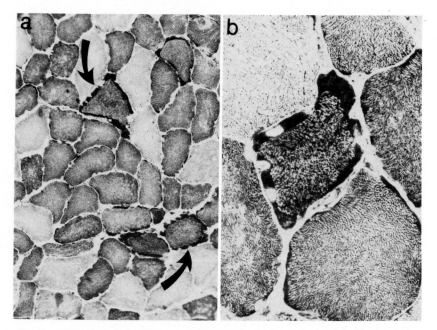

FIG. 7.   Light micrographs of a cross section of a brachial biceps muscle from a patient with
"mitochondrial myopathy," stained for a mitochondrial enzyme (NADH-tetrazolium reductase).
Note the almost normal histological appearance. The low power micrograph shows some ragged-red
fibers (arrows); the high magnification reveals that most of the reaction product is subsarcolemmal,
but also the staining pattern in the center of the affected fiber differs from that of adjacent "normal"
fibers. The different staining intensity indicate different fiber types. a, ×130; b, ×520.

Lipid droplets should not be confused with *residual bodies,* i.e., telelysomes
(lipochrome, lipofuscin, lipopigment), the incidence of which is increased in all
diseased muscles and which are often associated with subsarcolemmal clusters of
mitochondria. These bodies differ in electron density and often show one or
several electron-translucent centers and a peripheral rim or cap of heavily osmiophilic material (Fig. 6). They probably consist of a variety of lipids, including less soluble phosphatides.

In "mitochondrial myopathies," *glycogen granules* often seem to be more
numerous than in normal muscle fibers. This is difficult to ascertain by
morphology alone. An increased amount of glycogen has been documented
biochemically in a few cases only (for references, see DiMauro, 1979).

In several patients with "mitochondrial myopathy," *nemaline structures* or
rods were present (D'Agostino *et al.,* 1968; Lapresle *et al.,* 1972; Kamieniecka,
1976). These rods are derived from Z-line material; they are probably unspecific

changes that occur in denervated muscle fibers and in fibers released from their tendon attachments as well (Engel *et al.*, 1966; Shafiq *et al.*, 1969; Karpati *et al.*, 1972).

G. ABNORMAL MITOCHONDRIA IN NONMUSCLE CELLS IN PATIENTS WITH "MITOCHONDRIAL MYOPATHIES"

Abnormal mitochondria were present, not only in skeletal muscle fibers, but also in extraocular muscles in patients in whom these muscles were clinically affected (Zintz, 1966; Zintz and Villiger, 1967; Lessell *et al.*, 1969; Schneck *et al.*, 1973; Saraux *et al.*, 1974; Croft *et al.*, 1977). Since "mitochondrial myopathy" often is part of a multiorgan syndrome (oculocraniosomatic disease, Section VI,A,1,d), it is tempting to speculate that mitochondria of nonmuscle cells are also affected.

Brain tissue (cerebellum) of a patient who died with the cerebral symptoms associated with the fully manifest syndrome (Adachi *et al.*, 1973; Schneck *et al.*, 1973) contained intramitochondrial crystalloids that resembled those in the extraocular muscles from the same patient. In the mitochondria of the eccrine sweat glands in a skin biopsy from another patient, Karpati *et al.* (1973) found circular cristae and some plate-like inclusions resembling those shown by Reznik and Hansen (1969) in necrotizing muscles and by Hanzlíková and Schiaffino (1977) in ischemic muscles. Typical intramitochondrial crystalloids were absent. Identical plate-like inclusions had been found in a liver biopsy, but were considered as unspecific, since they occurred in liver biopsies from patients with alcoholism and diabetes as well (Gonatas *et al.*, 1967). Other authors described large mitochondria with tubular cristae (Okamura *et al.*, 1976), "many" mitochondria, and mitochondria with myeline figures in liver cells (Scarlato *et al.*, 1978). Intramitochondrial crystalloids were absent, but were abundant in muscle biopsies from these patients. Heart biopsies (McComish *et al.*, 1976; Harati *et al.*, 1977; Hart *et al.*, 1978) showed "abnormal collections of mitochondria," but no crystalloids.

Little is known about the normal structure of mitochondria and its unspecific variations in human eccrine glands. In liver cells and in heart muscle cells, giant mitochondria with bizarre internal structure were demonstrated in patients with diabetes or cardiac insufficiency and in rats under various experimental conditions (Trump and Ericsson, 1965; Reale, 1973). Since in patients with "mitochondrial myopathy" no intramitochondrial crystalloids and no mitochondria with typical concentric cristae were found, we do not feel that sufficient morphological evidence has been presented to conclude that mitochondrial abnormalities comparable to those in skeletal muscle fibers occur in heart, liver, and skin.

H. Comments on Abnormal Muscle Mitochondria

Diagnostically, intramitochondrial crystalloids are most important, since they can be identified even in nonoptimally prepared biopsies. Concentric cristae are a safe criterion as well, whereas in poor preparations, it is difficult to distinguish rarified cristae, "large" mitochondria, mitochondrial inclusions, or "mushy" looking mitochondria from artifacts. The size of subsarcolemmal clusters of mitochondria are also difficult to evaluate, because they vary considerably in normal muscle fibers.

The mechanism causing the formation of abnormal muscle mitochondria is unknown. A common property of all changes is that something occurs in excess: mitochondria are more numerous or larger in size, cristae are abundant, the amount of matrix increases, abnormal crystalline structures occur in the intracristal space, size and number of triglyceride droplets in the neighborhood of mitochondria increase. One may assume that the common cause is that the sequence of metabolic steps is disturbed and that the changes are due either to accumulation of substances that can no longer be metabolized or to attempts to compensate for the defect. This hypothesis is supported by the fact that abnormal mitochondria have been produced experimentally by ischemia or by uncoupling of oxidative phosphorylation.

Cytochemistry of crystalloids from patients did not reveal activity of cytochrome c oxidase or carnitine acetyltransferase (Bonilla et al., 1975). For the plate-like inclusions produced by ischemia, Hanzlíková and Schiaffino (1977) suggest that they derive from polymerization of enzymes such as mitochondrial creatine kinase present in the intracristal space. When the oxidative phosphorylation in mitochondria is uncoupled by 2,4-dinitrophenol, new mitochondrial membranes are formed and plate-like inclusions occur. Chloramphenicol, an inhibitor of mitochondrial protein synthesis, prevents the formation of new membranes, but not that of plate-like inclusions (Melmed et al., 1975). Hence, it is unlikely that plate-like inclusions result from excess production of an enzyme protein.

V. The Identification of Muscle Fibers Containing Abnormal Mitochondria by Light Microscopy

In routine histological preparations, the muscle biopsy from a patient with "mitochondrial myopathy" may appear normal. Some biopsies reveal increased variation of fiber diameters, few internally placed myonuclei, and, rarely, necrotic fibers. If the amount of intracellular lipid is increased, fibers show small vacuoles.

The incidence of fibers with abnormal mitochondria varies from 1% (Olson et *al.*, 1972) to more than 50% (Hyman *et al.*, 1977; Morgan-Hughes *et al.*, 1977); these fibers are randomly distributed and may be missed in preparations for electron microscopy.

To screen a muscle by light microscopy, different histological methods and criteria can be used.

1. Because of the large number of mitochondria, affected fibers stain more intensely for oxidative enzymes than normal fibers rich in mitochondria (Fig. 7). In serial sections, most fibers with abnormal mitochondria stain weakly for ATPase at pH 9.4 ("myofibrillar" ATPase, type I), but also fibers that stain intensely for ATPase at pH 9.4 (type II) may be affected (Price *et al.*, 1967; Schellens and Ossentjuk, 1969; Birnberger *et al.*, 1973; Karpati *et al.*, 1973; Morgan-Hughes and Mair, 1973; Carrier *et al.*, 1974; Iannaccone *et al.*, 1974; Tamura *et al.*, 1974; Kamieniecka, 1976; Morgan-Hughes *et al.*, 1977; Hyman *et al.*, 1977). Around these fibers, the number of capillaries is increased (Carrier *et al.*, 1974; Carpenter *et al.*, 1978).

2. Since the normal framework of mitochondria is lacking in affected fibers and the size of subsarcolemmal clusters of mitochondria is increased, the architecture of the mitochondrion is changed. This can be seen after staining for mitochondrial enzymes or by a modification of Gomori's trichrome stain (Engel and Cunningham, 1963). In sections stained by the Gomori method, affected fibers display a peripheral rim of granular bright-red material that is absent in fibers of normal skeletal muscles in man ("ragged-red" fibers; Olson *et al.*, 1972) (Fig. 7).

3. Affected fibers are rich in lipids, both in the form of triglyceride droplets and in the form of phosphatides contained in membranes, which can be identified by a lipid stain (Sudan black, Oil red).

VI. The Clinical Classification of "Mitochondrial Myopathies"

Since more and more laboratories use histochemistry and electron microscopy to study muscle biopsies from patients with neuromuscular disorders, the incidence of "mitochondrial myopathies" has increased. In a series of 343 patients, of whom 225 had a myogenic disorder, we found ragged-red fibers in 23 patients. Twenty-one of these biopsies were studied by electron microscopy, and abnormal mitochondria were found in 13. In eight biopsies, ragged-red fibers were not contained in the small samples embedded for electron microscopy.

The clinical findings suggest that the diagnosis "mitochondrial myopathy" comprises several disorders. Many cases have been reported, but only in a few of them has an attempt been made to localize the defect biochemically. In the following section, those cases that were studied biochemically are classified according to their clinical and biochemical properties. If one assumes that the subgroups obtained in that way represent *all* subgroups or nosological entities covered by "mitochondrial myopathy," one should be able to arrange those cases that had not been studied biochemically by their clinical properties alone into one of these subgroups (Table I).

A. Patients in Whom a Biochemical Defect Was Detected

1. *Loosely Coupled Oxidative Phosphorylation*

a. *Luft's Disease.* The first patient with a myogenic disorder with abnormal mitochondria was observed by Ernster *et al.* (1959), Ernster and Luft (1963), and Luft *et al.* (1962). A female, aged 35 years, suffered from hypermetabolism; the thyroid function was normal. The clinical impairment of skeletal muscle was moderate; electron microscopy showed a vast increase in the number and size of mitochondria and a variety of structural changes including intramitochondrial crystalloids. Biochemically, the oxidative phosphorylation was loosely coupled. The combination of hypermetabolism, normal thyroid function, and loosely coupled oxidative phosphorylation of mitochondria in skeletal muscle fibers was denominated "Luft's disease." The disorder is rare; up to now only one other patient has been observed (Haydar *et al.*, 1971; Afifi *et al.*, 1972; DiMauro *et al.*, 1976; Bonilla *et al.*, 1977).

b. *Congenital Generalized Muscle Weakness with Loose Coupling of Mitochondrial Phosphorylation without Hypermetabolism.* Schellens and Ossentjuk (1969) observed a patient in whom the biochemical defect resembled that in the patients with Luft's disease. Hypermetabolism was, however, absent. The patient was slender and complained mostly of fatigability.

c. *Pleoconial Myopathy.* An 8-year-old boy, that appeared floppy at birth, suffered from periods of paralysis lasting for several weeks. The muscle biopsy showed that 20% of the fibers contained abnormal mitochondria, many of them with concentric cristae. The oxidative phosphorylation was loosely coupled but, at variance with findings in the patients described above, only when α-glycerophosphate was offered as substrate. The basic metabolic rate was normal. The authors speculated that the periods of paralysis were related to abnormalities of the distribution of electrolytes (Shy *et al.*, 1966; Spiro *et al.*, 1970a). No similar patient has been reported.

d. *"Ophthalmoplegia Plus" Syndrome (Drachman, 1968, 1975).* This subgroup comprises patients that had ophthalmoplegia and a variety of other symptoms ("plus"). In some of these patients, loose coupling of mitochondrial phosphorylation was found (DiMauro *et al.*, 1973; Spiro *et al.*, 1970b; Harati *et al.*, 1977); in others, mitochondrial phosphorylation was not investigated, but the serum lactate level was increased at rest or increased more than normal after exercise (Reske-Nielsen *et al.*, 1976; Lou and Reske-Nielsen, 1976; Scarlato *et*

TABLE I

CLINICAL CLASSIFICATION OF MITOCHONDRIAL MYOPATHIES[a]

Groups	Clinical symptoms	Patients with known biochemical defect				Patients with no known biochemical defect			
		Onset in childhood		Onset in adults		Onset in childhood		Onset in adults	
		F	M	F	M	F	M	F	M
1. Luft's disease (Section VI,A,1,a)	Hypermetabolism, normal thyroid function, moderate weakness of skeletal muscles	2 (38,18)[b]							
2. Loosely coupled phosphorylation without hyper-metabolism (Section VI,A,1,b)	Congenital, generalized muscle weakness, fatigability		1 (49)			1 (13)			
3. Pleoconial myopathy (Section VI,A,1,c)	Congenital periodic paralysis		1 (53)						
4. Ophthalmoplegia plus (oculocraniosomatic disease and "formes frustes") (Section VI,A,1,d)	Ophthalmoplegia, retinitis pigmentosa, ptosis. (Deafness, dysphagia, heart block, hormonal defects, muscle weakness)	3 (45,48)	4 (17,26,45,57)	1 (29)	3 (29,58)	17[c]	12[d]	12[e]	9 (15,39,42,47,59)
						28 (Onset and sex not reported) (1,18a,36,43,46)			
5. Facioscapulohumeral type of muscle impairment (Section VI,A,1,e)	Muscle weakness, familial	2 (27)	1 (27)			3 (33)		1 (33)	
6. Limb-girdle type of muscle impairment (Section VI,A,1,e)	Muscle weakness	3 (28,51,53)	3 (6,40,63)			3 (14,33)	3 (2,3,22)	1 (33)	

(continued)

TABLE I (continued)

| Groups | Clinical symptoms | Patients with known biochemical defect | | | | Patients with no known biochemical defect | | | |
| | | Onset in childhood | | Onset in adults | | Onset in childhood | | Onset in adults | |
		F	M	F	M	F	M	F	M
7. Carnitine deficiency, muscular form (Section VI,A,2,a)	Skeletal muscles weakness, eye muscles unaffected. Excess lipid in muscle fibers	1 (20)			1 (31)				
8. Carnitine deficiency, generalized (Section VI,A,2,b)	Muscle weakness including eye muscles. Heart, liver, kidneys, brain affected. Excess lipid in muscle *and* in other organs	1 (6)	2 (21,35)						
9. Different lipid storage myopathies not due to carnitine deficiency (Section VI,A,2,c)									
(a) Decreased activity of cytochrome oxidase, lack of cytochromes	Familial, fatal within months		2 (16,62)						
(b) Acid maltase deficiency (glycogenosis)	Proximal muscle weakness				1 (19)				

(c) Lack of stimulation of glycolysis by adrenaline, decreased activity of enzymes of glycolysis	Ophthalmoplegia, muscle weakness	4F (56) (Onset not reported)		
10. Unclassified lipid storage myopathy with abnormal muscle mitochondria	Muscle weakness	2 (32,52)	1 (44)	2 (7,25)

[a] The patients in whom a biochemical defect was described are grouped according to biochemical and clinical properties; patients in whom no biochemical defect is known are grouped according to their clinical symptoms alone (Section VI). Whether or not these groups represent nosological entities is unknown. Patients in Group 1 to 6 showed, when studied biochemically, loosely coupled mitochondrial phosphorylation; hypermetabolism was present only in the patients in Group 1. Several of the patients of Group 4 and 6 had lactacidosis. All patients in Group 7 to 10 had lipid storage myopathy. F, female; M, male.

[b] References: (1) Bakouche et al. (1976); (2) Bender and Engel (1976); (3) Benke and Engel (1976); (4) Birnberger et al. (1973); (5) Black et al. (1975); (6) Boudin et al. (1976); (7) Bradley et al. (1972); (8) Butler and Gadoth (1976); (9) Carrier et al. (1974); (10) Carroll et al. (1976); (11) Castaigne et al. (1972); (12) Castaigne et al. (1977); (13) Coleman et al. (1967); (14) D'Agostino et al. (1968); (15) Denis et al. (1976); (16) DiMauro et al. (1978); (17) DiMauro et al. (1973); (18) DiMauro et al. (1976); (18a) Dubowitz and Brooke (1973); (19) Engel and Dale (1968); (20) Engel and Siekert (1972); (21) Engel et al. (1977); (22) Fisher and Danowski (1969); (23) Gadoth (1977); (24) Gérard et al. (1974); (25) Gullotta (1974); (26) Harati et al. (1977); (27) Hudson et al. (1972); (28) Hülsmann et al. (1967); (29) Hyman et al. (1977); (30) Iannaccone et al. (1974); (31) Isaacs et al. (1976); (32) Jerusalem et al. (1973); (33) Kamieniecka (1976); (34) Karpati et al. (1973); (35) Karpati et al. (1975); (36) Leshner et al. (1978); (37) Lessell et al. (1969); (38) Luft et al. (1962) (39) Morgan-Hughes and Mair (1973); (40) Morgan-Hughes et al. (1977); (41) Okamura et al. (1976); (42) Olson et al. (1972); (43) Poloni et al. (1978); (44) Price et al. (1967); (45) Reske-Nielsen et al. (1976); (46) Santa et al. (1975); (47) Saraux et al. (1974); (48) Scarlato et al. (1978); (49) Schellens and Ossenjuk (1969); (50) Schneck et al. (1973); (51) Schotland et al. (1976); (52) Sengers et al. (1976); (53) Shy et al. (1966); (54) Shy et al. (1967); (55) Simopoulos et al. (1971); (56) Sluga and Moser (1970); (57) Spiro et al. (1970b); (58) Sulaiman et al. (1974); (59) Tamura et al. (1974); (60) Toppet et al. (1977); (61) Vallat et al. (1975); (62) Van Biervliet et al. (1977); (63) Van Wijngaarden et al. (1967); (64) Völpel (1979); (65) Zintz and Villiger (1967).

[c] (10,12,12,18a,34,42,50,54,55,60,64,65).

[d] (4,8,10,23,33,41,42,55,61,65).

[e] (9,24,30,33,37,39,42,59).

341

al., 1978; Hyman *et al.*, 1977). One patient resembled the patient with pleoconial myopathy in that loose coupling of phosphorylation could only be demonstrated with α-glycerophosphate as substrate (DiMauro *et al.*, 1973); two
showed, in addition, a reduction in the cytochrome b content of mitochondria
(Spiro *et al.*, 1970b) and in the respiratory rate of mitochondria (Harati *et al.*,
1977).

In muscle biopsies, the number and size of mitochondria was increased, and a
variety of mitochondrial abnormalities was found. Clinically, other symptoms
("plus") were observed in different combinations: ptosis, retinitis pigmentosa,
deafness, dysphagia, weakness of skeletal muscle, cardiac block, cerebral and
cerebellar defects, short stature, hormonal dysfunction. The fully developed
syndrome was named "oculocraniosomatic disease" (Olson *et al.*, 1972),
"Kearns-Sayre syndrome," or "Kearns-Shy syndrome" (Kearns and Sayre,
1958; Kearns, 1965; Karpati *et al.*, 1973). Most cases are "formes fruste":
ptosis, ophthalmoplegia, and retinitis pigmentosa are found regularly, whereas
other symptoms may be absent. This disorder is more frequent than Luft's
disease or pleoconial myopathy. In most patients, symptoms occur early in life.
The clinical picture may be dominated by "plus" symptoms not related to the
muscular system and in some cases the diagnosis is probably not established.

e. *Facioscapulohumeral and Limb-Girdle Type of Muscle Impairment.* In
some patients, in all of whom loose coupling of oxidative phosphorylation in
muscle mitochondria was found, the weakness of skeletal muscle dominated the
clinical picture.

In one family, six members had a facioscapulohumeral distribution of muscle
weakness and wasting, and six members were "asymptomatically affected." In
none of these patients were eye muscles involved. In four patients, the serum
creatine phosphokinase was moderately increased; two had diabetes mellitus.
Prominent intramitochondrial crystalloids were present in biopsies obtained from
two patients and from one "asymptomatically affected" relative; both revealed
loose coupling of oxidative phosphorylation (Hudgson *et al.*, 1972).

In five unrelated patients (female, 2 years: Hülsmann *et al.*, 1967; male, 11
years: Van Wijngaarden *et al.*, 1967; female, 30 years: Black *et al.*, 1975;
female, 37 years: Schotland *et al.*, 1976; male, 38 years: Morgan-Hughes *et al.*,
1977), proximal muscles were mainly affected. Symptoms had started shortly
after birth or in childhood, and the affection was slowly progressing. Ragged-red
fibers and abnormally large mitochondria were found in biopsies from all patients, but intramitochondrial crystalloids were present in only two of them
(Schotland *et al.*, 1976; Morgan-Hughes *et al.*, 1977).

2. *Defects in the Metabolism of Lipids That Are Associated with Abnormal
   Mitochondria*

In several patients with muscle weakness, the muscle biopsy showed excessive
storage of intracellular lipids and of glycogen, together with abnormal mitochon-

dria. Different biochemical defects were reported. In most of these patients, defects of the carnitine metabolism were the cause of lipid storage; nevertheless, abnormal mitochondria were not found in all patients with disturbances of the carnitine metabolism.

a. *Benign Progressive Myopathy with Carnitine Deficiency.* Two patients with generalized (both proximal and distal) muscle weakness without involvement of the external eye muscles (Engel and Siekert, 1972; Engel and Angelini, 1973; Isaacs *et al.*, 1976) showed lipid storage in muscle fibers and mitochondria with intramitochondrial crystalloids. The muscle tissue contained less carnitine than normal, but the serum carnitine level was similar to control levels. [The biochemical role of carnitine for the metabolism of fatty acids in muscle fibers has been reviewed by Engel *et al.* (1974), Wilner *et al.* (1978), and DiMauro (1979).] Benign cases of carnitine deficiency without distinct morphological changes of mitochondria were reported by Di Donato *et al.* (1978), Markesbery et al. (1974), VanDyke *et al.* (1975), Smyth *et al.* (1975), Engel *et al.* (1974), and Jerusalem *et al.* (1975).

Another lipid storage myopathy is caused by lack of carnitine palmityl-transferase (Patten *et al.*, 1978; Hostetler *et al.*, 1978); it is characterized by recurrent muscle pain and myoglobinuria. No morphological abnormalities of mitochondria were found.

b. *Malignant Progressive Myopathy with Carnitine Deficiency.* In three children, a lipid storage myopathy was observed that differed clinically from the benign form. Muscle weakness was generalized and included the external eye muscle. In addition, liver, heart, and kidneys were affected. Muscle mitochondria contained typical crystalloids (Karpati *et al.*, 1975; Boudin *et al.*, 1976; Engel *et al.*, 1977). Two cases were fatal (Boudin *et al.*, 1976; Engel *et al.*, 1977). The level of carnitine was decreased, not only in muscle, but also in serum and liver (Scarlato and Pellegrini, 1979).

Two fatal cases of carnitine deficiency without morphologically abnormal muscle mitochondria were observed by Cornelio *et al.* (1977) and Hart *et al.* (1978).

c. *Progressive Myopathies with Abnormal Mitochondria, Storage of Lipid and Glycogen and Different Biochemical Defects.* A patient of Engel and Dale (1968) complained of muscle weakness of proximal muscles. The muscle biopsy showed storage of lipid and glycogen. The activity of acid maltase was decreased, and the patient was classified as having glycogenosis.

A patient with ophthalmoplegia, storage of fat and glycogen, and mitochondrial crystalloids was described by Sluga and Moser (1970), who attribute the metabolic abnormality to lack of stimulation of glycolysis by adrenaline.

Two fatal cases of muscular lipid and glycogen storage with increased size and number of mitochondria (without crystalloids) were observed in newborns. Cytochrome c oxidase activity was reduced (DiMauro *et al.*, 1978) and cytochromes were absent (Van Biervliet *et al.*, 1977). One of these patients had had two siblings who suffered from the same clinical symptoms and died shortly after birth (Van Biervliet *et al.*, 1977).

B. The Classification of Patients with "Mitochondrial Myopathies" in Whom Mitochondrial Metabolism Was Not Studied

1. *Patients Who Fit into Subgroups Established by Combined Biochemical and Clinical Parameters*

In most patients with "mitochondrial myopathy," the biochemical defect was not defined. In Table I, we try to classify these cases in the same way as those with known metabolic defects. Since several patients were studied before the concepts "mitochondrial myopathy" or "oculocraniosomatic syndrome" became commonplace, this classification may vary from the clinical diagnosis given in the original report. Patients who do not fit into one of the subgroups are described in Section VI, B, 2.

TABLE II. Unclassified

Reference	Clinical diagnosis	Age at first invest. (years)	Sex	Onset (years)	Pro-gression	Impair-ment of skeletal muscles	Ptosis	Ophthal-moplegia
Ketelsen *et al.* (1978)	Lactic acidosis	10	F	5	Slow	Yes (gen)	No	No
Sengers *et al.* (1975)	Mitochondrial myopathy	11	F	Birth	Slow	Yes	n.r.	n.r.
McLeod *et al.* (1975)	Epilepsy	27	F	Childhood	Slow	Yes (gen)	No	No
Crosby and Chou (1974)	Leigh's disease	10	M	5	Fatal	Yes	Yes	Yes
Shapira *et al.* (1975)	Familial poliodystrophy	4	F	4	Fatal	Yes (prox)	Yes	n.r.
Tsairis *et al.* (1973)	Familial myoclonic epilepsy	n.r.	F	n.r.	n.r.	Yes (gen)	n.r.	n.r.
		n.r.	F	n.r.	n.r.	Yes (gen)	n.r.	n.r.
		n.r.	F	n.r.	n.r.	Yes (prox)	n.r.	n.r.
		n.r.	M	n.r.	n.r.	Yes (prox)	n.r.	n.r.
Lapresle *et al.* (1972)	Congenital distal myopathy	63	M	Childhood	Slow	Yes (dist)	No	No
Dobkin and Verity (1976)	Spinal muscular atrophy	32	M	Childhood	Slow	Yes (gen)	Yes	No
Shafiq *et al.* (1967)	Spinal muscular atrophy	53	F	13	Slow	Yes (gen)	No	No
Schlote *et al.* (1979)	Friedreich's disease	26	M	7	Slow	Yes (dist)	No	No
Bastiaensen *et al.* (1977)	Friedreich's disease	34	F	Childhood	Fatal	Yes	Yes	Yes
Spalke *et al.* (1975)	Charcot-Marie-Tooth disease	38	F	10	Slow	Yes (dist)	No	Yes
		44	F	5	Slow	Yes (dist)	No	Yes
Norris and Panner (1966)	Hypothyroid myopathy	61	M	45	Slow	Yes (gen)	No	No
Godet-Guillain and Fardeau (1970)	Hypothyroid myopathy	63	F	48	Slow	Yes (prox)	n.r.	n.r.

[a] Eighteen patients in whom abnormal muscle mitochondria were found and who also had symptoms not related to skeletal muscle. These patients did not conform to the oculocraniosomatic syn-

2. Patients Who Resist Classification

a. *Oculocraniosomatic Syndrome without Affection of Eye Muscles?* In several patients with disorders of the central and peripheral nervous system and of endocrine organs, ragged-red muscle fibers and abnormal muscle mitochondria, including intramitochondrial crystalloids, were found. In some of these patients, the cerebral symptoms, and in others, muscle weakness and fatigability dominated the clinical picture. In two patients with hypothyroidism (Norris and Panner, 1966; Godet-Guillain and Fardeau, 1970), the muscle symptoms were thought to be secondary. In several of these patients, the blood lactate and pyruvate levels were increased as in patients with oculocraniosomatic syndrome. Ophthalmoplegia and ptosis were absent or developed late, or ptosis alone was found. Some of these cases probably belong to the group "oculocraniosomatic syndrome," even though fully developed ocular symptoms were absent; in others, it is doubtful whether this classification is justified. Since we believe that it is premature to include or exclude any of these cases, they are listed in Table II.

MITOCHONDRIAL MYOPATHIES[a]

Retinitis pigmentosa	Affection of muscles innervated by cranial nerves	Affection of peripheral nerves	Cerebral symptoms	Cerebellar symptoms	Spinal symptoms	Endocrine symptoms	Short stature	Heart block	Relatives affected	Blood lactate and pyruvate increased
No	Deafness	No	n.r.	n.r.	No	n.r.	Yes	No	No	Yes
n.r.	n.r.	n.r.	n.r.	n.r.	n.r.	n.r.	n.r.	Cardiomyopathy	Yes	Yes
Yes	Deafness	No	Yes	Yes	No	Yes	Yes	Yes	No	No
Yes	Yes	No	Yes	Yes	n.r.	n.r.	Yes	No	No	No
n.r.	Yes	n.r.	Yes	Yes	n.r.	Yes	Yes	No	Yes	Yes
n.r.	Yes	n.r.	Yes	Yes	n.r.	n.r.	n.r.	n.r.	Yes	Yes
n.r.	Yes	n.r.	Yes	Yes	n.r.	n.r.	n.r.	n.r.	Yes	Yes
n.r.	No	n.r.	Yes	No	n.r.	n.r.	n.r.	n.r.	Yes	Yes
n.r.	No	n.r.	Yes	No	n.r.	n.r.	n.r.	n.r.	Yes	No
No	No	No	No	n.r.	No	Yes	n.r.	No	Yes	No
n.r.	Yes	Yes	No	No	Yes	n.r.	No	Yes	Yes	n.r.
n.r.	n.r.	Yes	n.r.	n.r.	Yes	n.r.	n.r.	No	Yes	n.r.
No	n.r.	Yes	n.r.	Yes	Yes	n.r.	n.r.	n.r.	Yes	Yes
Yes	Yes	Yes	Yes	Yes	Yes	n.r.	n.r.	No	Yes	n.r.
No	No	Yes	No	No	No	n.r.	n.r.	n.r.	Yes	n.r.
No	No	Yes	No	No	No	n.r.	n.r.	n.r.	Yes	n.r.
n.r.	n.r.	No	Yes	n.r.	n.r.	Yes	n.r.	n.r.	No	n.r.
n.r.	Yes	No	n.r.	n.r.	n.r.	Yes	n.r.	n.r.	n.r.	n.r.

drome, but some probably represent "formes frustes" of this syndrome. n.r., Not reported; prox, proximal; dist, distal; gen, generalized.

b. *Thyrotoxic Hypokalaemic Periodic Paralysis.* A 29-year-old man (Schutta and Armitage, 1969) experienced an acute attack of hypokalaemic paralysis. A new attack was provoked by glucose, insulin, and prednisone. The patient suffered from severe weight loss due to thyrotoxicosis. Muscle biopsy showed vacuolar myopathy, intramitochondrial crystalloids, and cylindrical inclusions consisting of concentric membranes. These inclusions were identical with those described in Luft's disease (Luft *et al.*, 1962). Luft *et al.* (1962) assumed that they were derived from mitochondria, an assumption not confirmed by Schutta and Armitage (1969).

This patient had hypermetabolism, as did the patients with Luft's disease, but the hypermetabolism was due to thyrotoxicosis (the diagnosis was based on a 2-fold increase in P.B.I.). With respect to periodic paralysis, he was similar to the patient with pleoconial myopathy. The patient refused treatment of the thyrotoxicosis.

c. *Malignant Hyperthermia.* A male (22 years) and his father (48 years) (own unpublished observations) were investigated for malignant hyperthermia; halothane and caffeine tests were positive. Muscle biopsies showed central nuclei and mitochondria with typical crystalloids, but no ragged-red fibers. The relation between hypersensitivity to anesthetics and this mitochondrial abnormality is obscure. In a case with "ophthalmoplegia plus" with crystalloids in muscle mitochondria, Lessel *et et al.* (1969) observed apnea of 60 minutes duration after succinylcholine exposure.

d. *Late-Onset Myopathy with Generalized Affection of Skeletal Muscle.* Two patients (60-year-old male: Shibasaki *et al.*, 1973; 67-year-old male: Kamieniecka, 1976) complained of progressive weakness of limb muscles, one also of facial muscles. The creatine kinase level was moderately increased; biopsies showed about 15% ragged-red fibers with intramitochondrial crystalloids. These patients differed from those with the facioscapulohumeral form of "mitochondrial myopathy" in that no familial disposition was present, and from the limb-girdle form in that the symptoms started late in life.

e. *Familial Myopathy with Generalized Weakness and Muscle Cramps.* A 61-year-old female and her 35-year-old son suffered from generalized muscle weakness, moderate wasting, and painful muscle cramps since the age of 24 and 21, respectively (Buscaino *et al.*, 1970).

f. *Polymyositis.* Five patients (male, 48 years, female, 42 years: Shafiq *et al.*, 1967; male, 68 years: Chou, 1969; male, 37 years, male, 50 years: Kamieniecka and Schmarlbruch, 1978) were classified as polymyositis on the basis of a high sedimentation rate and muscle pain and tenderness. The biopsy showed inflammatory infiltration in three cases, abnormal mitochondria in all, and intramitochondrial crystalloids in three cases. Two patients responded to prednisone; two did not respond; and one was not treated appropriately. Intramitochondrial crystalloids were found in one patient with dermatomyositis

(Chew *et al.*, 1977) and in four patients (all males, 60 to 84 years of age) with "inclusion body myositis" (Carpenter *et al.*, 1978). None of the patients with "inclusion body myositis" responded to prednisone.

VII. Discussion

A. Neurogenic or Myogenic?

In most of the patients with abnormal muscle mitochondria, no obvious involvement of the peripheral nervous system was found. The pattern of branching of the distal motor axons (terminal innervation ratio) was normal in three cases of "mitochondrial myopathy" (Coërs *et al.*, 1976). When a biochemical defect responsible for muscle weakness and fatigability was demonstrated, it was localized in muscle mitochondria. Experimental denervation does never produce mitochondrial abnormalities as in "mitochondrial myopathy." In young muscle fibers grown aneurally in tissue culture from a biopsy from a patient with "mitochondrial myopathy," Askanas *et al.* (1978) found abnormal mitochondria. These observations suggest that the lesion of the muscle fiber is primarily myogenic rather than neurogenic.

Nevertheless, abnormal muscle mitochondria, including intramitochondrial crystalloids, were found in one fiber in a patient with amyotrophic lateral sclerosis (De Recondo *et al.*, 1966), in two patients classified as spinal muscular atrophy (Dobkin and Verity, 1976; Shafiq *et al.*, 1967), in one patient with Charcot-Marie-Tooth's disease (Spalke *et al.*, 1975), and in two patients with Friedreich's disease (Bastiaensen *et al.*, 1977; Schlote *et al.*, 1979). Both cases with spinal muscular atrophy were familial; in one patient, ptosis, bulbar symptoms, and cardiac block were present. A patient described by Shy *et al.* (1967) and Gonatas *et al.* (1967) showed symptoms of a fully developed oculocraniosomatic syndrome including cerebral and cerebellar symptoms. Muscle weakness was mainly proximal. The motor nerve conduction velocity was decreased. Histologically, no denervated muscle fibers were found. Schwann cells contained numerous lipid inclusions, but no intramitochondrial crystalloids. The findings in this patient suggest that the lesion of the muscle and of the peripheral nervous system were independent manifestations of the same pathological process rather than that the lesion of the muscle was due to denervation. The same explanation may apply to those cases in whom mitochondrial abnormalities were found in disorders of the peripheral nervous system (spinal muscular atrophy: Dobkin and Verity, 1976; Shafiq *et al.*, 1967), in particular in those in whom other organs than the nervous system were affected as well.

In a patient with Leigh's disease and signs of a myopathy with abnormal muscle mitochondria, ophthalmoplegia occurred during the final stage of the disease (Crosby and Chou, 1974). Whether ophthalmoplegia in this patient was of myogenic or neurogenic origin is unknown. Since both central nervous system and muscle fibers are affected in the oculocraniosomatic syndrome, it may be asked whether all cases of ophthalmoplegia may not be of neurogenic (nuclear) origin. Histological findings in extraocular muscle were assumed to suggest a myogenic origin of chronic progressive external ophthalmoplegia (CPEO: Kiloh and Nevin, 1951). Nevertheless, histology or electrophysiology are of little value in these cases because the criteria for skeletal muscle may not apply to extraocular muscles (Martinez *et al.*, 1976; Rosenberg *et al.*, 1968). In patients with ophthalmoplegia and involvement of skeletal muscles, the affection of skeletal muscles is always myogenic both electrophysiologically and histologically.

B. "Mitochondrial Myopathies" Differ from Classical Neurological Disorders

The majority of patients in whom abnormal muscle mitochondria were found had clinical symptoms that do not conform to classical neurological disorders. Only in a few papers were intramitochondrial crystalloids described in well recognized disorders (paramyotonia Eulenburg: Garcin *et al.*, 1966; amyotrophic lateral sclerosis: De Recondo *et al.*, 1966; myotonic dystrophy: Fardeau, 1970). Mitochondrial changes were also reported in polymyositis, dermatomyositis, or "inclusion body myositis" (Chou, 1969; Kamieniecka, 1976; Shafiq *et al.*, 1967; Chew *et al.*, 1977; Carpenter *et al.*, 1978). It is striking that nine of ten patients were males, and that only in two of eleven patients was treatment with prednisone beneficial. This distinguishes these cases from other types of polymyositis, in which females predominate and more than half of the patients respond to prednisone treatment (Pearson and Currie, 1974).

From the observations described, we conclude that abnormal mitochondria observed in muscle fibers from a patient with established neurological disorder justify a reconsideration of the diagnosis, though in a few well documented cases that obviously cannot be classified as "mitochondrial myopathy," abnormal mitochondria were found as well.

C. How Specific Are the Morphological Changes of Mitochondria in "Mitochondrial Myopathy"?

Different nosological entities are covered by the term "mitochondrial myopathy" that cannot be subgrouped by morphological criteria alone. Thus the mitochondrial abnormalities are "unspecific." Nevertheless, in experimental animals, only a few of these changes were reproduced, but not intramitochondrial crystalloids. Clinical and experimental evidence suggests that mitochondrial abnormalities are the expression of a metabolic defect at the level of the mitochondrion; the basic metabolic error need not be the same in all cases. In most patients, the unknown defect or defects caused an increase in the blood lactate or pyruvate level. This is, however, a gross symptom that does not reveal the site of the metabolic error (DiMauro, 1979). On the level of the single muscle fiber, metabolic defects that result in abnormal mitochondria may occur in all conditions, possibly even in normal subjects. This could account for the sporadic observations of abnormal mitochondria in classical neurological disorders. This "unspecificity" does not, however, devaluate the criterion "abnormal mitochondria," because the number of observations in well established neuromuscular disorders is negligible when compared to the number in disorders that are difficult to classify.

In several patients with "mitochondrial myopathy," loose coupling of mitochondrial phosphorylation was found (Section VI,A,1). The same phenomenon was observed in subsarcolemmal mitochondria of normal rat muscle (Hülsmann et al., 1968), whereas mitochondria of the internal framework appeared normal. Loose coupling was attributed to aging. The only observation of intramitochondrial crystalloids in animal muscles was by A. M. Kelly (personal communication, 1978) in a 3-year-old mouse. Most patients in whom abnormal mitochondria were found in myositis were elderly males; most patients with ophthalmoplegia, ptosis, and subclinical involvement of skeletal muscles, but without other symptoms of the oculocraniosomatic syndrome, were elderly females. From these findings one may speculate that, at least in some patients, the biochemical and morphological abnormalities were signs of premature aging of mitochondria.

D. What Kind of "Mitochondrial Myopathies" Can Be Distinguished?

Several patients reveal metabolic defects that allow distinct nosological entities to be established: Luft's disease (Section VI,A,1,a), the different forms of carnitine deficiency (Section VI,A,2,a,b), lack of cytochromes in newborns (Section VI,A,2,c). These entities appear, at least at present, well defined.

The group of patients in whom *loosely coupled mitochondrial phosphorylation* without generalized hypermetabolism was found, is heterogeneous. The only case with pleoconial myopathy (Section VI,A,1,c) suffered from periodic paralysis, possibly related to abnormalities in the distribution of electrolytes. The same symptom was present in a number of the other patients as well. Within the group of patients who complained of muscle weakness and fatigability, some suffered from a facioscapulohumeral form of weakness, others from a limb-girdle or more generalized form.

Shy et al.(1966) observed the first patient with limb-girdle type of muscle weakness, onset in childhood, and giant muscle mitochondria with intramitochondrial crystalloids. They assumed that this patient represented a new nosological entity, which was called "megaconial myopathy."

Nevertheless, some of the patients who complained of muscle weakness and fatigability, and also relatives of these patients, had symptoms that occurred in patients with the oculocraniosomatic syndrome (short stature, dementia, diabetes mellitus, ataxia, epilepsy) (D'Agostino *et al.*, 1968; Hudgson *et al.*, 1972; Kamieniecka, 1976). It is tempting to speculate that the "mitochondrial myopathies" with facioscapulohumeral or limb-girdle type of involvement are in reality "formes fruste" of the oculocraniosomatic syndrome. If this were so, the heredity should be the same. The information available, though sparse, seems to disprove this speculation.

The familial cases of the oculocraniosomatic syndrome indicated either autosomal recessive (Tamura *et al.*, 1974; Scarlato *et al.*, 1978) or autosomal dominant (Iannaccone *et al.*, 1974; Shapira *et al.*, 1975; Kamieniecka, 1976; Leshner *et al.*, 1978) inheritance. All patients with facioscapulohumeral impairment were from three families (Hudgson *et al.*, 1972; Kamieniecka, 1976). In one family with six affected and six "asymptomatically affected" members (Hudgson *et al.*, 1972), women outnumbered men and the trait was passed from mother to children. The authors suggest that the mitochondrial defects are not inherited in the Mendelian mode, but through cytoplasmic, that is maternal, inheritance.

The term "oculocraniosomatic syndrome" covers a great variety of phenotypes, but possibly only one entity. Within the same family, a spectrum from ptosis alone over oculopharyngeal syndrome up to a complete oculocraniosomatic syndrome was observed (Saraux *et al.*, 1974; Tamura *et al.*, 1974; Leshner *et al.*, 1978). Ocular symptoms are most consistent, but apparently not obligatory. Several patients obviously had an oculocraniosomatic syndrome, though ptosis and ophthalmoplegia were lacking (Table II).

One patient with "mitochondrial myopathy" and symptoms of an oculocraniosomatic syndrome died from Leigh's disease documented at necropsy (Crosby and Chou, 1974); another patient had a sister who died from Leigh's disease (Schlote *et al.*, 1979). The relation between this autosomal recessive necrotizing encephalomyelopathy and the oculocraniosomatic syndrome remains obscure. In several related patients who died from Leigh's disease, the activity of pyruvate dehydrogenase was decreased in fibroblasts (Blass *et al.*, 1976); in one patient, deficiency of cytochrome c oxidase and loosely coupled mitochondrial phosphorylation was found (Willems *et al.*, 1977). It is striking that intramitochondrial crystalloids resembling those in skeletal muscle fibers were demonstrated solely in brain tissue (oculocraniosomatic syndrome: Schneck *et al.*, 1973; Adachi *et al.*, 1973).

Some patients with ophthalmoplegia suffered from moderate or subclinical involvement of the skeletal muscles only. No other organs were affected and none of these cases was familial. They are listed as "ophthalmoplegia plus" (Table I). Nevertheless, there is evidence that these patients form a distinct group, i.e., are not a "forme fruste" of the oculocraniosomatic syndrome. Clinically, they were classified as "chronic progressive external ophthalmoplegia (CPEO)" (Kiloh and Nevin, 1951), but differed from the majority of these patients by their abnormal muscle mitochondria (Morgan-Hughes and Lambert, 1974; Danta et al., 1975). We observed seven females, 57 to 70 years of age, with late-onset ophthalmoplegia and subclinical or minimal involvement of nuchal and shoulder girdle muscles. Three of these patients were admitted with the diagnosis "myasthenia gravis" because of fatigability. In all, ragged-red fibers were present in biopsies from the brachial biceps muscle.

Several patients described in Section VI do not fit into the subgroups mentioned above. Biochemically, the patient of Schellens and Ossentjuk (1969) resembled patients with Luft's disease, but lacked hypermetabolism (Section VI,A,1,b). The patient of Engel and Dale (1968) revealed a decrease of activity of acid maltase and was classified as glycogenosis (Section VI,A,2,c). Other patients were different because they had unusual clinical symptoms (cramps) (Section VI,B,1,e) or the time of onset of symptoms was unusual late (Section VI,B,1,d). Two patients showed hypothyroidism and the muscular changes were thought to be secondary (Table II); one had hypermetabolism and periodic paralysis secondary to hyperthyroidism and hypokalaemia (Section VI,B,1,b). These inconsistencies may be due to (1) normal variation of the clinical picture, (2) the fact that not the entire "spectrum" of "mitochondrial myopathy" was covered by the subgroups established by biochemical and clinical parameters, (3) incomplete clinical information, or (4) "unspecific" occurrence of abnormal mitochondria (Section VII,C).

It is still unknown whether the morphological abnormalities of muscle mitochondria are a direct manifestation of the metabolic defect responsible for muscular and nonmuscular symptoms. Independent of that, the concept of "mitochondrial myopathy" may help to elucidate the etiology of several obscure multisystem syndromes, many of them with involvement of the central nervous system (Tanabe et al., 1975). Those disorders that are associated with abnormal muscle mitochondria have now been transferred into one group of disorders with evidence of a metabolic defect ("mitochondrial myopathy") and will be transferred increasingly to subgroups with a well-defined metabolic error. We believe that here rests the true advantage of the concept of "mitochondrial myopathy" based on predominantly morphological criteria.

ACKNOWLEDGMENTS

We wish to thank Dr. M. Lennox-Buchthal for correcting the English, and Mrs. E. Fischer for typing the manuscript. The authors' own investigations had been supported by the Danish Medical Research Council.

REFERENCES

Adachi, M., Torii, Z., Volk, B. W., Briet, W., Wolintz, A., and Schneck, L. (1973). *Acta Neuropathol.* **23**, 300–312.

Afifi, A. K., Ibrahim, M. Z., Bergman, R. A., Haydar, N. A., Mire, J., Bahuth, N., and Kaylani, F. (1972). *J. Neurol. Sci.* **15**, 271–290.

Askanas, V., Engel, W. K., Britton, D. E., Adornato, B. T., and Eiben, R. M. (1978). *Arch. Neurol. (Chicago)* **35**, 801–809.

Bakouche, P., Lamotte-Barrillon, S., and Lagarde, P. (1976). *Sem. Hop.* **52**, 1013–1016.

Bastiaensen, L.A.K., Jaspar, H.H.J., Stadhouders, A. M., Egberink, G.J.M., and Korten, J. J. (1977). *Acta Neurol. Scand.* **56**, 483–507.

Beckett, E. B., and Bourne, G. H. (1973). *In* "The Structure and Function of Muscle" (G. H. Bourne, ed.), Vol. 4, pp. 289–358. Academic Press, New York.

Bender, A. N., and Engel, W. K. (1976). *J. Neuropathol. Exp. Neurol.* **35**, 46–52.

Benke, B., and Szendröi, M. (1971). *Virchows Arch. B.* **9**, 145–152.

Birnberger, K. L., Weindl, A., Struppler, A., Schinko, I., and Pongratz, D. (1973). *Z. Neurol.* **205**, 323–340.

Black, J. T., Judge, D., Demers, L., and Gordon, S. (1975). *J. Neurol. Sci.* **26**, 479–488.

Blass, J. P., Cederbaum, S. D., and Dunn, H. G. (1976). *Lancet* **1**, 1237–1238.

Bonilla, E., Schotland, D. L., DiMauro, S., and Aldover, B. (1975). *J. Ultrastruct. Res.* **51**, 404–408.

Bonilla, E., Schotland, D. L., DiMauro, S., and Lee, C-P. (1977). *J. Ultrastruct. Res.* **58**, 1–9.

Boudin, G., Mikol, J., Guillard, A., and Engel, A. G. (1976). *J. Neurol. Sci.* **30**, 313–325.

Bradley, W. G., Jenkison, M., Park, D. C., Hudgson, P., Gardner-Medwin, D., Pennington, R.J.T., and Walton, J. N. (1972). *J. Neurol. Sci.* **16**, 137–154.

Bubenzer, H.-J. (1966). *Z. Zellforsch. Mikrosk. Anat.* **69**, 520–550.

Burke, R. E., Levine, D. N., Zajac, F. E., III, Tsairis, P., and Engel, W. K. (1971). *Science* **174**, 709–712.

Burke, R. E., Levine, D. N., Tsairis, P., and Zajac, F. E., III. (1973). *J. Physiol. (London)* **234**, 723–748.

Buscaino, G. A., DeGiacomo, P., and Mazzarella, L. (1970). *In* "Muscle Diseases" (J. N. Walton, N. Canal, and G. Scarlato, eds.), pp. 112–115. Excerpta Med. Found., Amsterdam.

Butler, I. J., and Gadoth, N. (1976). *Arch. Intern. Med.* **136**, 1290–1293.

Carpenter, S., Karpati, G., Heller, I., and Eisen, A. (1978). *Neurology* **28**, 8–17.

Carrier, H., Garde, A., Tommasi, M., Kopp, N., and Savet, J.-F. (1974). *Acta Neuropathol.* **30**, 295–303.

Carroll, J. E., Zwillich, C., Weil, J. V., and Brooke, M. H. (1976). *Neruology* **26**, 140–146.

Castaigne, P., Laplane, D., Fardeau, M., Dordain, G., Autret, A., and Hirt, L. (1972). *Rev. Neurol.* **126**, 81–96.

Castaigne, P., Lhermite, F., Escourolle, R., Chain, F., Fardeau, M., Hauw, J. J., Curet, J., and Flavigny, C. (1977). *Rev. Neurol.* **133**, 369–386.

Chew, E. C., Araoz, C., Sun, C. N., and White, H. J. (1977). *Ann. Clin. Lab. Sci.* **7**, 29–34.

Chou, S. M. (1969). *Acta Neuropathol.* **12**, 68–89.

Close, R. (1967). *J. Physiol. (London)* **193**, 45–55.

Coërs, C., Telerman-Toppet, N., Gérard, J. M., Szliwowski, H., Bethlem, J., and Van Wijngaarden, G. K. (1976). *Neurology* **26**, 1046–1053.

Coleman, R. F., Nienhuis, A. W., Brown, W. J., Munsat, T. L., and Pearson, C. M. (1967). *J. Am. Med. Assoc.* **199**, 118–124.

Cornelio, F., Di Donato, S., Peluchetti, D., Bizzi, A., Bertagnolio, B., D'Angelo, A., and Wiesmann, U. (1977). *J. Neurol., Neurosurg. Psychiatry* **40**, 170–178.

Croft, P. B., Cutting, J. C., Jewesbury, E. C. O., Blackwood, W., and Mair, W.G.P. (1977). *Acta Neurol. Scand.* **55**, 169-197.

Crosby, T. W., and Chou, S. M. (1974). *Neurology* **24**, 49-54.

D'Agostino, A. N., Ziter, F. A., Rallison, M. L., and Bray, P. F. (1968). *Arch. Neurol. (Chicago)* **18**, 388-401.

Danta, G., Hilton, R. C., and Lynch, P. G. (1975). *Brain* **98**, 473-492.

DeHaan, E. J., Groot, G.S.P., Scholte, H. R., Tager, J. M., and Wit-Peeters, E. M. (1973). *In* "The Structure and Function of Muscle" (G. H. Bourne, ed.). Vol. 3, pp. 417-469. Academic Press, New York.

Denis, B., Morena, H., Rossignol, B., Machecourt, J., Sebag, M., Stoebner, P., and Martin-Noel, P. (1976). *Arch. Mal. Coeur Vaiss.* **69**, 747-753.

DeRecondo, J., Fardeau, M., and Lapresle, J. (1966). *Rev. Neurol.* **114**, 169-192.

Di Donato, S., Cornelio, F., Balestrini, M. R., Bertagnolio, B., and Peluchetti, D. (1978). *Neurology* **28**, 1110-1116.

DiMauro, S. (1979). *In* "Handbook of Clinical Neurology" (P. J. Vinken and G. W. Bruyn, eds.), Vol. 41. North-Holland Publ., Amsterdam (in press).

DiMauro, S., Schotland, D. L., Bonilla, E., Lee, C.-P., Gambetti, P., and Rowland, L. P. (1973). *Arch. Neurol. (Chicago)* **29**, 170-179.

DiMauro, S., Schotland, D. L., Bonilla, E., Lee, C. P., DiMauro, P. M., and Scarpa, A. (1974). *In* "Exploratory Concepts in Muscular Dystrophy II" (A. T. Milhorat, ed.), pp. 506-515. Excerpta Med. Found., Amsterdam.

DiMauro, S., Bonilla, E., Lee, C. -P., Schotland, D. L., Scarpa, A., Conn, H., Jr., and Chance, B. (1976). *J. Neurol. Sci.* **27**, 217-232.

DiMauro, S., Mendell, R., Sahenk, Z., Bachman, D., and Scarpa, A. (1978). *Proc. Int. Congr. Neuromusc. Dis., 4th, 1978* Abstract 348.

Dobkin, B. H., and Verity, M. A. (1976). *Neurology* **26**, 754-763.

Drachman, D. A. (1968). *Arch. Neurol. (Chicago)* **18**, 654-674.

Drachman, D. A. (1975). *In* "Handbook of Clinical Neurology" (P. J. Vinken and G. W. Bruyn, eds.), Vol. 22, pp. 203-216. North-Holland Publ., Amsterdam.

Dubowitz, V., and Brooke, M. H. (1973). "Muscle Biopsy: A. Modern Approach," pp. 242-250. Saunders, Philadelphia.

Eisenberg, B. R., and Kuda, A. M. (1976). *J. Ultrastruct. Res.* **54**, 76-88.

Engel, A. G., and Angelini, C. (1973). *Science* **179**, 899-902.

Engel, A. G., and Dale, A.J.D. (1968). *Proc. Staff Meet. Mayo Clin.* **43**, 233-279.

Engel, A. G., and Siekert, R. G. (1972). *Arch. Neurol. (Chicago)* **27**, 174-181.

Engel, A. G., Angelini, C., and Nelson, R. A. (1974). *In* "Exploratory Concepts in Muscular Dystrophy II" (A. T. Milhorat, ed.), pp. 601-617. Excerpta Med. Found., Amsterdam.

Engel, A. G., Banker, B. Q., and Eiben, R. M. (1977). *J. Neurol., Neurosurg. Psychiatry* **40**, 313-322.

Engel, W. K., and Cunningham, G. G. (1963). *Neurology* **13**, 919-923.

Engel, W. K., Brooke, M. H., and Nelson, P. G. (1966). *Ann. N.Y. Acad. Sci.* **138**, 160-185.

Ernster, L., and Luft, R. (1963). *Exp. Cell Res.* **32**, 26-35.

Ernster, L., Ikkos, D., and Luft, R. (1959). *Nature (London)* **184**, 1851-1854.

Fardeau, M. (1970). *In* "Muscle Diseases" (J. N. Walton, N. Canal, and G. Scarlato, eds.), pp. 98-108. Excerpta Med. Found., Amsterdam.

Fawcett, D. W., and McNutt, N. S. (1969). *J. Cell Biol.* **42**, 1-45.

Fisher, E. R., and Danowski, T. S. (1969). *Am. J. Clin. Pathol.* **51**, 619-630.

Folkow, B., and Halicka, H. D. (1968). *Microvasc. Res.* **1**, 1-14.

Gadoth, N. (1977). *Isr. J. Med. Sci.* **13**, 159-160.

Garcin, R., Legrain, M., Rondot, P., and Fardeau, M. (1966). *Rev. Neurol.* **115**, 295-311.

Gérard, J. M., Rétif, J., Telerman-Toppet, N., Demols, E., and Coërs, C. (1974). *Acta Neurol. Belg.* **74**, 284–296.

Godet-Guillain, J., and Fardeau, M. (1970). In "Muscle Diseases" (J. N. Walton, N. Canal, and G. Scarlato, eds.), pp. 512–515. Excerpta Med. Found., Amsterdam.

Gollnick, P. D., and King, D. W. (1969). *Am. J. Physiol.* **216**, 1502–1509.

Gonatas, N. K., Evangelista, I., and Martin, J. (1967). *Am. J. Med.* **42**, 169–178.

Green, D. E., Asai, J., Harris, R. A., and Penniston, J. T. (1968). *Arch. Biochem. Biophys.* **125**, 684–705.

Gruner, J. E. (1963). *C.R. Seances Soc. Biol. Ses. Fil.* **157**, 181–182.

Gullotta, F., Payk, T. R., and Solbach, A. (1974). *Z. Neurol.* **206**, 309–326.

Hall-Craggs, E.C.B. (1968). *J. Anat.* **102**, 241–255.

Hanzlíková, V., and Schiaffino, A. (1977). *J. Ultrastruct. Res.* **60**, 121–133.

Harati, Y., Patten, B. M., Sheehan, M., Judge, D., and Wood, J. M. (1977). *Proc. World Congr. Neurol., 11th, 1977* p. 318.

Hart, Z. H., Chang, C.-H., DiMauro, S., Farooki, Q., and Ayyar, R. (1978). *Neurology* **28**, 147–151.

Haydar, N. A., Conn, H. L., Afifi, A., Wakid, N., Ballas, S., and Faway, K. (1971). *Ann. Intern. Med.* **74**, 548–558.

Henneman, E., and Olson, C. B. (1965). *J. Neurophysiol.* **28**, 581–598.

Hostetler, K. Y., Hoppel, C. L., Romine, J. S., Sipe, J. C., Gross, S. R., and Higginbottom, P. A. (1978). *N. Engl. J. Med.* **298**, 553–557.

Hudgson, P., Bradley, W. G., and Jenkison, M. (1972). *J. Neurol. Sci.* **16**, 343–370.

Hülsmann, W. C., Bethlem, J., Meijer, A.E.F.H., Fleury, P., and Schellens, J.P.M. (1967). *J. Neurol., Neurosurg. Psychiatry* **30**, 519–525.

Hülsmann, W. C., de Jong, J. W., and Van Tol, A. (1968). *Biochim. Biophys. Acta* **162**, 292–293.

Hyman, B. N., Patten, B. M., and Dodson, R. F. (1977). *Am. J. Ophthalmol.* **83**, 362–371.

Iannaccone, S. T., Griggs, R. C., Markesbery, W. R., and Joynt, R. J. (1974). *Neurology* **24**, 1033–1038.

Isaacs, H., Heffron, J.J.A., Badenhorst, M., and Pickering, A. (1976). *J. Neurol., Neurosurg. Psychiatry* **39**, 1114–1123.

Jerusalem, F., Angelini, C., Engel, A. G., and Groover, R. V. (1973). *Arch. Neurol. (Chicago)* **29**, 162–169.

Jerusalem, F., Spiess, H., and Baumgartner, G. (1975). *J. Neurol. Sci.* **24**, 272–282.

Kamieniecka, Z. (1976). *Acta Neurol. Scand.* **55**, 57–75.

Kamieniecka, Z., and Schmalbruch, H. (1978). *Muscle & Nerve* **1**, 413–415.

Karpati, G., Carpenter, S., and Eisen, A. A. (1972). *Arch. Neurol. (Chicago)* **27**, 237–251.

Karpati, G., Carpenter, S., Larbrisseau, A., and Lafontaine, R. (1973). *J. Neurol. Sci.* **19**, 133–151.

Karpati, G., Carpenter, S., Melmed, C., and Eisen, A. A. (1974). *J. Neurol. Sci.* **23**, 129–161.

Karpati, G., Carpenter, S., Engel, A. G., Watters, G., Allen, J., Rothman, S., Klassen, G., and Mamer, O. A. (1975). *Neurology* **25**, 16–24.

Kearns, T. P. (1965). *Trans. Am. Ophthalmol. Soc.* **63**, 559–625.

Kearns, T. P., and Sayre, G. P. (1958). *Arch. Ophthalmol.* **60**, 280–289.

Kelly, A. M., and Zacks, S. I. (1969). *J. Cell Biol.* **42**, 135–153.

Ketelsen, U. -P., Beckmann, R., and Nolte, J. (1978). *J. Neurol. Sci.* **35**, 275–290.

Kiloh, L. G., and Nevin, S. (1951). *Brain* **74**, 115–143.

Kuner, J. M., and Beyer, R. E. (1970). *J. Membr. Biol.* **2**, 71–84.

Lapresle, J., Fardeau, M., and Godet-Guillain, J. (1972). *J. Neurol. Sci.* **17**, 87–102.

Leshner, R. T., Spector, R. H., Seybold, M., Romine, J., Sipe, J., and Kelts, A. (1978). *Neurology* **28**, 364–365.

Lessell, S., Kuwabara, T., and Feldman, R. G. (1969). *Am. J. Ophthalmol.* **68**, 789–796.

Lou, H. C., and Reske-Nielsen, E. (1976). *Arch. Neurol. (Chicago)* **33**, 455–456.

Luft, R., Ikkos, D., Palmieri, G., Ernster, L., and Afzelius, B. (1962). *J. Clin. Invest.* **41**, 1776-1804.

McComish, M., Compston, A., and Jewitt, D. (1976). *Br. Heart J.* **38**, 526-529.

McLeod, J. G., Baker, W. De C., Shorey, C. D., and Kerr, C. B. (1975). *J. Neurol. Sci.* **24**, 39-52.

Markesbery, W. R., McQuillen, M. P., Procopis, P. G., Harrison, R., and Engel, A. G. (1974). *Arch. Neurol. (Chicago)* **31**, 320-324.

Martinez, A. J., Hay, S., and McNeer, K. W. (1976). *Acta Neuropathol.* **34**, 237-253.

Mayr, R. (1971). *Tissue & Cell* **3**, 433-462.

Melmed, C., Karpati, G., and Carpenter, S. (1975). *J. Neurol. Sci.* **26**, 305-318.

Miledi, R., and Slater, C. R. (1968). *J. Cell Sci.* **3**, 49-54.

Morgan-Hughes, J. A., and Lambert, C. D. (1974). *Trans. Am. Neurol. Assoc.* **99**, 35-38.

Morgan-Hughes, J. A., and Mair, W.G.P. (1973). *Brain* **96**, 215-224.

Morgan-Hughes, J. A., Darveniza, P., Kahn, S. N., Landon, D. N., Sherratt, R. M., Land, J. M., and Clark, J. B. (1977). *Brain* **100**, 617-640.

Norris, F. H., and Panner, B. J. (1966). *Arch. Neurol. (Chicago)* **14**, 574-589.

Okamura, K., Santa, T., Nagae, K., and Omae, T. (1976). *J. Neurol. Sci.* **27**, 79-91.

Olson, W., Engel, W. K., Walsh, G. O., and Einaugler, R. (1972). *Arch. Neurol. (Chicago)* **26**, 193-211.

Patten, B. M., Wood, J., Howell, R. R., Heferan, P., and DiMauro, S. (1978). *Neurology* **28**, 399.

Pearson, C. M., and Currie, S. (1974). *In* "Disorders of Voluntary Muscle" (J. N. Walton, ed.), pp. 614-652. Churchill, London.

Penniston, J. T., Harris, R. A., Asai, J., and Green, D. E. (1968). *Proc. Natl. Acad. Sci. U.S.A.* **59**, 624-631.

Poloni, M., Cosi, V., Piccolo, G., Scelsi, R., Marchetti, C., and Moglia, A. (1978). *Proc. Int. Congr. Neuromusc. Dis., 4th, 1978* Abstract 426.

Price, H. M., Gordon, G. B., Munsat, T. L., and Pearson, C. M. (1967). *J. Neuropathol. Exp. Neurol.* **26**, 475-497.

Reale, E. (1973). *In* "Grundlagen der Cytologie" (G. C. Hirsch, H. Ruska, and P. Sitte, eds.), pp. 305-343. Fischer, Jena.

Reske-Nielsen, E., Lou, H. C., and Lowes, M. (1976). *Acta Ophthalmol.* **54**, 553-573.

Reznik, M., and Hansen, J. L. (1969). *Arch. Pathol.* **87**, 601-608.

Rosenberg, R. N., Schotland, D. L., Lovelace, R. E., and Rowland, L. P. (1968). *Arch. Neurol. (Chicago)* **19**, 362-376.

Santa, T., Hosokawa, S., Kuroiwa, Y., Okamura, K., Tamura, K., and Omura, I. (1975) *Adv. Neurol. Sci.* **19**, 49-51 (in Japanese, cited from MDA Abstr., Vol. 20, p. 76).

Saraux, H., Offret, H., Nou, B., and Mikol, J. (1974). *Bull. Soc. Ophthalmol. Fr.* **74**, 305-317.

Scarlato, G., and Pellegrini, G. (1979). *Zentralbl. Allg. Pathol. Pathol. Anat.* **123** (in press).

Scarlato, G., Pellegrini, G., and Veicsteinas, A. (1978). *J. Neuropathol. Exp. Neurol.* **37**, 1-12.

Schellens, J.P.M., and Ossentjuk, E. (1969). *Virchows Arch. B* **4**, 21-29.

Schiaffino, S., Hanzlíková, V., and Pierobon, S. (1970). *J. Cell Biol.* **47**, 107-119.

Schlote, W., Meyer, H. J., and Roos, W. (1979). *Zentralbl. Allg. Pathol. Pathol. Anat.* **123** (in press).

Schmalbruch, H. (1971). *Z. Zellforsch. Mikrosk. Anat.* **119**, 120-146.

Schmalbruch, H. (1976). *Tissue & Cell* **8**, 673-692.

Schmalbruch, H. (1979a). *In* "Muscle Regeneration" (A. Mauro, ed.), pp. 217-229. Raven Press, New York.

Schmalbruch, H. (1976b). *Cell Tiss. Res.* **204**, 187-200.

Schneck, L., Adachi, M., Briet, P., Wolintz, A., and Volk, B. W. (1973). *J. Neurol. Sci.* **19**, 37-44.

Schotland, D. L., DiMauro, S., Bonilla, E., Scarpa, A., and Lee, C. -P. (1976). *Arch. Neurol. (Chicago)* **33**, 475-479.

Schutta, H. S., and Armitage, J. L. (1969). *J. Neuropathol. Exp. Neurol.* **28**, 321-336.

Sengers, R. C. A., ter Haar, B. G. A., Trijbels, J. M. F., Willems, J. L., Daniels, O., and Stadhouders, A. M. (1975). *J. Pediatr.* **86**, 873-880.

Sengers, R.C.A., Stadhouders, A. M., Jaspar, H.H.J., Trijbels, J.M.F., and Daniels, O. (1976). *Neuropaediatrie* **7**, 196-208.

Shafiq, S. A., Milhorat, A. T., and Gorycki, M. A. (1967). *Arch. Neurol. (Chicago)* **17**, 666-671.

Shafiq, S. A., Gorycki, M. A., Asiedu, S. A., and Milhorat, A. T. (1969). *Arch. Neurol. (Chicago)* **20**, 625-633.

Shapira, Y., Cederbaum, D., Cancilla, P. A., Nielsen, D., and Lippe, B. M. (1975). *Neurology* **25**, 614-621.

Shibasaki, H., Santa, T., and Kuroiwa, Y. (1973). *J. Neurol. Sci.* **18**, 301-310.

Shy, G. M., Gonatas, N. K., and Perez, M. (1966). *Brain* **89**, 133-158.

Shy, G. M., Silberberg, D. H., Appel, S. H., Mishkin, N. M., and Godfrey, E. H. (1967). *Am. J. Med.* **42**, 163-168.

Simopoulos, A. P., Delea, C. S., and Bartter, F. C. (1971). *J. Pediatr.* **79**, 633-641.

Sjöstrand, F. S., and Bernhard, W. (1976). *J. Ultrastruct. Res.* **56**, 233-246.

Sluga, E., and Moser, K. (1970). *In* "Muscle Diseases" (J. N. Walton, N. Canal, and G. Scarlato, eds.), pp. 116-119. Excerpta Med. Found., Amsterdam.

Smyth, D.P.L., Lake, B. D., MacDermot, J., and Wilson, J. (1975). *Lancet* **1**, 1198-1199.

Spalke, G., Heene, R., and Herold, D. (1975). *J. Neurol.* **209**, 9-29.

Spiro, A. J., Prineas, J. W., and Moore, C. L. (1970a). *Arch. Neurol. (Chicago)* **22**, 259-269.

Spiro, A. J., Moore, C. L., Prineas, J. W., Strasberg, P. M., and Rapin, I. (1970b). *Arch. Neurol. (Chicago)* **23**, 103-112.

Stern, L. Z., Gruener, R., Kirkpatrick, J. B., and Nemeth, P. (1972). *Exp. Neurol.* **36**, 530-538.

Strock, P. E., and Majno, G. (1969). *Surg., Gynecol. Obstet.* **129**, 1213-1224.

Sulaiman, W. R., Doyle, D., Johnson, R. H., and Jennett, S. (1974). *J. Neurol., Neurosurg. Psychiatry* **37**, 1236-1246.

Tamura, K., Santa, T., and Kuroiwa, Y. (1974). *Brain* **97**, 665-672.

Tanabe, H., Nozawa, T., Shiozawa, R., and Tomonaga, M. (1975). *Adv. Neurol. Sci.* **19**, 36-38 (in Japanese, cited from *MDA* Abstr., Vol. 20, p. 196).

Toppet, M., Telerman-Toppet, N., Szliwowski, H. B., Vainsel, M., and Coërs, C. (1977). *Am. J. Dis. Child.* **131**, 437-441.

Trump, B. F., and Ericsson, J.L.E. (1965). *In* "The Inflammatory Process" (B. W. Zweifach, L. Grant, and R. T. McCluskey, eds.), 1st ed., pp. 35-120. Academic Press, New York.

Tsairis, P., Engel, W. K., and Kark, P. (1973). *Neurology* **23**, 408.

Vallat, M., Julien, J., Vallat, J. -M., Vital, C., and Faussier, P. (1975). *Arch. Ophtalmol.* **35**, 509-520.

Van Biervliet, J.P.G.M., Bruinvis, L., Ketting, D., De Bree, P. K., van Der Heiden, C., and Wadman, S. K. (1977). *Pediatr. Res.* **11**, 1088-1093.

VanDyke, D. H., Griggs, R. C., Markesbery, W., and DiMauro, S. (1975). *Neurology* **25**, 154-159.

Van Wijngaarden, G. K., Bethlem, J., Meijer, A.E.F.H., Hülsmann, W. C., and Feltkamp, C. A. (1967). *Brain* **90**, 577-592.

Völpel, M. C. (1979). *Zentralbl. Allg. Pathol. Pathol. Anat.* **123**, (in press).

Walsh, G., DeVivo, D., and Olson, W. (1971). *Arch. Neurol. (Chicago)* **24**, 83-93.

Willems, J. L., Monnens, L.A.H., Trijbels, J.M.F., Veerkamp, J. H., Meyer, A.E.F.H., van Dam, K., and van Haelst, U. (1977). *Pediatrics* **60**, 850-857.

Williamson, J. R. (1979). *Annu. Rev. Physiol.* **41**, 485-506.

Wilner, J. H., Ginsburg, S., and DiMauro, S. (1978). *Neurology* **28**, 721-724.

Worsfold, M., Park, D. C., and Pennington, R. J. (1973). *J. Neurol. Sci.* **19**, 261-274.

Zintz, R. (1966). *In* "Progressive Muskeldystrophie, Myotonie, Myasthenie" (E. Kuhn, ed.), pp. 109–114. Springer, Berlin and New York.

Zintz, R., and Villiger, W. (1967). *Ophthalmologica* **153**, 439–459.

NOTE ADDED IN PROOF

L. A. K. Bastiaensen [(1978). "Chronic Progressive External Ophthalmoplegia," pp. 5–385. Stafleu, Alphen-Rhyn, Netherlands] has reviewed several hundred patients with external ocular myopathy and concludes that patients who do not have a familial oculopharyngeal dystrophy have an oculocraniosomatic syndrome or one of its abortive forms. Most of these patients have discrete plus-symptoms, and no well-documented cases with normal muscle mitochondria are known. Familial cases are frequent but usually missed because little affected relatives are overlooked.

Subject Index

neurogenic or myogenic, 347–348
what kind of mitochondrial myopathies can be distinguished?, 349–351
Neuronal cells, contractile proteins in, 239–243
Nucleolus, ribonucleic acid synthesisin, 258–266

O

Osteoclastic bone removal, similarities to calcium carbonate excavation
boring sponges as model system, 317
modes of action by boring sponges, 316–317
modes of action of bone resorption, 315–316

P

Placenta, estrogen transport into, 98
Plasma membrane, contractile proteins in, 223–227
Plasma proteins, role in transport of steroid hormones, 51–52
evidence for transfer of free steroids across membrane, 52–53
other effects on target tissues, 58–59
regulation of gradient across membrane, 54–58
Prostate, androgen transport into, 88–92

R

Ribonucleic acid, extranucleolar synthesis
further distribution and, 275–290
visualization of transcription sites, 266–275
Ribonucleic acid, nucleolar synthesis
conclusion, 266
further distribution within nucleolus, 263–266
sites of transcription, 258–262
Ribonucleic acid, transcription and mitosis
conclusion, 292
distribution of perichromosomal RNA, 290–291
localization of transcription sites, 291–292

S

Smooth muscle-like cells, contractile proteins in, 238–239

Steroid(s)
binding to membrane proteins
high affinity, 104–108
low affinity, 102–104
possible role of proteins, 108–109
transport, into rat liver cells, 94–98
Steroid hormones
permeability of cell membranes to
effects of changes in membrane composition, 81–84
methods of study, 73–75
permeability coefficients, 75–79
rates of transfer into cells and tissues, 79–81
role of plasma proteins in transport, 51–52
evidence for transfer of free steroids across membrane, 52–53
other effects on target tissues, 58–59
regulation of gradient across membrane, 54–58
transport into target tissues
androgens into prostate, 88–92
estrogens into placenta, 98
estrogens into uterus, 84–88
glucocorticoids into cultured cells, 98–102
glucocorticoids into thymocytes, 93–94
steroid transport into rat liver cells, 94–98
Steroid molecules, physicochemical characteristics of, 59–61
electrical charges, 70–71
energetic considerations, 61–62
mass, configuration and conformation, 62–66
orientation and movement in lipid membranes, 71–73
presence of polar groups, 66–70
Sugar acceptor, glycosyltransferases and, 4
Sugar donor, glycosyltransferases and, 3–4

T

Thrombocytes, contractile proteins in, 237–238
Thymocytes, glucocorticoid transport into, 93–94
Tight junctions, opening of, 133–137
Transendothelial channels, blood-brain barrier and, 143–144
Tropomyosin, antibodies to, 208–210

Contents of Previous Volumes